I0874452

Jörg Sander | Lars Kirmaier | Mircea Manea | Dmitry Shchukin | Ekaterina Skorb

Anticorrosive Coatings

Fundamentals and New Concepts

Jörg Sander et al.: Anticorrosive Coatings

ISBN 978-3-86630-911-1

Cover: Simon Cataudo - sxc and Sergey Kolesnikov - Foltolia.com

Sander, Jörg; Kirmaier, Lars; Manea, Mircea; Shchukin, Dmitry; Skorb, Ekaterina
Anticorrosive Coatings
Hanover: Vincentz Network, 2010
European Coatings Tech files
ISBN 3-86630-911-2
ISBN 978-3-86630-911-1

Please ask for our book catalogue
Vincentz Network, Plathnerstr. 4c, 30175 Hanover, Germany
T (202) 684-6630, F (202) 380-9129
E-mail: books@american-coatings.com, www.american-coatings.com

Layout: Vincentz Network, Hannover, Germany

ISBN 3-86630-911-2
ISBN 978-3-86630-911-1

European Coatings Tech Files

**Jörg Sander | Lars Kirmaier | Mircea Manea |
Dmitry Shchukin | Ekaterina Skorb**

Anticorrosive Coatings

Fundamentals and New Concepts

Jörg Sander et al.: Anticorrosive Coatings

ISBN 978-3-86630-911-1

Foreword

Metals, in particular steel and aluminium, are among the most important construction materials to be met in everyday life. However, these versatile materials are prone to corrosion, resulting in safety impairments, aesthetic failures, and, on the bottom line, economic damage. A major part of surface engineering of metals therefore is focused on corrosion protection.

Organic coatings have been used on any substrate for design purposes as well as to preserve the outward appearance. When applied to metals, corrosion protection becomes their most important technical feature. Though, in this respect, the primary effect of a coating is to form a physical barrier, there is no simple rule of "the thicker, the better". Effective corrosion protection provided by an organic coating requires proper preparation of the substrate surface, expert formulation of treatment and coating chemicals and appropriate application processes, as well as adaptation to different uses and service environments.

A lot of literature has been published on single aspects of corrosion protective coatings. However, corrosion protection by organic coatings is a truly cross-functional issue. Whereas, to the lead author's opinion, a unified approach to this task is lacking that highlights the role of all disciplines involved in the creation and use of corrosion protection coatings for metals.

The intention of this book is to provide this missing synopsis. It features an up-to-date picture of the quality and chemistry of a substrate surface, its proper preparation by conversion treatment, the function of resins and anticorrosive pigments in paints, and novel concepts for corrosion protection. It is addressed to all parties involved in metal surface and coatings engineering, both the supplier and the user, both the expert as well as the student in any of the single disciplines. It is intended to contribute to a better understanding of the mutual roles and responsibilities in corrosion protective organic coatings, and pave the way to a more durable and sustainable preservation of our valuables and resources.

Velbert, Germany, April 2010

Jörg Sander

Jörg Sander et al.: Anticorrosive Coatings

ISBN 978-3-86630-911-1

Contents

1	**Introduction**	**13**
1.1	Why corrosion-protective coatings	13
1.2	Literature	14
2	**Corrosion protection coatings**	**15**
2.1	Principles of function	15
2.1.1	Electrochemistry of corrosion inhibition	15
2.1.2	Metal oxide formation	16
2.1.3	Cathodic protection	17
2.1.4	Passivation and conversion coating	18
2.2	Design of organic coating systems	19
2.2.1	Diffusion barrier features – humidity uptake and electrolyte permeation	19
2.2.2	Active pigments	20
2.3	Function of individual coating layers	21
2.4	Literature	22
3	**Surface preparation**	**25**
3.1	Industrial cleaning	25
3.1.1	Importance of cleaning process	25
3.1.2	Contaminants	25
3.1.3	Surface energy and tension	26
3.2	Mechanical cleaning	27
3.3	Chemical cleaning	28
3.3.1	Plasma and corona processes	28
3.3.2	Solvent cleaning	29
3.3.3	Chemistry of aqueous cleaners	29
3.3.3.1	Mechanism: alkalinity, saponification and metal dissolution	29
3.3.3.2	Ingredients of aqueous cleaners	31
3.3.3.2.1	General considerations	31
3.3.3.2.2	Surfactants	32
3.3.4	Physics of aqueous cleaning, bath life and rinsing	34
3.4	Literature	37
4	**Organic coating materials**	**39**
4.1	Ingredients of organic coating materials	40
4.2	Resins	40
4.2.1	Alkyd resins	41
4.2.1.1	Alkyd resins manufacturing process	43
4.2.1.2	Decay of alkyd resins	43
4.2.1.3	Composition of alkyd resins	44
4.2.1.4	Curing of alkyd resins	45
4.2.2	Chlorinated rubber	46

Jörg Sander et al.: Anticorrosive Coatings

ISBN 978-3-86630-911-1

4.2.3 Polyvinyl chloride ... 48
4.2.4 Epoxy resins ... 49
4.2.4.1 Raw materials for epoxy resins ... 50
4.2.4.2 Manufacturing process for epoxy resins ... 50
4.2.4.3 Cross-linking of the epoxy resins ... 52
4.2.5 Epoxy esters ... 53
4.2.6 Acrylic resins ... 55
4.2.6.1 Manufacturing of acrylic resins ... 56
4.2.6.2 Thermoplastic and thermosetting acrylic resins ... 56
4.2.7 Polyurethanes ... 57
4.2.7.1 Reactivity of isocyanate group ... 59
4.2.7.2 Waterborne polyurethanes ... 60
4.2.7.3 Isocyanate free polyurethanes ... 61
4.2.8 Polyaspartics ... 62
4.2.8.1 Polyurea systems ... 63
4.2.8.2 Polyaspartic systems ... 64
4.2.9.1 Nomenclature of the silicon chemistry ... 66
4.2.9.2 Manufacturing of alkyl silicates ... 68
4.2.9.3 Reactions of alkyl silicates ... 69
4.2.10 Polysiloxanes ... 70
4.2.10.1 Reactions of siloxanes ... 70
4.2.10.2 Manufacturing of siloxanes ... 72
4.3 Pigments - Introduction ... 74
4.3.1 Corrosion protection pigments ... 75
4.3.1.1 Lead and chromate pigments ... 76
4.3.1.1.1 Lead pigments ... 76
4.3.1.1.2 Chromate pigments ... 76
4.3.1.2 Phosphate based pigments ... 79
4.3.1.2.1. Zinc phosphate ... 79
4.3.1.2.2 Modified orthophosphates ... 80
4.3.1.2.3 Modified polyphosphates ... 82
4.3.1.3 Inorganic/organic synergies ... 82
4.3.1.4 Wide spectrum anticorrosives ... 83
4.3.1.5 Phosphites and phosphides ... 84
4.3.1.5.1 Zinc hydroxyphosphite ... 84
4.3.1.5.2 Iron phosphide ... 84
4.3.1.6 Borates ... 85
4.3.1.6.1 Barium metaborate ... 85
4.3.1.6.2 Zinc borate ... 85
4.3.1.6.3 Calcium borosilicate ... 85
4.3.1.7 Molybdates ... 85
4.3.1.8 Ion-exchange pigments ... 86
4.3.1.9 Zinc cyanamide ... 86
4.3.1.10 Hybrid anticorrosive inhibitors ... 86
4.3.2 Barrier pigments ... 87
4.3.2.1 Micaceous iron oxide ... 87
4.3.2.2 Aluminium flakes ... 88
4.3.2.3 Zinc flakes ... 88
4.3.3 Sacrificial pigments ... 88
4.3.3.1 Zinc dust ... 88
4.3.3.2 Magnesium ... 89

4.3.4 Colouring agent 89
4.3.4.1 White pigment, titanium dioxide 89
4.3.4.2 Red pigments 90
4.3.4.3 Yellow pigments 90
4.3.4.4 Green pigments 90
4.3.4.5 Blue pigments 90
4.3.4.6 Black pigments 90
4.4 Extender pigments 91
4.4.1 Carbonates 91
4.4.2 Sulphates 91
4.4.3 Silicas 91
4.4.4 Silicates 91
4.4.4.1 Talc 91
4.4.4.2 Kaolin 92
4.4.4.3 Wollastonite 92
4.4.4.4 Mica 92
4.5 Other additives 92
4.6 Solvents 94
4.7 Raw materials for powder coatings 95
4.8 Literature 97

5 Film formation 101
5.1 Physical drying 101
5.2 Chemical curing 102
5.2.1 Thermal cross-linking: chemistry, mechanism, imparted properties 102
5.2.2 Radiation curing 103
5.2.2.1 Chemical principles and intrinsic properties 103
5.2.2.2 Applications 107
5.2.2.3 Equipment 108
5.3 Literature 110

6 Mechanism of protection and properties of organic coatings 111
6.1 Measurement of physical properties and influence 111
6.2 Dry film thickness 111
6.3 Adhesiveness 112
6.3.1 Role of adhesion and factors of influence 112
6.3.2 Measurement of adhesion and elasticity 113
6.3.2.1 Industrial methods 113
6.3.2.2 Laboratory methods 114
6.4 Permeation through organic coatings 115
6.5 Corrosion protective performance 115
6.5.1 Titanium and zirconium fluoro complex based pretreatments 115
6.5.2 Weldable corrosion protection primer for automotive sheet 118
6.5.3 Thermally curing 2-in-1 primer-pretreatment 120
6.5.4 Chromous based pretreatments - chromiting 120
6.5.5 Active pigments, ion exchangers and scavengers 121
6.5.6 UV-curable primer-pretreatment 122
6.6 Degradation and ageing 122
6.6.1 Weathering 122
6.6.2 Electrochemical degradation 124
6.6.2.1 Cathodic disbonding: oxygen reduction 124

6.6.2.2 Anodic disbonding: filiform corrosion ... 125
6.7 Literature ... 126

7 Testing of organic coatings ... 129
7.1 Performance testing ... 129
7.2 Accelerated corrosion tests ... 129
7.2.1 Overview ... 129
7.2.2 Constant climate tests ... 129
7.2.2.1 Salt spray ... 129
7.2.2.2 Constant climate, humidity ... 130
7.2.2.3 Filiform corrosion ... 130
7.2.2.4 Condensation ... 131
7.2.2.5 Boiling, water soak ... 131
7.2.3 Cyclic climate tests ... 131
7.2.3.1 Cyclic humidity ... 131
7.2.3.2 Prohesion ... 131
7.2.3.3 VDA test ... 131
7.2.3.4 UV test, weathering ... 131
7.3 Electrochemical testing ... 132
7.3.1. General remarks ... 132
7.3.2. Electrochemical potential ... 132
7.3.2.1 Standard potential ... 132
7.3.2.2 Cyclovoltammetry ... 133
7.3.3 Electrochemical impedance spectroscopy ... 133
7.3.4. Electrochemical techniques with high spatial resolution ... 134
7.3.4.1 Scanning vibrating electrode ... 134
7.3.4.2 Height-regulated Scanning Kelvin Probe ... 134
7.3.4.2.1 General technique ... 134
7.3.4.2.2 Blister test ... 135
7.4 Outdoor exposure tests ... 136
7.5 Literature ... 137

8 Chemical conversion treatment ... 139
8.1 Substrates ... 139
8.2 Chemicals for pretreatment ... 141
8.2.1 General remarks ... 141
8.2.2 Alkaline passivation ... 142
8.2.3 Phosphating ... 143
8.2.3.1 Iron phosphating ... 143
8.2.3.2 Zinc phosphating ... 143
8.2.4 Chromating ... 144
8.2.5 Anodising of aluminium ... 144
8.2.6 Chromiting ... 145
8.2.7 Chromium-free pretreatment ... 145
8.2.7.1 Titanium and zirconium fluoro-complex technology ... 145
8.2.7.2 Other chromium-free pretreatments ... 148
8.2.8 Hybrid pretreatment coatings ... 148
8.2.8.1 Silane/siloxane coatings ... 148
8.2.8.2 Combined thermal processes for primer-pretreatment ... 148
8.2.9 Surface preparation of other substrates – copper alloys, white metal, magnesium, stainless steel ... 149

8.2.10 Environmental considerations ... 149
8.3 Application of pretreatments ... 151
8.3.1 Immersion and Spray Treatment ... 151
8.3.2 Pretreatment application for coil: spray/squeeze, spray-cell, roll-coating.. 152
8.4 Literature ... 153

9 Organic coatings for maintenance ... 157
9.1 Surface tolerant coatings ... 157
9.1.1 General considerations ... 157
9.1.2 Surface tolerant coating materials ... 157
9.2 Organic coatings on residual rust and old coatings ... 158
9.3 Literature ... 159

10 New corrosion protection concepts ... 161
10.1 Thin films ... 161
10.1.1 Self-assembling monolayers ... 163
10.1.2 Conducting polymers ... 167
10.1.3 Biopolymers ... 175
10.2 Nanomaterials ... 179
10.2.1 Nanocomposites ... 179
10.2.2 Sol-gel derived ceramic and hybrid coatings ... 182
10.3 Self-healing coatings ... 184
10.3.1 Self-repairing polymer films ... 185
10.3.2 Inhibitor release ... 189
10.4 Conclusions ... 194
10.5 Literature ... 195

11 Standards and guidelines ... 201
11.1 General information ... 201
11.2 General norms ... 201
11.2.1 Norms on mechanical testing of organically coated metallic work pieces. 202
11.2.2 Norms on corrosion testing of organically coated metallic work pieces ... 202
11.3 Selected European legislation on environmental protection ... 203
11.4 Literature ... 204

Authors ... 205

Index ... 206

1 Introduction

Jörg Sander

1.1 Why corrosion-protective coatings

Man has used metal since about 6,500 B.C. Dating back to this time long past, copper items were found in the neolithic settlement of Çatal Höyük (Catalhoyuk) in Anatolia, where their appearance coincided with the obvious decline in the use of stone tools [1, 2]. While copper and its alloys served for decoration, everyday objects, cutlery, tools and weapons of their era, widespread use of metal only started with the establishing of iron making in the second half of the second millennium B.C. [3].

Iron has dominated history, culture, engineering sciences and industrial manufacture grace to its properties, combining availability and convenient refining, versatile workability by casting, forging, rolling and machining, and, last but not least, structural strength and ductility. From Damascus steel scimitar blades to Stephenson's Rocket, from the common passenger car to the Sears Tower Building, from a washing machine to luxury cruise ships – the applications of iron and steel are innumerable.

However, since the discovery of iron and the invention of its relative, steel, man has also been confronted with the degeneration and decay of this so very useful material by corrosion. The corrosion of iron and steel is a very fast process that destroys valuable goods every day and causes a lot of economic damage. The prevention of corrosion therefore is of paramount importance.

Organic coatings offer a very attractive road to the corrosion protection of metals. Their purposes and applications span from the temporary protection of surfaces during storage, transport and onward processing, over the role as protective primers that may provide a number of additional features like surface structuring, biostatic finish, conductor properties, (dry) lubrication etc., cosmetic coatings for colour, haptic and gloss design of surfaces, and finally surface finishes that provide anti-graffiti, anti-fingerprint, or wear resistant properties. Organic coatings provide a unique combination of aesthetic appearance and protection against corrosive decay for man's most important construction material – metal.

A lot of literature has been issued on corrosion prevention, and a lot of scientific and engineering effort as well as public and institutional funding are spent to understand corrosion and deterioration phenomena and to find ways for their prevention. The protective coating of metals primarily intended to retard the start of corrosion damage has often been discussed. In particular, the organic coating of metals has been described in basic reference works by representatives of the paint industry, in papers by the scientific community, in guides to surface engineering, and in monographs dealing with single aspects of use. All authors had their specialist perspectives on metal and metal finishing, for example in steel making, on passivation and surface treatment, on the vast number of uses and their respective anticorrosive measures, and the expertise to formulate and apply coatings. However, it has proven difficult to obtain the big picture that would demonstrate the interdependence and the interplay of all parties.

The intention and the purpose of the present book is to provide a synopsis on all aspects of corrosion protection coatings for metal, from the substrate quality and its chemical proper-

Jörg Sander et al.: Anticorrosive Coatings

ISBN 978-3-86630-911-1

ties, the importance of proper cleaning and surface preparation by conversion treatment, the role of paint ingredients, in particular novel resins and anticorrosive pigments, surface engineering techniques, and the ongoing research for novel trends that will lead the way into the future of corrosion-protective coatings.

Five authors, each of them with a good record in the industrial and scientific community, have joined in this work.

1.2 Literature

[1] Zohar, M., Catal Hüyük, Residential Structures, Mortuary Customs, Material Culture, Artistic Expression, Catal Hoyuk, a Neolithic Town in Anatolia, Net Industries, Kingston ON, Canada 2010 (www.jrank.org/history/pages/5978/Catal-H%C3%BCy%C3%BCk.html) (28-03-2010; 15:35 h)

[2] Weipert, H., Palästina in vorhellenistischer Zeit (Palestine in Pre-hellenistic Times) in: Hausmann, U. (ed.), Handbuch der Archäologie (Handbook of Archeology - Middle East) 2, 1st Vol., Beck, Munich 1988, p. 120

[3] Coghlan, H. H., Prehistoric Iron Prior to the Dispersion of the Hittite Empire, Man 41, Royal Anthropological Institute of Great Britain and Ireland, 1941, pp. 74 ff

2 Corrosion protection coatings

Jörg Sander

2.1 Principles of function

2.1.1 Electrochemistry of corrosion inhibition

Corrosion is an electrochemical process that destroys the surface of metals by dissolution reactions and formation of corrosion products. On a coated metal surface corrosion takes place at the interface between substrate and coating. Consequently, a coating must provide sufficient protection to retard an onset and propagation of corrosion reactions. Strategies for corrosion prevention therefore involve a number of possible measures that either improve electrochemical stability of metal surface or restrict access of corrosive media to surface and along metal/coating interface. Common preventive strategies are summarised below, that have been discussed elsewhere [1, 2].

Corrosion prevention strategies

- Conversion coatings
- Cathodic protection
  - Galvanising
  - Zinc dust primers
- Sealing of pores
  - Multi-layer coatings
- Barrier effect
- Dielectric properties of a coating
- Reduced moisture and gas uptake
- Interception of corrosive agents
  - Active anions
  - Ion exchange pigments
- Improvement of substrate adhesion
  - Adhesive primers
  - Silane/siloxane coatings
- Increase of alkalinity
  - Cement coatings

The metallic state is defined as a feature of solid matter, where atoms are located, densely packed, at the sites of a crystal lattice, and one or more of their electrons (bonding electrons) released to be freely distributed across the entire macroscopic crystal [3]. Therefore, in principle all metals are prone to corrosion due to free availability of their electrons at the outer surfaces of the metal crystal.

One driving force for corrosion is the ease of electrons to be released, or in other terms, of the metal to be oxidised. This depends on the kind of metal, and is usually characterised by the metal's electrochemical potential. The chemical reactions that take place during atmospheric corrosion can be resolved into a pair of red-ox reactions, shown in Equations 2.1 and 2.2.

Jörg Sander et al.: Anticorrosive Coatings

ISBN 978-3-86630-911-1

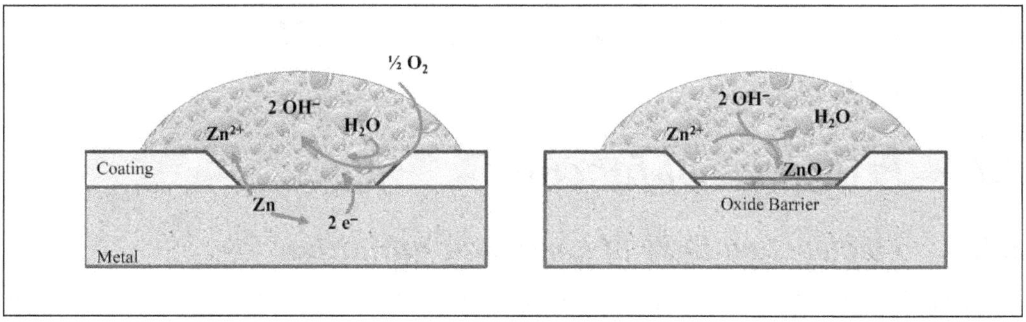

Figure 2.1: (a) Atmospheric corrosion at a punctured coating over a zinc substrate; (b) Formation of a barrier layer by precipitation of zinc oxide (native oxide barrier)

Equation 2.1 $Zn \rightarrow Zn^{2+} + 2\ e^-$ Anodic reaction

Equation 2.2 $1/2\ O_2 + H_2O + 2\ e^- \rightarrow 2\ (OH)^-$ Cathodic reaction

The metal serves as the anode, i.e. the source of free electrons which are released upon the dissolution of metal atoms. These are oxidised to metal cations (e.g. Zn^{2+}) in the first reaction, and the corresponding equivalent amount of electrons (e^-) is made available for the second reaction. In case of atmospheric corrosion, these electrons pass through the interface into the surrounding solution, where they are consumed for the reduction of oxygen. Figure 2.1 shows a schematic view of this situation [4, 5].

This description, in fact, leads directly to the major strategic principle of any anticorrosive coating: If either of the reaction branches is slowed down effectively, the onset and development of corrosion can be largely retarded. Corrosion inhibition (passivation) hence can be achieved by the formation of a physical barrier that either insulates electrically, i.e. prevents the transition of electrons from the metal to surrounding electrolyte, or mechanically, i.e. blocks the direct access of electrolyte or atmospheric oxygen to the metal surface.

2.1.2 Metal oxide formation

A passivation or conversion coating, therefore, usually involves formation of a dense coating of metal oxide, in order to display this barrier effect. The metal oxide, most often incorporating dissolved ions of the substrate metal itself, precipitates on the substrate surface and forms a physical barrier. This barrier can only be penetrated by oxygen or ions through diffusion, thereby slowing down the access of these materials to the base metal.

An effect of metal oxide coatings is sometimes also described in terms of a band model. Many oxides of transition and higher main group metals have semiconductor properties. For example, zinc oxide (ZnO) displays characteristics of an n-semiconductor, which means that electrons in the valence band of the oxide can be promoted to the conducting band by relatively low energy input. The energy gap (i.e. the distance of energy levels between the valence and conducting bands) is a characteristic feature for each substance. The band gap of magnesium oxide, MgO, for example, is much wider, which makes the compound an insulator. Oxides of alloys, like the species $MgZn_2$, display energy gap values between those observed for oxides of the pure elements. Additionally, defects in the oxide coating result in a reduction of the energy gap. A better resistance of an oxide coating against corrosion is obtained with the widening of the energy gap. Though $MgZn_2$ is the prevailing component in the new Zn-Mg galvanising coatings, the situation in technical Zn-Mg coatings is complicated by the additional presence of aluminium. The effect of the various corrosion products in this model is also not yet entirely understood [6].

One approach for a rationalisation of this effect is the respective stability of corrosion products (oxides and hydroxides) in alkaline conditions, as occur under the regime of oxygen reduction. MgO and $Mg(OH)_2$ are insoluble in alkalis, and therefore form stable barrier layers. Consequently, zero-corrosion current densities are observed for lower (more negative) electrode potentials and higher alkalinities, the more Mg is present in the overall metallic coating composition (cf. Figure 2.2) [7]. In coherence, micrographic and photo electron spectroscopic (XPS) studies have shown mainly Zn corrosion products in defective areas with high dissolution, while areas with a low corrosion aspect were found to be covered by an $MgO/Mg(OH)_2$ film.

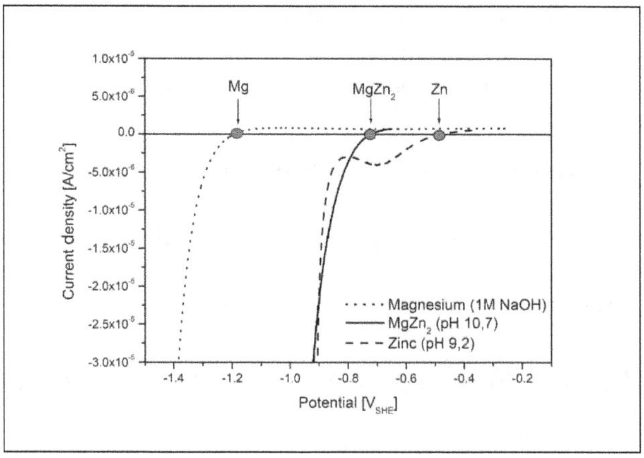

Figure 2.2: Current/potential curves of passive electrodes in aerated solutions. Passive layers of Zn, $MgZn_2$ and Mg consist of ZnO, a mixed Zn-Mg hydroxide ([Mg/Zn] = 0.2) and $Mg(OH)_2$, respectively Source: Elsevier Ltd. [7]

Native oxide coatings of this kind develop on zinc (Zn), aluminium (Al), magnesium (Mg), or chromium (Cr), sometimes also incorporating carbonate ions from atmospheric carbon dioxide (cf. Chapter 3.3.3.1).

Most oxides and hydroxides are instable when contacted with chloride containing electrolytes. Chloride can replace oxygen in the crystal lattice, so that mixed oxy/hydroxy chlorides are formed that are known as intermediates in the dissolution process which often leads to local (pitting) corrosion. Hydrotalcite pigments are reported to act as chloride scavengers by virtue of their ion-exchange properties [8]. Calcium compounds are also known to intercept residual chloride in coatings for polyalkylene packaging foils [9]. A similar effect can be expected from calcium compounds in organic coatings for metals.

2.1.3 Cathodic protection

The native oxide layer of ferrous materials is too porous to withstand further corrosive attack. Iron (Fe) and steel are often saved from corrosion by a coating with less noble metals, e.g. with a Zn layer in galvanising. Further to its less penetrable native oxide barrier, Zn corrodes preferentially on cut edges or scratches, serving as a "sacrificial anode", thus keeping the base metal from destruction. This behaviour is driven by the different electrochemical potentials. While the potential of Fe is E_{pot} = -0.44 V, (cf. Chapter 7.3.2.1, Table 7.2), that of Zn is lower, i.e. E_{pot} = -0.76 V. Protection is even active at some distance from the border of the Zn layer, and may be attributed to a local increase in electron density in front of the Zn border (cf. Chapter 7.3.4.1). This long-range effect is usually called cathodic protection.

Cathodic protection can also be made a function of the primer. Zinc pigmented primers have been used as anti-corrosion barrier for various artisan and industrial coating applications, in particular in heavy machinery, architecture and industrial construction, bridge engineering, water engineering, shipbuilding, etc. For these purposes, primer coats are used with gauges of 50 µm and more [10-13]. The usefulness of zinc-rich primers is often attributed to the cathodic protection provided by zinc particles to otherwise unprotected steel surfaces. Zinc dust primers have also been reported as protective coatings under powder coatings.

As the authors did not observe any influence of the zinc content in various primer formulations, they conclude that the major contributions to the improved performance are made by the mechanical barrier effect and the adhesion promotion that is brought about by the extra primer layer [14].

The principle of cathodic protection has also been used in corrosion protection primers (CPP) that are applied on sheet for car manufacture [15, 16]. By their use, improved corrosion protection is achieved particularly in critical areas of a car body, like seams, box sections or flanges, where a pretreatment is geometrically difficult to apply and may therefore be faulty, or the cathodic electro-dip primer is only insufficiently deposited due to shielding effects.

2.1.4 Passivation and conversion coating

Native oxide layers may serve well as efficient anticorrosive barriers for bare bulk metal, but from the perspective of technical surfaces, it is often desired to create a more uniform, controlled passivation layer, to better preserve the surface aspect and features. To this end, it is necessary to remove the native oxide film (cf. Chapter 3.3.3.1), and replace it by a similar film under controlled conditions, with improved features, for instance, uniform thickness, lower porosity, better transparency, or higher electrical resistance.

Chromate containing chemicals have been quite commonly used for the generation of passivating layers, due to the very high oxidative power of the chromate ion in acid systems. This results in a very fast superficial red-ox reaction (cf. Chapter 8.2.4, Equations 8.8 to 8.10), forming an insoluble, very dense mixed oxide layer of Cr^{3+} and the substrate metal, e.g. Zn. The cathodic branch of this reaction system is also described by Equation 2.3. Its standard potential is E_{pot} = 1.195 V [17].

Equation 2.3 $(CrO_4)^{2-} + 8\ H^+ + 3\ e^- \rightarrow Cr^{3+} + 4\ H_2O$ Cathodic reaction

Underneath a coating, the situation is more complicated. It is important to provide a good adhesion of the organic coating by means of a conversion pretreatment. A good conversion pretreatment has to form chemically bonded layers that are insoluble in water under changing pH conditions. Under a mild atmosphere, the reduction of oxygen tends to result in an increase of the pH, as hydroxyl (OH^-) ions are released in the course of reaction (Equation 2.2). While these ions may be consumed again by ongoing precipitation of oxides, locally and temporarily high alkalinity may occur.

The long-lasting corrosion resistance provided by chromate passivation is also due to the so-called self-healing effect. This is usually attributed to residual chromate ions that are incorporated in the oxide layer. In case of mechanical damage of the passivation layer that penetrates through the underlying substrate metal these chromate ions are available for a quick passivation reaction at the newly exposed metallic location. Other “active anions” that may also serve as strong oxidants for metals, like the oxo-anions ferrate, (per)manganate, molybdate, tungstate, or vanadate, are considered as possible direct replacements for chromate [18]. For example, Patent WO 0036182 [19] describes a pretreatment for aluminium and its alloys, based on an alkaline ferrate (VI) solution and additional oxo-anions like molybdate. The patent is valid in several European states, but no commercial application is known yet.

2.2 Design of organic coating systems

2.2.1 Diffusion barrier features - humidity uptake and electrolyte permeation

While cross-linking of a coating renders a macroscopically intact and closed network with the required mechanical and chemical features, on a molecular scale, any polymer film will remain penetrable to some extent by gases and electrolytes, due to diffusion and migration. Increasing the film thickness is one possible measure to improve on this situation. For instance, in coil coating, primer coatings are normally used at 5 µm dry-film thickness. For high anticorrosive requirements, e.g. certain appliances and marine architectural applications, coil primer coats of up to 30 µm are common.

The uptake of water or moisture is considered an important feature of organic coatings, as it is linked with changes in the density and dielectric properties of the polymer film. Absorption and incorporation of water molecules will lead to physical swelling (volume expansion) and results in a softening of the coating surface. Moreover, as water is an electrolyte carrier, its presence is related with the ease of dissolved ions to penetrate through a coating, where they may accumulate and, finally, cause degradation of the polymer film and corrosion of the metal substrate [20].

Water uptake is influenced by chemical and physical properties of the coating, and therefore it is determined by the molecular design of the polymer. Typical coating resins like epoxies or polyurethanes display a modest water uptake in humid atmosphere. However, the rate of incorporation is strongly accelerated when approaching the dew point. In principle, polar (hydrophilic) polymers will allow better wetting and hence also a higher water absorption [21].

The elasticity of a coating, on the other hand, plays a role in the diffusion transport of molecules through the interstices of the polymer film, which can be interpreted as the relative mobility of polymer moieties [22]. While this property changes with temperature, usually showing a jolt from the rigid to the elastic regime at the glass transition temperature (T_g), it will also be affected when the film is plasticised by incorporation of water [23]. The design of the polymer therefore includes the task of reconciling disparate features like flexibility (formability) vs. curing temperature vs. diffusion barrier properties, or water miscibility vs. chemical resistance vs. adhesion vs. wetting.

Another strategy involves the use of nanoscale particles like titanium dioxide (TiO_2) preparations, or natural or synthetic mineral pigments like clays and hydrotalcites. The latter two contain agglomerates of nanosized crystalline platelets that can be separated from the agglomerate and distributed in the bulk coating. The idea is that these platelets can form an additional barrier to penetrating media, as they could extend the diffusion path for any corrosive ion or media through the coating, cf. Figure 2.3. The observation of the expected effect was reported for oxygen, carbon dioxide and water vapour [24].

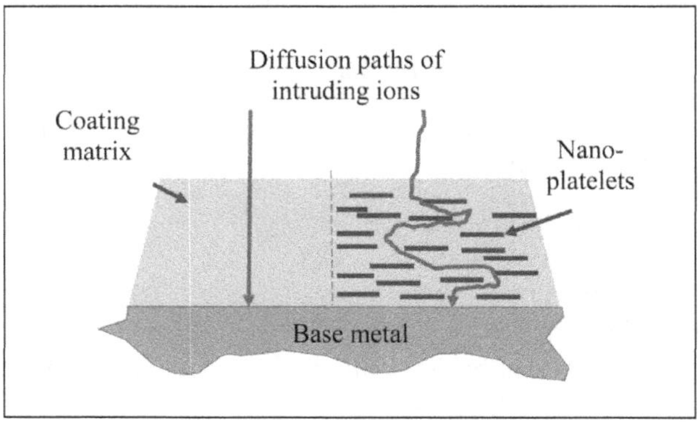

Figure 2.3: Schematic view of the extension of the diffusion path by a nano-platelet filled coating; according to Lewis [1]

2.2.2 Active pigments

Additives are also used to obtain a higher anticorrosive performance. Often active pigments are incorporated that interfere with penetrating corrosive agents (cf. also Chapter 4.3). Here again, chromates, in particular strontium chromate, and dichromates have been used, but due to their toxicity and pollution control requirements, are being replaced by successors that are free from hexavalent chromium (chromate) or entirely chromium-free, e.g. zinc phosphates.

Mineral or synthetically generated particles like hydrotalcites or zeolites can also be designed as carriers to host e.g. anticorrosive ions like molybdates or vanadates. Such carrier concepts are considered to be particularly useful, because they enable retarded release of the active agents that thus remain available for longer periods, instead of quickly bleeding out of a coating, depriving it of its properties. The release might even be controlled by the conditions of the surrounding medium, e.g. pH value, temperature or the presence of (corrosive) counter ions, so that the active ions are only available on demand (cf. Chapter 6.5.5) [25].

Other suitable anticorrosive additives comprise ion-exchange pigments like calcium-modified silicates that release calcium ions (Ca^{2+}) in the presence of acids [26] thereby reducing the corrosivity of the electrolyte, and precipitating corrosive ions like chlorides or sulphates. A number of products are commercially available.

Condensed conductive polymers like polyaniline (PAni) have been around for a number of years [27]. They have been shown to suppress delamination on steel. The mechanism of PAni is attributed mainly to an increase of the electrochemical potential of the substrate (ennoblement) thereby obstructing oxygen reduction [28]. PAni is rendered conductive by the formation of its protonated salts, and the reverse deprotonation process is triggered by the potential decrease during the delamination reaction. If counter-anions were suitably selected as to inhibit corrosion, they might be truly released on demand with the onset of a corrosion reaction. However, a controversial discussion is going on around this point. It has been commented that PAni can be useful when added in low amounts, but can be detrimental to corrosion protection in other conditions [29-31].

2.3 Function of individual coating layers

Industrial organic coating systems consist of a conversion coating and a single- or multi-layered paint coating. Any coating is designed to combine a set of varying features appropriate for the final use [32, 33]. In many cases, two or three-coat paint systems are used that comprise a primer layer, an optional intermediate coating and a topcoat. In the manufacture of cars, multi-coat systems are used. An electrocoat paint is applied as primer, followed by a filler, then the coloured basecoat and finally a clearcoat.

Features provided by an organic coating system on metals can be summarised as follows (cf. also Figure 2.4)

- Colour, gloss, structure
- Corrosion resistance
- Chemical resistance
- Adhesion
  - Substrate adhesion (fingerprints, condensation, etc.)
  - Intercoat adhesion
- Elasticity, formability
- Wear and scratch resistance

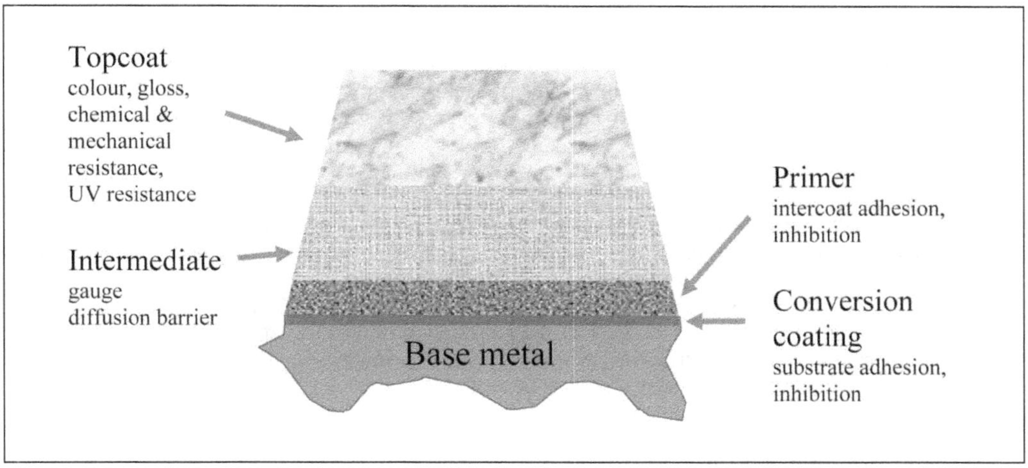

Figure 2.4: Schematic of a typical multi-coat paint system and functions of individual layers

- Soil and stain resistance
- Heat resistance
- UV resistance

The conversion coating is necessary to obtain the primary adhesion to the substrate and a sufficient electronic anticorrosive barrier (cf. Chapters 2.1 and 8). Usually, the primer provides an additional part of the corrosion resistance and flexibility, while the topcoat is used to give aesthetic appearance (colour, gloss) and mechanical and chemical resistance.

Primers therefore form a physical barrier that prevents access of moisture, water, electrolytes and reactive gases (e.g. oxygen) to the metal surface. This is achieved by choice of the resin chemistry. Resins that are commonly used in primers comprise acrylates, polyesters, epoxies and polyurethanes. For example, acrylates are used in primers for polyvinyl chloride (PVC) plastisol coatings. Polyesters usually display good adhesion to the substrate, and provide good corrosion resistance. Polyurethanes are known for their flexibility that allows deformation, resistance against mechanical wear, and a low sensitivity towards thermal stress, particularly by low temperatures in the environment. Chemical resistance is brought on by cross-linking of single polymer strands, which is also enabled by the suitable formulation of the paint.

Intermediate coatings are used to build up film thickness, and mediate the distribution of mechanical stress. They often contain platelet pigments, like micaceous iron oxide, in order to extend the diffusion path for intruding corrosive species (cf. Chapter 2.2.1).

Applications in particulary demanding corrosive environments may require a topcoat containing such a pigmentation which may restrict the obtainable colour and the level of gloss. It is also important that the topcoat is non-transparent to UV radiation.

2.4 Literature

[1] Lewis, O. D., A Study of the Influence of Nanofiller Additives on the Performance of Waterborne Primer Coatings, PhD Thesis, Loughborough Univ., Loughborough 2008, p. 78

[2] Vogelsang, J. A., Selbstheilende Beschichtungssysteme - Ein Überblick (Self-Healing Coating Systems...), Schiff und Hafen (8) 2009, pp. 28 ff

[3] Barrow, G. M., Physikalische Chemie (Physical Chemistry), vol. 3, 3rd ed., Bohmann-Vieweg, Vienna 1977, pp. 203 ff

[4] Meuthen, B., Jandel, A.-S., Coil Coating, 2nd ed., Vieweg, Wiesbaden 2008, p. 89 ff

[5] Fafilek, G., Kronberger, H., Thermodynamik, metallische Werkstoffe und Korrosion (Thermodynamics, Metallic Construction Materials and Corrosion), Experimental Exercises Script, Exercise TK, Inst. Chem. Technologies and Analytics, TU Vienna 2005, p. 6, www.tuwien.ac.at/echem/education/158076/pcvt_lu_TK_skriptum.pdf (15-03-2010; 18:42 h)
[6] Giza, M., In-situ Spectroscopic and Kelvin Probe Studies of the Modification of Solid Surfaces in Low Temperature Plasmas, Diss., University of Paderborn, 2008
[7] Hausbrand, R., Stratmann, M., Rohwerder, M., Corrosion of Zinc-Magnesium Coatings: Mechanism of Paint Delamination, Corr. Sci. 51, 2009, pp. 2107 ff
[8] Mahajanam, S. V. P., Buchheit, R. G., Characterization of Inhibitor Release from Zn-Al-$[V_{10}O_{28}]^{6-}$ Hydrotalcite Pigments and Corrosion Protection from Hydrotalcite-Pigmented Epoxy Coatings, Corrosion 64, 2008, pp. 230 ff
[9] Crass, G., Janocha, S., Bothe, L. (inv.), Biaxially oriented multilayered polyolefin film which is printable and of which both sides are sealable, its production and use, European Patent EP 0263963, Hoechst AG 1987
[10] Meyer, K., Schutz von Stahlkonstruktionen, Beispiele aus der Eisenindustrie und dem Bergbau (Protection of Steel Constructions…), Fette, Seifen, Anstrichmittel 69, 1967, pp. 90 ff
[11] Schröder, T., Erfahrungen mit Schutzbeschichtungen an Stahlwasserbauten der Wasser- und Schiffahrtsverwaltung im Bereich der Nord- und Ostsee (Experiences with Protective Coatings on Hydraulic Structures…), Materials and Corrosion 23, 1972, pp. 993 ff
[12] anon., Kanalüberführung Leine (Canal Flyover Leine River), Wasser- und Schifffahrtsverwaltung des Bundes - Neubauamt Hannover (Federal Waterways and Shipping Administration), Hannover 2009; www.nba-hannover.wsv.de/baumassnahmen/abgeschlossene_baumassnahmen/neubau_leinequerung.html (15-03-2010; 18:45 h)
[13] Vogelsang [2] p. 30
[14] Schütz, A., Kaiser, W.-D., Kein Rasten gegen das Rosten (No Resting versus Rusting), FARBE UND LACK 110, 2004, pp. 26 ff
[15] Meuthen, Jandel [4] p. 75 ff
[16] Schinzel, M., Advanced Corrosion Protection of Automotive Body Sheet - A Challenge for Coil Coating, ECCA Autumn Congress Brussels, Proc., European Coil Coating Association, Brussels 2009
[17] Weast, R. C., Astle, M. J. (ed.), Handbook of Chemistry and Physics, 60th ed., CRC Press, Boca Raton 1981, p. D-155
[18] Kendig, M. W., Buchheit, R. G., Corrosion Inhibition of Aluminum and Aluminum Alloys by Soluble Chromates, Chromate Coatings, and Chromate-Free Coatings, Corrosion 59, 2003, pp. 379 ff
[19] Minevski, Z., Eylem, C., Maxey, J. (inv.), Ferrate Conversion Coatings for Metal Substrates, WO 0036182, Lynntech Inc., 2001
[20] Goldschmidt, A., Streitberger, H.-J., Basics of Coating Technology, 2nd Edition, BASF Coatings AG, Münster 2007, p. 447
[21] Öchsner, W. P., Bergk, B., Fischer, E., Gaszner, K., Sorptionsisothermen für Wasser in organischen Beschichtungen und deren Einfluss auf die Beschichtungseigenschaften (Sorption Isotherms for Water in Organic Coatings…), FARBE UND LACK 111, 2005, pp. 42 ff
[22] Lewis [1] p. 87
[23] Goldschmidt, Streitberger [20] p. 401
[24] Lewis [1] p. 34
[25] Mahajanam, Buchheit [8] p. 231
[26] Vogelsang [2] p. 29
[27] Wessling, B., Corrosion prevention with an organic metal (polyaniline): Surface ennobling, passivation, corrosion test results, Materials and Corrosion 47, 1996, pp. 439 ff
[28] Holness, R. J., Williams, G, Worsley, D. A., McMurray, H. N., Polyaniline Inhibition of Corrosion-Driven Organic Coating Cathodic Delamination on Iron, J. Electrochem. Soc. 152, 2005, pp. B73 ff
[29] Rohwerder, M., Intelligent Corrosion Protection by Conducting Polymers, Smart Coatings II, ACS Symposium Ser. 1002, Am. Chem. Soc. 2009, pp. 274 ff
[30] Meine, D., Korrosionsschutz der Zukunft? (Future Corrosion Protection?), Conf. Report, Europ. Coatings. Conf., Vincentz Network 2007, FARBE UND LACK 113, 2007, pp. 33 f
[31] Saji, V. S., Thomas, J., Nanomaterials for corrosion control, Current Science 92, 2007, pp. 51 ff
[32] Kittel, H., Streitberger, H.-J. (ed.), Lehrbuch der Lacke und Beschichtungen (Coursebook of Paints and Coatings), Vol. 6, 2nd ed., Hirzel, Stuttgart 2008, pp. 485 ff
[33] Meuthen, Jandel [4] pp. 51, 79 ff

3 Surface preparation

Jörg Sander

3.1 Industrial cleaning

3.1.1 Importance of cleaning process

Anticorrosive metal coatings require a clean and well-prepared metal surface, if a long-lasting product with an aesthetic appeal and high durability is to be obtained. The efficiency of the surface preparation determines the endurance of any conversion, anodic, galvanic or organic coating between weeks and years. Quite often however, the difference between a clean and a contaminated surface does not become apparent by visual inspection. Thorough monitoring of the cleaning [1, 2] stage is therefore mandatory in order to ensure that both a high-class pretreatment and final coating are achieved.

Mechanical cleaning like brushing, grinding, shot or grit-blasting etc. can serve for a basic level of surface preparation. Much more often, chemical cleaning is a necessity. Several application methods are available for industrial cleaning operations, like immersion, spraying, or vapour cleaning. Prior to painting, metal surfaces are most efficiently and economically cleaned by aqueous spray cleaners, but the choice is made according to the size and shape of the work pieces and the type of contamination.

In many cases, e.g. job-coater facilities, mixed-metal car body manufacture or coil coating, a mix of different substrates must be cleaned and pretreated. Most industrial cleaners therefore must apply on different substrates. Aqueous cleaner solutions are used on steel, galvanised steel, aluminium, but also on die-cast alloys (Al, Zn, Mg), other non-ferrous metals and – in the automotive and appliance industries – prephosphated sheet.

3.1.2 Contaminants

Industrial cleaning deals with a great variety of soils and contaminants.

Rust, scale, residues, e.g. of old paints, galvanising layers, faulty conversion or anodic coatings, usually require mechanical precleaning or pickling.

Oils and fats usually originate from the generation process of the unmanufactured or semi-finished material. They stem from protective oils (temporary corrosion protection, often containing additional organic inhibitors, e.g. triethanol amine) or metal working fluids (drilling, cutting, stamping, drawing oils etc.).

Polishing agents like stearic or paraffinic waxes, usually contain additional grinding compounds like finely dispersed silica, lime, corundum, colloidal clay minerals etc.

Sweat or fingerprints contain fats, fatty acids, proteins etc.

Fines and metal abrasions originate from rolling, grinding and cutting processes, in particular also from forming (deep-drawing, flanging).

Pigment soil, graphite and carbonised oil often appear after assembly processes (grinding, joining, welding) or arise from annealing processes e.g. during coil production.

Jörg Sander et al.: Anticorrosive Coatings

ISBN 978-3-86630-911-1

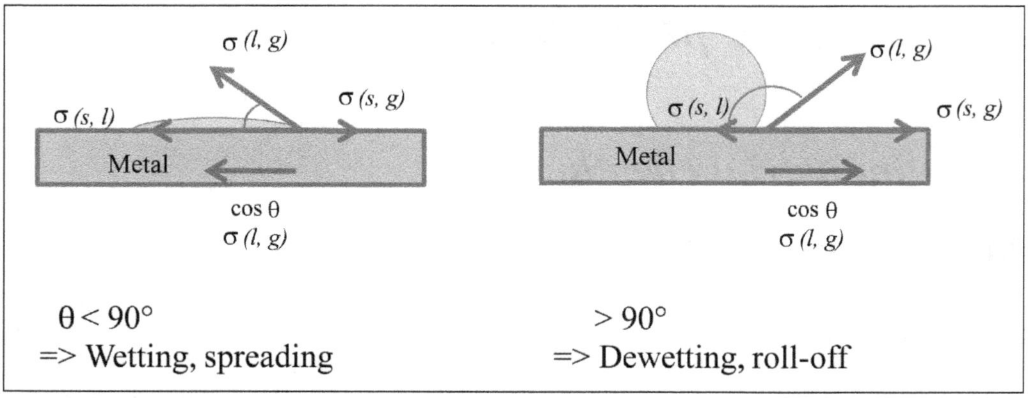

Figure 3.1: Schematic: Sessile drop showing the interfacial tension acting at the wetting front
The angle θ between the tension vectors σ for the solid-liquid (s, l) and the liquid-gas (l, g) interfaces is measured by microscopic inspection.

3.1.3 Surface energy and tension

"Chemical cleanliness" is often mistaken for a water-break-free surface, i.e. the complete wettability of a metal. While water-break-free rinsing gives a rough indication of the presence or absence of oils, fats etc., it depends on the surface tension of contaminating substances (e.g. self-emulsifying oils) and thickness of the water film, and can be affected by evaporation of water or the presence of surfactants. Native oxide layers may not result in particularly poor surface cleanliness, but they usually inhibit any further surface reaction that is necessary to build up a proper conversion coating prior to painting.

Surface tension occurs at any interface between non-miscible phases. It is brought about by differences of attractive forces that apply to molecules inside vs. between phases. This effect rules the form of droplets, micelles and foam bubbles, and it determines the adsorption of gases to solids, as it brings on capillarity phenomena like wetting and spreading of liquids on a metal surface [3]. The wettability of a solid by a liquid is characterised by the contact angle θ that liquid makes on the solid. The mathematical description is given by Young's equation:

Equation 3.1: $$\cos\theta = \frac{\sigma(s,g) - \sigma(s,l)}{\sigma(l,g)}$$

where: Θ is the contact angle of a sessile drop on a solid surface, σ_s represent the tension vectors for the solid-liquid (s, l), solid-gas (s, g) and the liquid-gas (l, g) interfaces (cf. Figure 3.1).

Table 3.1: Contaminants

Oils, fats, waxes	**Fines, abrasives**	**Pigments**
Rolling, casting oils, emulsifiers	Grinding, cutting, tool fines	Graphite, molybdenum disulphide
Corrosion protective oil		Grinding compounds
Drawing, pressing lubricants, soaps		Carbonised oil, welding, brazing residues
Drilling, cutting fluids		
Waxes		
Sweat, fingerprints		

The surface tension of liquids is usually measured by capillarity or sessile drop methods. For the latter, the contact angle, i.e. the angle that is enclosed by the tension vectors for the solid-liquid and liquid-gas interfaces (cf. Figure 3.1), is measured by microscopic inspection. Table 3.2 gives some exemplary values [4]. Contact angles <90° (i.e. cos θ > 0) indicate wetting, angles >90° (cos θ < 0) repulsion situations. Accordingly, good wetting is obtained when the interfacial tension between solid and liquid exceeds the tension at the solid-gas boundary. If the following process requires the surface to be wet by aqueous media, the interfacial tension should exceed 72 mN/m, which is the surface tension of water vs. air [5].

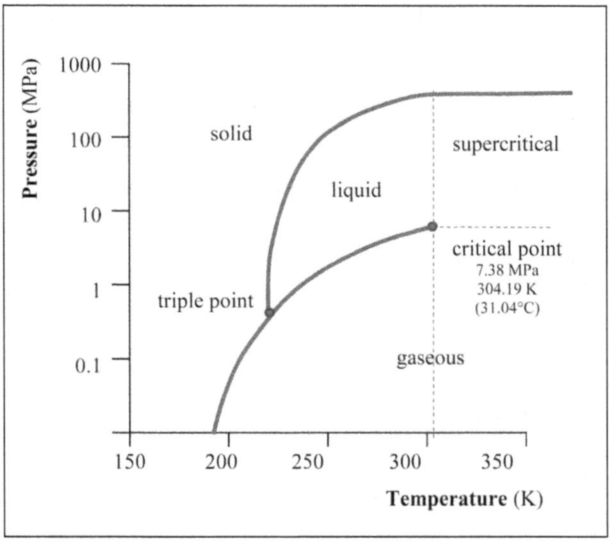

Figure 3.2: Phase diagram for carbon dioxide (CO_2) 0.1 MPa ≈ 1 atmosphere (Illustration according to Kukova [9])

To determine the surface tension of a solid surface, the contact angles of droplets of different liquids with known surface tensions are measured and the resulting overall surface tensions calculated [6]. A simple method for measuring surface tension uses prefabricated test inks [7].

3.2 Mechanical cleaning

Mechanical cleaning is necessary when thick scale or rust layers must be removed. For steel surfaces, this is quite often supported by acid pickling. Mechanical cleaning, however, does not give sufficiently clean surfaces for subsequent galvanising or conversion coating. Apart from brushing or grinding, shot or grit-blasting is often used to remove coarse contaminations like rust, scale, carbonised oil or old paint. Grit-blasting uses various abrasive particles (e.g. sand, slag, corundum, glass or plastic beads) that are shot at the target by pressurised air. For delicate surfaces, dry ice (solid CO_2) has become a familiar abrasive that is used to remove fats and oils, glues, paints and inks, corrosion products etc. Dry ice sublimates, i.e. evaporates directly from solid state, at a temperature of -78 °C at atmospheric pressure. Dry ice blasting therefore generates less waste than other grit-blasting processes. Figure 3.2 depicts the phase diagram of carbon dioxide, showing the solid, liquid and gaseous states [8, 9].

Table 3.2: Surface tension of liquids at different temperatures (in mN/m)

Liquid	20 °C	60 °C	100 °C
H_2O	72.25	66.18	58.85
C_2H_5OH	22.3	22.3	19.0
C_6H_6	28.9	23.7	
$(C_2H_5)_2O$	17.0		80.0
Hg	480[a)]		
Ag	800[b)]		
NaCl	94[c)]		
AgCl	125[d)]		

a) at 0°C; b) molten at 970°C; c) molten at 1080°C; d) molten at 452°C; cf. Barrow [4]

In cases when larger engineered constructions have to be cleaned on the site of their erection, the ultra-high pressure water jet technique has gained importance, to avoid dust and soil in the vicinity of the construction ground. Water pressures of 1,700 to 2,100 bar are sufficient to remove non-adhering rust and salt contaminations [10, 11].

Some discussion has been focused on supercritical CO_2 ($scCO_2$). The supercritical state is achieved when a real gas is heated and pressurised beyond its critical temperature and pressure. For CO_2, the respective data are 31.04 °C and 7.38 MPa. Beyond this point, there is no physical difference between the liquid and gaseous form, and gas obtains fluid and solvent features that enable its use as an efficient extractant. $scCO_2$ is used for high purity extraction of natural substances (e.g. decaffeination) and other food processing applications[12]. Further applications of $scCO_2$ comprise extractions of fragrances [13], synthesis and polymer production and processing (octene hydroformylation, fluorinated polymer and polycarbonate production) and powder production. $scCO_2$ has some potential use as fluid for cleaning of small metallic work pieces that can be handled in an enclosed, high-pressure cleaning tank[14].

Finally, ultrasonic cleaning [15] employs the conversion of DC voltage into mechanical waves above the audible frequency range (>20 kHz) by a piezo-electric generator. The vibrations create local vacua and overpressure in rapid succession (cavitation) which literally blast off soil from a contaminated surface. Ultrasonic cleaning achieves a high degree of efficiency at a very fast rate. It has been used on a wide variety of work pieces of different materials like glass, ceramics, or metal, regardless of dimension and shape.

3.3 Chemical cleaning

3.3.1 Plasma and corona processes

For finer cleaning that achieves a higher degree of cleanliness, plasma [16] treatments have gained interest. Plasma is generated by ionisation of vapour, and is sometimes called the fourth aggregate state of matter. Being accelerated against a solid surface, the energy of the plasma is transferred to this surface and contaminants are chemically destroyed and removed. Plasma cleaning is now available in atmospheric operations, saving expensive vacuum technique that used to be necessary. The equipment that is available nowadays can operate at a speed of several hundreds of mm/min, and plasma source arrays can span up to two metres width.

Atmospheric zero-potential plasma cleaning is technically used for applications where composite materials made from plastics, metal, glass and even ceramics are to be treated. The plasma beam makes no distinction between materials. Plasma cleaning has proven sufficient for removal of neat oil residues in stamping of painted lids on jam jars or air conditioning equipment parts in cars and trucks. The treated surfaces are sufficiently susceptible for labelling adhesives or printing inks. Applications on bare metal surfaces include a small aluminium coil coating operation and riveting in aircraft industry, where former solvent cleaning processes could be successfully replaced [17].

Plasma cleaning has been combined with deposition of materials from the gas or plasma phase, to form thin protective, in particular soil-repellent coatings [18]. High-voltage corona discharge treatment is commonly used to clean and activate plastics surfaces, especially of polyalkylenes [19]. It has also been used to clean and deposit thin coatings on metal [20].

3.3.2 *Solvent cleaning*

Fatty, oily and wax contaminations can be removed by immersion of work pieces into organic solvents or by degreasing in condensing solvent vapour. Halogenated hydrocarbons have been used to this end due to their inflammability and solubilising power towards most fatty contaminations. To remove pigment and soils, solvent cleaning must be supported by mechanical means, e.g. brushing or ultrasonic treatment. Soil that adheres by chemisorption, oxides and metal soaps cannot be removed by solvent cleaning [21]. Furthermore, the use of solvents is subject to legal regulation, i.e. the EU Directive 2008/1/EC [22], commonly referred to as the Integrated Pollution Prevention and Control (IPPC) Directive. This directive and related regulations, e.g. on emission of volatile organic compounds (VOCs) [23], restrict emission levels of volatiles and the use of certain solvents that pose a health hazard, i.e. due to carcinogenity or mutagenity. Hermetically closed cleaning systems are therefore mandatory.

3.3.3 *Chemistry of aqueous cleaners*

3.3.3.1 *Mechanism: alkalinity, saponification and metal dissolution*

In most metal coating lines, aqueous immersion or spray processes are used to clean the metal surface, and remove oil, solid contaminations and superficial scale. Typical aqueous cleaners contain alkalis for saponification of fats and oils, and for pickling of the metal surface, builders (e.g. phosphate, silicate) to disperse solid dirt particles in the solution after their removal from the surface, and surfactants to quickly wet the metal surface and to emulsify fatty and oily contaminations in a cleaner bath. Moreover, additives like defoamers etc. may be present.

The alkalinity of the cleaner is selected according to the degree of ageing and contamination of the feedstock. The alkalinity will change during the course of reaction, mainly due to consumption of the alkali by saponification and metal dissolution. However, ageing effects caused by entrapment of carbon dioxide (carbonation, see Equation 3.2) from the air must also be considered, as they leave less “free” alkalinity available for the main reaction paths (cf. Equations 3.3 to 3.8). The alkalinity therefore must be monitored manually or automatically, preferably by titration methods.

Equation 3.2: $2\ OH^- + CO_2 \rightarrow (CO_3)^{2-} + 2\ H_2O$ Carbonation reaction

The reactions of the alkaline cleaning process on galvanised surfaces were studied using an electro-chemical quartz crystal micro-balance (ECQM) array [24]. The principle of this method is that the oscillation of a piezo-electric quartz wafer is proportional to the wafer’s thickness. A galvanic-coated quartz wafer will therefore change its oscillating frequency in response to any chemical reaction that takes place on its galvanised surface.

Dissolution results in reduction of thickness, hence accelerated oscillation, while coating formation is supposed to cause an increase of the wafer thickness, thereby slowing down its motion. The sensitivity of the method allows monitoring of thickness changes on the atomic scale, so that molecular layers can be detected while forming, recording mass and charge simultaneously and in-situ. The schematic presentation of the experimental set-up of the ECQM for registration is given in Figure 3.3. Experimental results were obtained with a zinc-coated, special quartz crystal, that was mounted in a flow cell and then subjected to cleaner solutions at various flow rates, temperatures etc. Two different reactions can be identified when this experiment is carried out in different atmospheres, i.e.

Stage A: Dissolution of the native oxide/carbonate/hydrate layer (chemical)

Equation 3.3:	$ZnO + 2\ OH^- + H_2O$	→	$[Zn\ (OH)_4]^{2-}$
Equation 3.4:	$Zn\ (OH)_2 + 2\ OH^-$	→	$[Zn\ (OH)_4]^{2-}$
Equation 3.5:	$Zn_5\ (OH)_6\ (CO_3)_2 + 14\ OH^-$	→	$5\ [Zn\ (OH)_4]^{2-} + 2\ (CO_3)^{2-}$

Stage B: Dissolution of metallic zinc and zinc hydroxide layer formation (electrochemical)

Equation 3.6:	$Zn + 4\ OH^-$	→	$[Zn\ (OH)_4]^{2-} + 2\ e^-$	Anodic reaction
Equation 3.7:	$O_2 + 2\ H_2O + 4\ e^-$	→	$4\ (OH)^-$	Cathodic reaction

In an aerated solution, both electrochemical reactions are possible. However, in the absence of oxygen, the cathodic reaction is inhibited, so that only chemical dissolution is observed. While the chemical reaction is largely unaffected by process conditions, Zn dissolution is also profoundly influenced by aeration, alkalinity, "age" of the cleaning solution and kinetic effects. Bath ageing thus can be interpreted as the build-up of dissolved zinc ions and carbon dioxide absorption from the air. The results are illustrated in Figures 3.4 and 3.5.

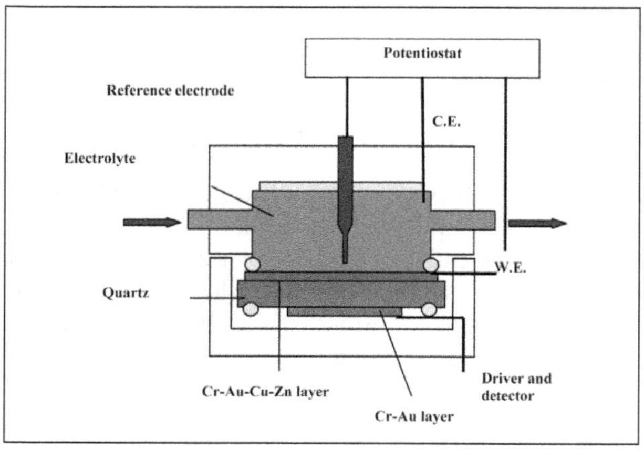

Figure 3.3: Experimental set-up: electro-chemical quartz micro-balance (ECQM). Source: [24]
W.E.: Working electrode; C.E.: Counter electrode

On aluminium surfaces, similar reactions take place. For this substrate, an acidic rinse is recommended after alkaline cleaning, to remove compounds of alloying elements that are otherwise inso-

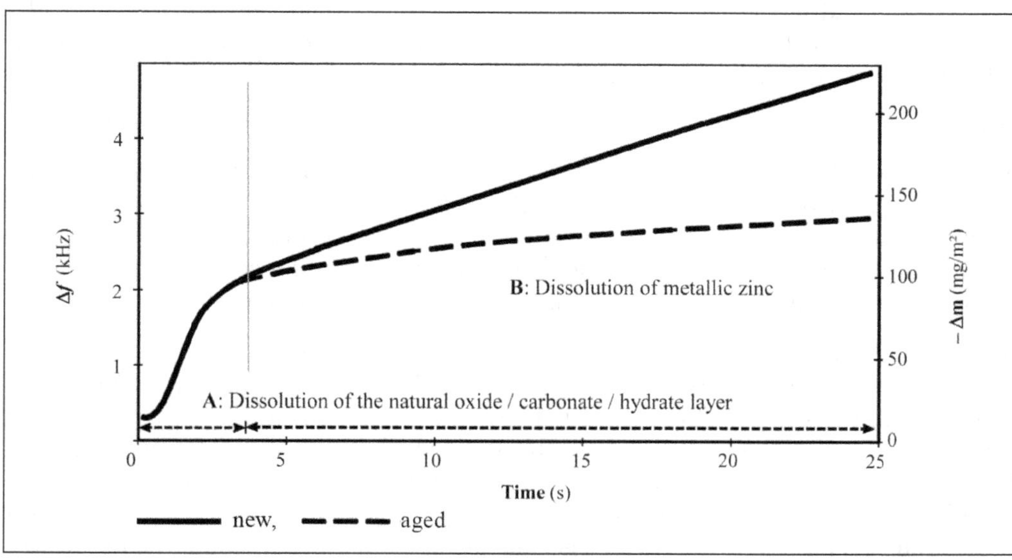

Figure 3.4: ECQM simulation: dissolution of zinc surfaces in alkaline cleaner solution with increasing age. *Source:* [24]

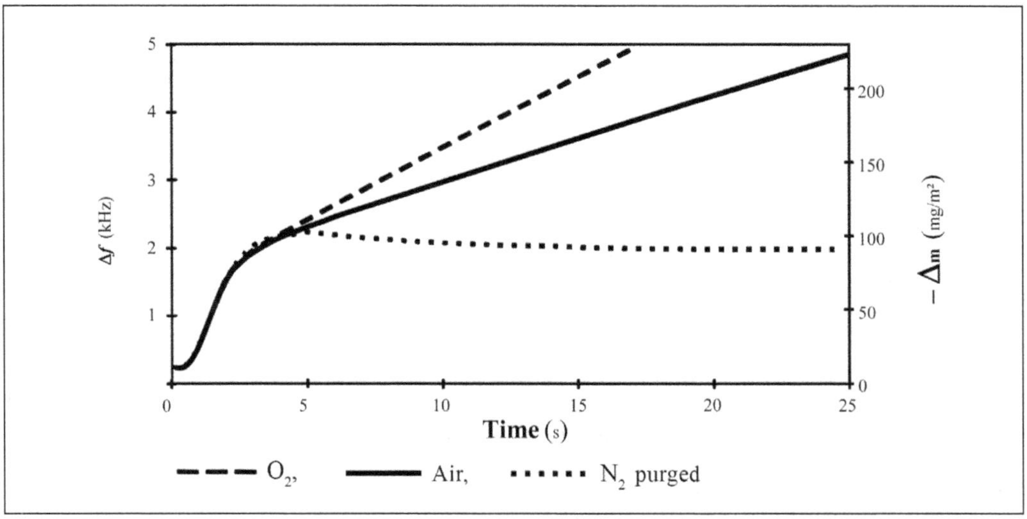

Figure 3.5: ECQM simulation: dissolution of zinc surfaces in alkaline cleaner solution with different aeration. *Source: [24]*

luble. As aluminium is amphoteric, which means it is also dissolved in acids, acid cleaning is common practice whenever a low contamination level allows (coil, can).

3.3.3.2 *Ingredients of aqueous cleaners*

3.3.3.2.1 *General considerations*

Major functions of an aqueous cleaner [25] comprise its ability to emulsify liquid contaminations, deflocculate solid soils and prevent their redeposition. These features are ruled by the interfacial tension between solution and soil, thus directly depending on the wetting power of the solution.

In general, fats, oils and waxes can be saponified, and the soap that is generated acts as a surfactant and emulsifier. Most industrial cleaners therefore are alkaline, in order to bring on the required saponification reaction (cf. Equation 3.8). Strong alkaline cleaners often contain caustic soda (sodium hydroxide), in case of liquid products also potassium hydroxide due to its higher solubility. To reduce the alkaline attack, carbonate may be present in the formulation.

Equation 3.8:

$$\begin{array}{l} H_2C-O-(CO)-(CH_2)_n-CH_3 \\ \quad | \\ HC-O-(CO)-(CH_2)_n-CH_3 \quad + \; 3\,OH^- \\ \quad | \\ H_2C-O-(CO)-(CH_2)_n-CH_3 \end{array}$$

Triglyceride (fat) + alkali

$$\rightarrow \begin{array}{l} H_2C-OH \\ \quad | \\ HC-OH \; + \; 3\,H_3C-(CH_2)_n-CO_2^- \\ \quad | \\ H_2C-OH \end{array}$$

Glycerol + carboxylate (soap)

Especially when used for metal mixes, alkaline cleaners may also contain silicates which display a high alkalinity themselves, however form a thin silica layer, once the bare metal is exposed, thereby inhibiting excessive pickling reaction.

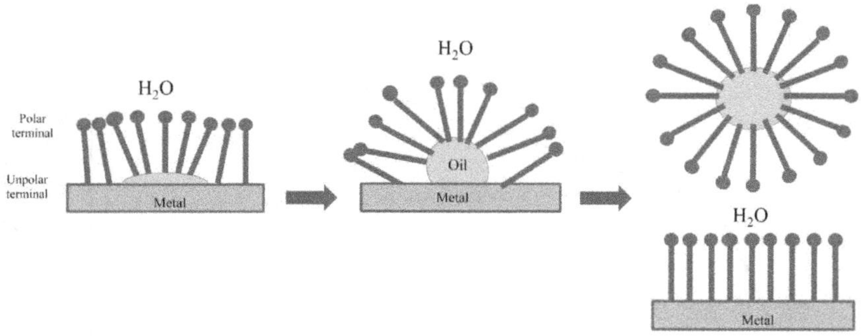

Figure 3.6: Schematic: removal of an oil droplet by surfactant action

For wrought and rolled alloyed aluminium (Mg, Mn containing alloys), alkaline cleaners usually contain complexants to help solubilise these elements. However, an acid rinse is often necessary to remove alkali-insoluble alloying elements and their oxides and hydroxides (e.g. MnO). In alkaline and acidic aqueous cleaner formulations, the task of dispersing solid contaminants is frequently taken by builder ingredients, i.e. salts. To remove fat, oil, and often entrapped solid contaminations from the metal surface, the aqueous products also contain surface-active substances (surfactants). Table 3.3 summarises the common ingredients of aqueous cleaners.

Table 3.3: Ingredients of industrial cleaners

Acidic cleaners	Alkaline cleaners	Neutral cleaners
Phosphoric acid, dihydrogen phosphate (Al also: sulphuric acid)	Alkali hydroxides	Surfactants, solubilisers, inhibitors
	Phosphates, carbonates	
Fluorides, fluoro complexes	Silicates	
Surfactants, solubilisers	Surfactants, solubilisers	
	Complexants, sequestrants	
	Inhibitors	

3.3.3.2.2 Surfactants

Surfactants act by virtue of their amphiphilic nature, i.e. they assemble at the interface of non-miscible phases, self-aligning according to the polarity of the two phases and to their own dipolar character. They can displace oily matter (and any solid that is entrapped in it) from a metal surface. Figure 3.6 illustrates this process.

The amphiphilic character of surfactants is caused by the molecular structure, having a polar moiety bonded to an unpolar, often longer hydrocarbon chain. Depending on the type of polar functionality, surfactants are classified as anionic, cationic or non-ionic. Figure 3.7 illustrates examples for the three surfactant classes.

Important anionic surfactant groups are salts of fatty acids (soaps), sulphuric acid semi-esters of long-chain fatty alcohols or alkyl benzene sulphonates. Historically, tetrapropylene benzene sulphonate (TPBS) played a particularly dominant role as a widely used surfactant in detergents. However, TPBS had to be replaced, as it was particularly poorly degradable in the environment. Cationic species are e.g. quaternary alkyl group containing ammonium bases. Typical representatives of the non-ionic classes are polyethoxylated alcohols or esters.

Anionic Surfactants	
$C_{17}H_{35} - CO_2^-$	carboxylate (soap)
$H - (CH_2CH[CH_3])_4 - C_6H_4 - SO_3^-$	alkyl benzene sulphonate
Cationic Surfactants	
$C_nH_{2n+1} - NH^+([C_2H_4O]_mH)_2$	ethoxylated ammonium salt
$C_nH_{2n+1} - CO - N(C_2H_4{-}NR_3^+)_2$	quaternary ammonium base
Non-ionic Surfactants	
$C_nH_{2n+1} - CO - O(C_2H_4O)_mH$	ethoxylated ester
$C_nH_{2n+1}O - (C_2H_4O)_mH$	ethoxylated fatty alcohol

Figure 3.7: Exemplary formulas of surfactants

A lot of interest has been focused on environmental effects of surfactants, since, being surface active substances they interfere with biological reactions which may make them toxic to aquatic species. In terms of waste water treatment in public sewage works, it is most important that surfactants will not be harmful towards aerobic bacteria, and neither are their metabolic successors that are formed by biodegradation reactions. According to OECD Screening Test specification of 1976, a surfactant is considered to be biodegradable if its concentration is reduced by at least 80 % during an incubation test with a small amount of sewage sludge within 19 days or less [26]. This requirement is codified by European law, for example the German Surfactant Directive (Tensidverordnung, version of June 1986) [27].

For domestic and industrial detergents, it is required for a surfactant to be even "readily biodegradable". Since the OECD method involves detection of anionic and non-ionic surfactants by their specific reactions with methylene blue or bismuth iodide, which may not be given by metabolites, more stringent requirements consider the total chemical oxygen demand (COD), as an appropriate indicator. Surfactants that pass the 60 % reduction level within 28 days are considered as "readily biodegradable" [28]. Other accepted tests are codified by EU Directives and Regulations, like the Dissolved Organic Carbon Test (DOC), Closed Bottle Test, or the Modified OECD Screening Test - DOC Analysis, and related ISO norms [29, 30].

By virtue of the surfactants features, the removed dirt particles are also reliably kept away from the substrate surface. In immersion cleaners, these dirt particles are often distributed homogeneously, i.e. emulsified in the diluent. On the contrary, in spray cleaners it is desirable to displace liberated oil droplets as fast as possible from the aqueous phase, to allow their removal from the system by skimming. Both - contradicting - tasks must be performed by suitably chosen surfactant blends in case of combined spray-dip cleaners.

As further important tasks, the surfactant mix must effectively allow the adjustment of the substrate surface wettability for subsequent pretreatment processes, and to control foam formation in case of spray cleaners.

The chemistry of the non-ionic surfactant class is particularly well suited to match the tasks described above. Molecules of such surfactants comprise an organic water-repellent part, to which a polar, hydrophilic non-ionic group is affixed. This polar group is usually formed by polyethylene oxide/polypropylene oxide chains of different length.

Any surfactant blend must match the physico-chemical features required for its intended use. As a rule, foaming behaviour, emulsifying power, wetting and detergency of a surfactant blend are laboratory tested in combination with the entire builder composition.

Finally, practical trialling is indispensable. Particularly with respect to their foaming characteristics, many factors that influence cleaning practice cannot be simulated by experiment. This includes bath volumes, type and array of spray nozzles, deflector sheets, power data and arrangement of pumps, as well as chemical parameters like water hardness, type and amount of contamination on work pieces and other sources of drag-in to the cleaning bath.

3.3.4 Physics of aqueous cleaning, bath life and rinsing

The action of cleaners is substantially affected by the availability of sufficient amounts of cleaner ingredients at any time, at the metal surface as the point of cleaning. As with any chemical reaction taking place at an interface, the cleaning action is decelerated when there is no relative motion between the work piece and the solution. The reaction becomes diffusion controlled, i.e. the ingredients are only transported to the point of reaction by random motion of molecules. At a given temperature, this means that the reaction speed cannot exceed a maximum value that is specific to the substance.

Apart from heating, acceleration is only possible by inducing relative motion. Whilst with spray cleaners, the spray jet impacting on the work piece brings on sufficient motion in the reaction zone (therefore, in principle, spray cleaners can be used at lower concentrations), with immersion cleaners, extra measures must be taken. Relative motion between work piece and solution can be induced by moving treated items (rotating, oscillating or pitch installations), or by bath agitation (stirring, convection, circulation, air or steam injection).

Typical conditions for spray cleaning are cleaner concentrations of 0.5 to 1 %, pressures of 1 to 2 bar and temperatures of 50 to 70 °C that are applied during a period of usually 5 to 60 s. Following the cleaning stage, excess cleaner solution, soil and reaction products are removed from the metal surface in a multi-stage rinsing section. The rinse stages are operated in a counter-flow cascade to achieve a most efficient use of water. The efficiency of the rinse increases exponentially with the number of rinse baths. The final rinse is often fed with deionised water (DI). A DI rinse is mandatory when a so-called no-rinse pretreatment is used (cf. Chapter 8), to avoid high electrolyte concentration and contamination by unwanted, potentially corrosive ions. Usually a maximum acceptable electrolyte concentration is specified for the last rinse, given in units of electrical conductivity. Recommended values range between 30 and 100 μS at room temperature. To minimise drag-out, the liquid film is generally removed in between individual stages, the drip-off supported by blowers, squeeze rollers etc., depending on the work piece geometry. For instance in coil operations, the loss of liquid can be reduced to 2 to 5 ml/m^2 with properly maintained squeeze rollers. In immersion processes, drag-out can be reduced by hot water spray nozzles that are mounted on both tank sides above the bath surface. It is particularly advantageous to operate those nozzles by valve control, spraying each rack with fresh hot water when emerging from the bath liquid.

However, a certain drag-out is often desirable to control the contamination level, and extend the bath service life. Independently, each tank should be equipped with an overflow sewer.

Increasing energy costs are also influential in cleaning as they are in other fields of the metal finishing process. Changing the process to aqueous, alkaline low-temperature cleaners is facilitated by use of optimised and increased surfactant contents, combined with highly efficient dispersants. Thus, the working temperature can be reduced to 30 °C under appropriate conditions. However, higher temperatures (50 to 60 °C) still apply when highly viscous fats and oils, pastes, and aged, hardened soils must be removed. The cleaner must never dry upon a work piece, so that long transfer times must be avoided.

Table 3.4: Typical data for drag-out and required dilution (of initial concentration)
(a) Drag-out *(b) Required dilution*

Shape	Drag-out (ml/m²)	Process	Dilution
Flat, single	100	Cleaning, hot	35 to 100
Flat, strip	10	electrolytic	100 to 200
Corrugated	200	Acidic rinse	100 to 200
Bowl-shaped	1000	Caustic pickle (Al)	500 to 1000
		Galvanising	1000 to 100000[a)]
		Pretreatment	2000 to 5000

a) High figures apply for environmentally critical elements like nickel and chromium

The increase of service life of a cleaner bath by overflow operation can be appreciated by simple model calculations: The result of such a calculation is that a cleaner bath can achieve substantially longer service life when, at a given set of equipment and operational conditions, a part of the cleaner solution (overflow) is continually discarded and replenished. The bath life becomes shorter when the overflow amount falls below the calculated level. The required overflow rate is dependent on the ratio between drag-in and the acceptable maximum concentration of the contaminant that reduces bath life. In practice, of course, several contaminants must be considered simultaneously.

The principle applies for any type of process where residual drag-out material must be reduced by a rinsing operation. Tables 3.4 and 3.5 show some considerations about expected drag-outs, as well as required dilution ratios for various work piece geometry and treatment processes [31]. Operational savings in a cleaning process with an overflow system do not only substantiate in lower water demand, but in reduced downtime otherwise necessary for discontinuous dumping and make-up of the bath. Overflow operation is therefore a mandatory measure when working with large bath volumes.

Table 3.5: Rinse water consumption
Required rinse water volume (l/h) per litre of initial electrolyte drag-in

Required dilution	10,000	5,000	1,000	200
Rinse tanks				
1	10,000	5,000	1,000	200
2	100	71	32	14
3	22	17	10	6
4	10	8	6	4
5	6	5	4	3

The most efficient use of rinse water can be achieved when rinsing is performed via a counter-flow bath cascade. This principle can be employed regardless of whether immersion or spray technique is used. When a number of similar bath tanks are connected in-line through a pump or overflow system, and fresh water is fed (preferably via a separate spray riser) into the final tank, the concentration of rinsed-off substances (excess cleaner, by-products, contaminants) is reduced drastically from bath to bath along the line. The last tank contains virtually fresh water. The fresh water demand is calculated according to equation 3.9 [32].

Equation 3.9:
$$V = \sqrt[N]{[\text{requ. dilution}]} \times [\text{throughput}] \times [\text{drag} - \text{out}]$$

where: N is the number of rinse tanks employed, the throughput is given in m² surface area and drag-out is given in l/m²

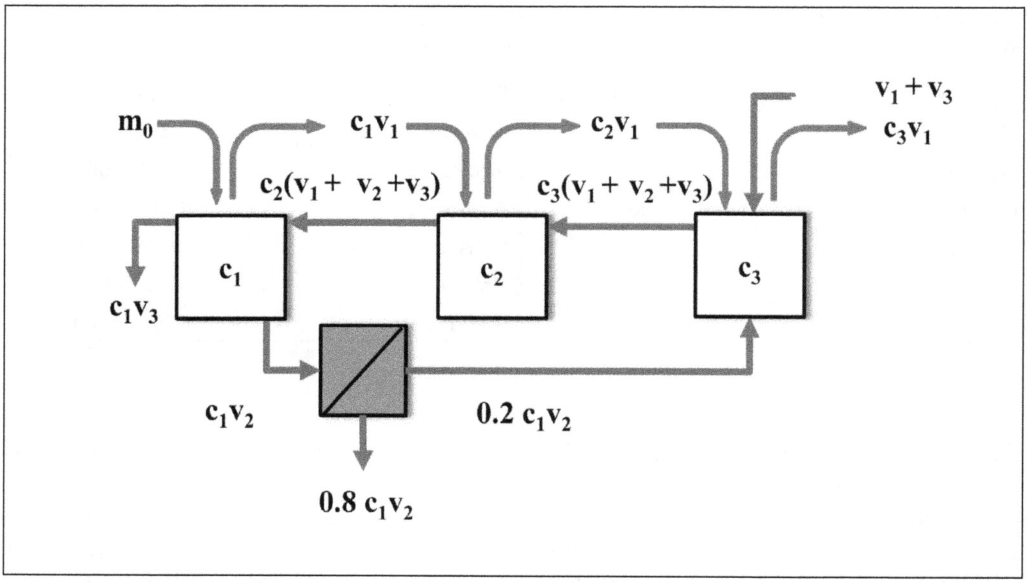

Figure 3.8: Schematic for a counterflow rinse cascade with additional filter system (efficiency 80 %) for the reduction of contaminants from a circulated rinse bath; m_0 = contaminant mass flow; v_{1-3} = drag-out and overflow volumes; c_{1-3} = local instantaneous concentrations of contaminant in tanks 1 to 3

The required number of tanks depends on the ratio between the amount of substances dragged in and the counter-flowing water volume (stationary state). A tank number of three are considered state-of-the-art according to the European Reference Documents on Best-available Techniques (BREF Standards) [33]. Table 3.5 also shows that the economic gain with higher numbers of tanks is only small. After some time, the first rinse bath may obtain a similar composition as the cleaner bath itself, and can be used to replenish the working bath. Preferably, this first rinse water is continuously deoiled (by demulsification, skimming, and ultra-filtration) and reused to replenish the main bath, together with an appropriate builder/surfactant package. Figure 3.8 shows a schematic for a cascade and bath maintenance layout.

3.4 Literature

[1] Sander, J., Praktische Fragestellungen beim Einsatz technischer Reinigungsmittel auf Metalloberflächen (Practical Considerations Regarding the Use of Industrial Cleaners...), Seminar Proc., Techn. Akad. Wuppertal, 1988

[2] Wichelhaus, W., Buetfering, L., Reinigung und chemische Vorbehandlung von Metallen (Cleaning and Chemical Pretreatment...), Galvanotechnik 103, 2005, pp. 2712 ff

[3] Asthana, R., Sobczak, N., Wettability, Spreading, and Interfacial Phenomena in High-Temperature Coatings, J. Metals e-Version, 52 (1), 2000, www.tms.org/pubs/journals/JOM/0001/Asthana/Asthana-0001.html (07-06-2010;18:55h)

[4] Barrow, G. M., Physikalische Chemie (Physical Chemistry), vol. 3, 3rd ed., Bohmann-Vieweg, Vienna 1977, p. 74

[5] anon., Plasma technology: Determination and measuring of the surface tension of starting materials, Press Release, Plasmatreat GmbH, Steinhagen 2007; www.plasmatreat.co.uk/measuring_surface_tension_determination.html; (15.03.2010; 18:49 h)

[6] Gerstenberg, K. W., Netzung, Oberflächenenergie und Young'sche Gleichung (Wetting, Surface Energy and Young's Equation), TIGRES Publication 01, Dr. Gerstenberg GmbH, Rellingen 2006, pp. 5 ff; www.tigres.de/publikationen.php; (15-03-2010; 18:50 h)

[7] anon., Test inks: Determination of Surface Tension with the Aid of Test Liquids, Press Release, Plasmatreat GmbH, Steinhagen 2007; www.plasmatreat.co.uk/test_inks_surface_tension.html; (15-03-2010; 18:55 h)

[8] anon.: What is Supercritical CO_2? Press Release, Applied Separations Inc., Allentown PA, USA 2008; www.appliedseparations.com/supercritical/Supercritical_CO2.asp (15-03-2010; 18:57 h)

[9] Kukova, E., Phasenverhalten und Transporteigenschaften binärer Systeme aus hochviskosen Polyethylen-glykolen und Kohlendioxid (Phase Behaviour ... of Binary Systems of... Polyethylene Glycols and Carbon Dioxide), Diss., Univ. Bochum 2003

[10] Wilds, N., Surface Tolerant Coatings for Offshore Maintenance, NACE Corrosion Congress 2004, New Orleans LA, USA, Proc., NACE International, Houston TX, USA 2004

[11] Mühlberg, K., Surface-Tolerant Coatings - Some Experiences, J. Prot. Coat. Lin. Prot. Coat. Europe, Technology Publ. 2001, pp. 13 ff, 18

[12] Raventos, M., Duarte, S., Alarcon, R., Application and Possibilities of Supercritical CO_2 Extraction in Food Processing Industry: An Overview, Food Science and Technology International 8, 2002, pp. 269 ff

[13] Reverchon, E., Della Porta, G., Gorgoglione, D., Supercritical CO_2 fractionation of jasmine concrete, The Journal of Supercritical Fluids 8 (1), 1995, pp. 60 ff

[14] Wright, W. (ed.), Carbon Dioxide Can Be Part of the Solution - Supercritical CO_2: A Green Solvent, Press Release 0908, SRI Consulting, Menlo Park CA, USA 2009; www.sriconsulting.com/SRIC/Public/NewsEventsArt/PR_Articles/PR0908SupercriticalCO2.html; (15-03-2010; 18:59 h)

[15] anon., Ultrasonic Cleaning of Automotive Parts, Factsheet 2, National Center for Remanufacturing & Resource Recovery, Rochester Inst. Techn., Rochester NY, USA 2000; www.cims.rit.edu/ne/pubs/NC3R_2ultrasonic_web.pdf; (15-03-2010; 19:01 h)

[16] anon., Industrial applications: Surface pretreatment with plasma for the most diverse applications in industry, Press Release, Plasmatreat GmbH, Steinhagen 2007; www.plasmatreat.co.uk/surface_treatmeant_industrial_applications.html; (15-03-2010; 19:04)

[17] Melamies, I. A., Es geht auch umweltfreundlich - Kaltes Plasma ersetzt chemische Vorreinigung beim Coil Coating (Cold Plasma Displaces the Use of Chemicals...), Coating Int. (10), 2009, p. 16 ff

[18] Klages, C.-P., Grischke, M., Höing, T., Thyen, R., Beschichtungen mit definierter Oberflächenenergie. Herstellung und Bedeutung für die Medizintechnik (Coatings with Defined Surface Energy...), Proc., FDS-Workshop Dresden, Fraunhofer IST, Braunschweig 1996

[19] Kiesow, A., Meinhardt, J., Heilmann, A., Coronabehandlung von Polymerfolien (Corona Treatment of Polymer Films), Coating 37, 2004, pp. 34 ff

[20] Thyen, R., Weber, A., Klages, C.-P., Einsatz der Corona-Entladung für die Abscheidung dünner Schichten (Use of Corona Discharge for the Deposition of Thin Coatings), Platin-Plasmatechnologie-Initiative NRW, Proc., PLATIN-Seminar Wuppertal, Fraunhofer-IST, Braunschweig 1997

[21] Blecher, A., Grasme, D., Hinüber, H., Husmeier, F., Meyer, B. D., Röhricht, G., Sander, J., Beschichten von Aluminiumguß - Lackieren (Coating of Cast Aluminium - Painting), Worksheet, Dtsch. Forschungsges. f. Oberflächenbehandlung DFO, Düsseldorf 1991, p. 15

[22] anon., Directive 2008/1/EC of the European Parliament and of the Council of 15 January 2008 concerning integrated pollution prevention and control (Codified version) (1), Official Journal of the European Union 51, 2008, L 24; eur-lex.europa.eu/...

[23] anon., Council Directive 1999/13/EC of 11 March 1999 on the limitation of emissions of volatile organic compounds due to the use of organic solvents in certain activities and installations, Official Journal of the European Union 42, 1999, L 85; eur-lex.europa.eu/...

[24] Androsch, F. M., Stellnberger, K.-H., Wolpers, M., Jandel, L., Drescher, D., Sander, J., Seidel, R., Chromate-free Coil Coating and One Year of Production Experience, ECCA General Meeting Monte Carlo, Proc., European Coil Coating Association, Brussels 1999

[25] Wichelhaus, W., Buetfering, L., [2] pp2714ff

[26] Karsa, D. R., Porter, M. R. (eds.), Biodegradability of surfactants, Blackie, Glasgow 1995, pp. 1946 ff

[27] anon., BGBl. (Fed. Law Gazette) I (25), Art. 1.2, Bundesanzeiger, Bonn 1986, p. 851

[28] Karsa, D. R., Porter, M. R. (eds.), Biodegradability of surfactants, Blackie, Glasgow 1995, p. 1954

[29] anon., Regulation (EC) No 1272/2008 of the European Parliament and of the Council of 16 December 2008 on classification, labelling and packaging of substances and mixtures, amending and repealing Directives 67/548/EEC and 1999/45/EC, and amending Regulation (EC) No 1907/2006, Chapter 4.1.2.9. Rapid degradability of organic substances, Official Journal of the European Union L 353, 2008, pp. 1 ff; eur-lex.europa.eu/…

[30] anon., 2007/506/EC: Commission Decision of 21 June 2007 establishing the ecological criteria for the award of the Community eco-label to soaps, shampoos and hair conditioners (notified under document number C(2007) 3127, Appendix "Environmental Criteria", Part 3.a) Aerobic Biodegradability of Surfactants, Official Journal of the European Union L 186 , 2007, pp. 36 ff; eur-lex.europa.eu/…

[31] Fresner, J., Zero Emission Retrofitting Method for Existing Galvanising Plants (ZERMEG), Austrian Fed. Min. Traffic, Innovation, Technology (ed.), Vienna 2003, pp. 48 ff; www.nachhaltigwirtschaften.at/nw_pdf/0321_zermeg.pdf; (15-03-2010; 19:07 h)

[32] Fresner [31] p. 54

[33] Bosse, K. (ed.), Entwurf des deutschen Beitrags zu den besten verfügbaren Techniken bei der Behandlung metallischer und nichtmetallischer Oberflächen mit chemischen und elektrochemischen Verfahren. (Draft … Best Available Techniques for the Treatment of Metallic … Surfaces with Chemical … Processes), AG BREF Oberflächentechnik, Umweltbundesamt (Fed. Environment Agency), Berlin 2003, pp. 49 ff; www.bvt.umweltbundesamt.de/archiv/oberflaechenbehandlungvonmetallen.pdf; (15-03-2010; 19:08)

4 *Organic coating materials*

Mircea Manea, Lars Kirmaier, Jörg Sander

Coatings are defined as liquid, paste or powder materials which enable transparent and/or opaque coverings exhibiting decorative, protective and/or specific technical properties [1]. Paints may be divided in accordance with different principles (Table 4.1) [6–8].

A coating system selection must be in direct accordance with the specific technical, decorative and functional requirement, as well as with specific application conditions, substrate, and curing. All these requirements are convergent to achieve the target of an optimum performance [4, 5].

All coatings have in common a range of properties which secure the functional performance of the coating:

- Ready to use delivery form
- Easy and fairly quick manufacturing
- Good substrate wetting
- Good formation of a closed film
- Proper rheological behaviour
- Workability
- Good final mechanical and chemical properties
- Good storage stability, small degree of syneresis and sedimentation, redispersibility on stirring
- Fast curing without chalking
- Sufficient diffusion into substrate for good adhesion and surface hardening
- Good and uniform colour retention on the substrate
- Good weather stability
- Low energy consumption
- Environmental compliance

Table 4.1: Coatings with different principles according to the specific technical, decorative and functional requirement

Function	Layer in coating system	Purpose	Environmental compatibility	Chemistry of the film forming agent	Curing conditions
clear coat	primer	car paint	water borne	alkyd paint	baking
metallic paint	surfacer	decorative paint	solvent borne	acrylic paint	oxidative curing
solid paint	topcoat	industrial paint	radiation curing	cellulose derivatives	physical curing
		wood paint	high solids	polyurethane	radiation curing
			powder paint		urethane curing

Jörg Sander et al.: Anticorrosive Coatings

ISBN 978-3-86630-911-1

4.1 Ingredients of organic coating materials

An organic coating is a disperse system comprising a range of materials specifically chosen to accomplish a certain role in the manufacturing or application process or in the service of the final product.

An organic coating is a system comprising organic film formers, solvents (in case of liquid coatings), pigments, fillers and a range of additives.

4.2 Resins

An organic coating must indispensably contain a film forming agent which is capable to solidify as a result of physical and/or chemical processes. A film forming agent is an oligomeric up to a polymeric compound which controls most of the properties and performance of the coating:

- hardness
- strength
- adhesion
- elongation
- chemical properties
- outdoor resistance, etc.

The word polymer derives from πολυ ("poly"=many) and μέρος ("meros"=part), and defines a substance built from many molecules, having a large molecular mass. The name polymer has been introduced by *Berzelius* in 1833. The repeating parts are structural units (monomers) bound together by chemical covalent bonds.

Pioneering work on cellulose derivatives by *Braconnot* and later, the modification of natural rubber brought polymers into daily industrial life. The first synthetic polymer "Bakelite" was reported by *Baekeland in* 1907 and the work by *Carothers and Staudinger* in the 1920s made significant inroads in rational synthesis of polymers and helped the understanding of polymer chemistry.

Polymers use different types of monomers or combinations thereof to generate specific families used in the coating industry and related industries. Referring to polymers, four different terms are commonly used in the coating industry:

- **Resin** as defining any kind of polymeric material in its original state.
- **Binder** as defining the polymer in a paint composition capable of binding by dispersed pigment/filler materials.
- **Vehicle** as defining the property of the polymeric material to carry the pigments in dispersion by a subsequent multistage manufacturing process.
- **Organic film former** as the polymer is capable to form a closed solid film which is the backbone of the organic coating.

The term of **organic film former** is more suitable for the concept of anticorrosion coatings. The meaning is directly referring to the capability of a polymeric coating applied to metallic substrates to provide a barrier against corrosive species. The polymeric film is not purely impermeable and due to exterior factors the polymer will decay in time (as further explained by humidity and UV irradiation) but also due to defects or mechanical damage. Pigments having lamellar or plate like shapes, provide further stability and protection by

- increased length of the diffusion path for the corrosive species and

- decreased permeability of the organic coating when an orientation of pigments in the coating parallel to the surface is achieved.

A requirement is that the pigments involved in the coating formulation exhibit a high compatibility with the polymer matrix, for example by forming anticorrosive complexes with the binder.

The choice of binders in anti corrosive formulations is mainly related to the hydrophobic properties of the binders but also to their light stability. The hydrophobic properties of a binder are related to the capability of the organic film to repel humidity. Under UV irradiation the organic film may undergo changes which will yield a hydrophilic surface. Recent studies explain the behaviour of a superhydrophobic surface to become superhydrophylic by UV irradiation and back to superhydrophobic by heating [59].

These organic film formers are either natural substances and/or modified natural substances or intentionally produced by industrial synthesis processes.

The organic film formers derived from natural substances apparently lost some of their importance. They still are out in the market and their technical relevance is now mostly related to the concepts of **"renewable raw materials"** and **"neutral carbon balance"**.

The organic film formers may be grouped in two categories:

- Higher and **high molecular weight polymers** which undergo no further chemical reaction in the film forming process but form organic films by physical drying, eventually releasing solvent emissions
- Low **and lower molecular weight polymers** which purposely are designed having a certain chemical functionality and are capable of forming organic films by a chemical reaction initiated by light, heat, catalyst, etc.

The liquid organic film formers are produced and supplied in a solvent or solvent blends but the latter group of polymers generally requires less solvent. However the addition and selection of solvents is made according to the application processing conditions, in order to achieve a trouble-free film forming process.

4.2.1 Alkyd resins

Alkyds are B-stage polymers that exhibit by far lower molecular weight compared to thermoplastic polymers. This feature, however, is compensated by the presence of functional groups (such as double bonds, hydroxyl groups, and carboxylic groups) to a large extent. Alkyd resins have been regarded as backbones containing ester bonds and having a modification with a fatty acid.

The etymology of the word alkyd is derived from AL-cohol and a-CID later changed to ALKYD. For the first time alkyds were mentioned in 1929 in the Chemical Abstracts under the chapter "Resins". Although a very strong and well established market segment, the alkyd resin business has been revived due to the interest in new drying chemistry and environmental concerns, targeting high solids and waterborne products [16-19]. To a large extent, alkyd resins are considered as being renewable raw materials in the coating industry due to their large content of natural oils and polyols.

Alkyd resins undoubtedly represent a large segment in the coating market due to some reasons listed below:

- well known technologybroad application range
- one-component application

- broard application range
- good price-quality
- large content of renewable raw materials
- long shelf life
- high build per application
- adhesion for multiple substrates
- predictable drying times
- acceptable odour
- less surface preparation
- independence with respect to humidity

Alkyd resins may be categorised according to the oil length as follows:

- very long oil alkyds (oil content higher than 70 %)
- long oil alkyds (oil content higher than 55 %)
- medium long oil alkyds (oil content of 45 to 55 %)
- short oil alkyds (oil content of less than 45 %)

Further classification of alkyd resins is related to the film forming mechanism:

- air drying alkyds
- non-air-drying alkyds
- oven drying alkyds

Alkyds may be modified in order to improve certain properties, like the hydrophobic behaviour and the light fastness, as previously discussed. Modifications done on an alkyd binder backbone are meant to increase water repellency, improve UV resistance by consuming the double bonds of the oil by vinylation or by cyclisation through different processes. The stabilisation of the backbone of the alkyd binder may be done by hydrolytically stable phenol modification or by introduction of silicone moieties.

A classification according to the chemical modification may be considered as:

- **vinylated alkyd resins** (modified with acrylic monomers employing unsaturation of the fatty acid)
- **phenol modified alkyd resins** (modified with phenol-formaldehyde resins employing unsaturation of the fatty acid)
- **urethane modified alkyd resins** (modified with urethanes when isocyanates are employed as difunctional building blocks replacing a diacid)
- **silicone modified alkyd resins** (modified with siloxanes when siloxanes are employed as functional building blocks in a transesterification reaction)
- **maleic anhydride modified alkyd resins** (modified with maleic anhydride employing unsaturation of the fatty acids by a Diels-Alder reaction)

Properties of the alkyd resins are strongly related to the composition of raw materials, achieved molecular weight and molecular weight distribution. Other preliminary reactions of double bonds, e.g. etherification, dehydration and undesired cyclisation through hydroxyl and carboxyl groups, affect the final properties of the polymer.

An alkyd may be schematically represented by the following formula. The alkyd resin is a condensation product which is the result of the reaction between functional groups [20, 21]:

a-A-a + b-B-b → a-A-B-b +a-b

In most cases the reactive group **a** is a carboxyl group and the reactive group **b** is a hydroxyl group in which case the condensation product **a-b** is water, but some alkyd binders are urethane modified and in the respective case the **a** group is an isocyanate group.

The alkyd resins may be considered as polyesters modified by natural oil. The final properties of the alkyds are a result of the quality and property of the raw materials involved in the manufacturing process.

The choice of raw materials for alkyd formulation is made with respect to the target application and final parameters of the alkyd:

- solubility
- viscosity
- drying properties
- compatibility
- coating hardness
- acid number and hydroxyl functionality
- **water resistance**
- **chemical resistance**
- choice of ingredients
- commercial production and cost
- availability
- environmental issues
- hazards to health
- hazards to the environment
- processing
- issues imposed by equipment

Hence, the wide latitude of formulations and choices of raw materials, apart from oils, comprises polyols, polyacids and hydroxylated acids (mixed functionality). A range of modified polyols is also available aiming to improve the drying properties of the binders.

4.2.1.1 *Alkyd resins manufacturing process*

Alkyd resins manufacturing processes are summarised as follows (see Table 4.2):

4.2.1.2 *Decay of alkyd resins*

The decay of alkyd resins may occur in the polyester backbone by two different mechanisms but also in the pendent fatty acids chains:

Table 4.2: Manufacturing processes of alkyd resins

Alcoholysis process	Fatty acid process	Fatty acid/oil process	Acidolysis process
two step process	one step process	one step process	two step process
cost and availability of oil recommend the process	no alcoholysis step in alkyd synthesis	oil is integrated in the polymer by interchange esterification (oil at most 1/3 of total furnish of fatty acids)	high melting dibasic acids like isophthalic acid and terephthalic acid are used
distinct phase in alkyd synthesis. Polyacid is added in a second stage			distinct phase in alkyd synthesis. Polyol is added in the second stage
reaction between oil and polyol results in redistributed fatty acid	alkyd results in a more linear backbone	higher viscosity products due to seedy polycondensates between polyol and polyacids	reaction between oil and polyacid results in redistributed fatty acid

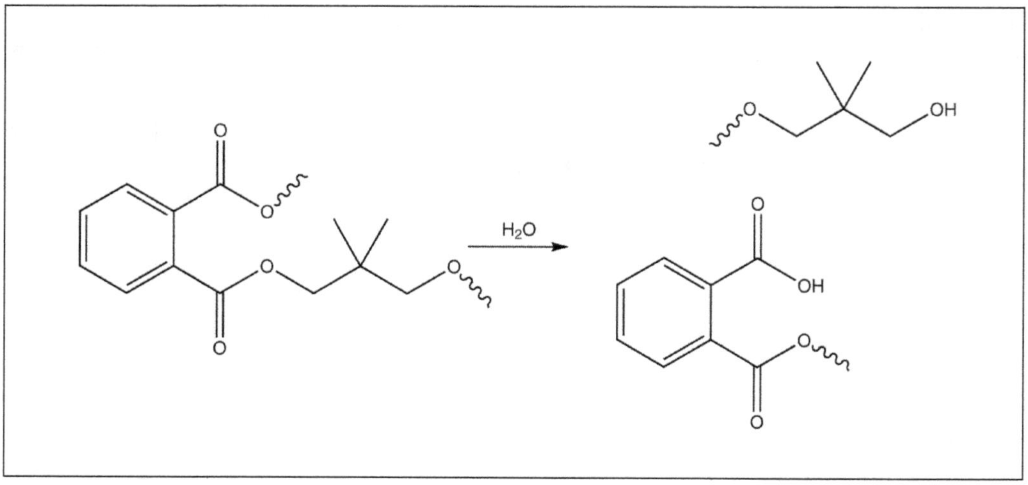

Figure 4.1: Hydrolytical decay of the polyester backbone

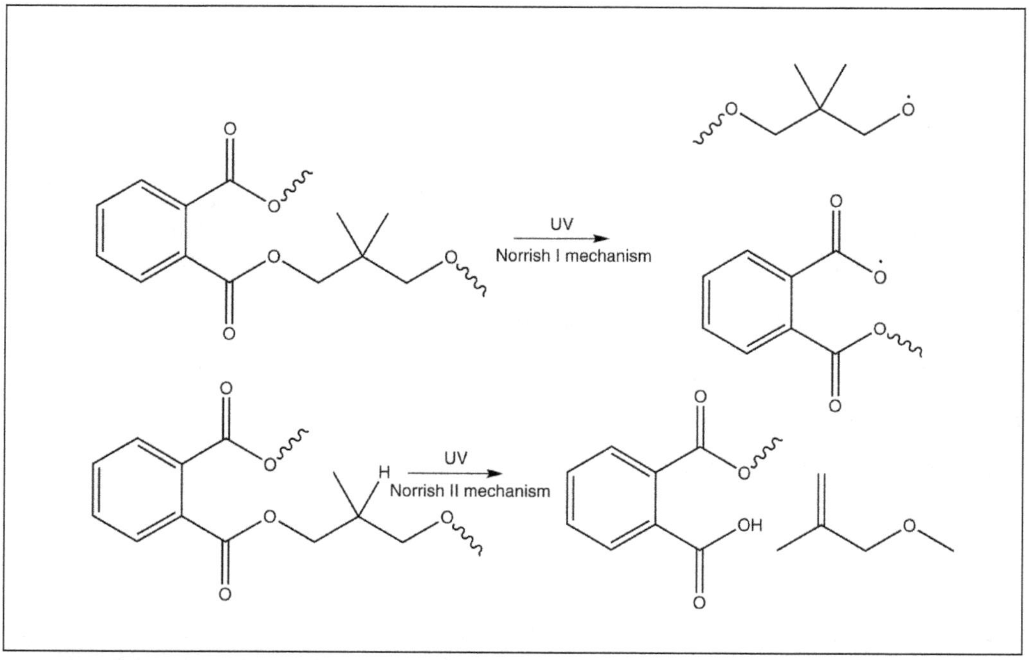

Figure 4.2: Light induced decay of the polyester backbone

- hydrolytical decay of the polyester backbone
- UV initiated decay
- Decay in the pendent fatty acid chain (indoor yellowing of alkyd resins) [4]

4.2.1.3 Composition of alkyd resins

- To a large extent natural oils are the working horses in the alkyd resin composition. The oil importance is increasing as a result of the impact on the alkyd resin performance and as well due to the fact that the natural oils are renewable sources of raw materials [21].
- The backbone of the alkyd resin is a result of condensation of polyols and polyacids. In most cases the backbone is hydroxyl functional, in which case the fatty acid is pendent on the backbone through an esterification reaction on the exceeding hydroxyl groups.

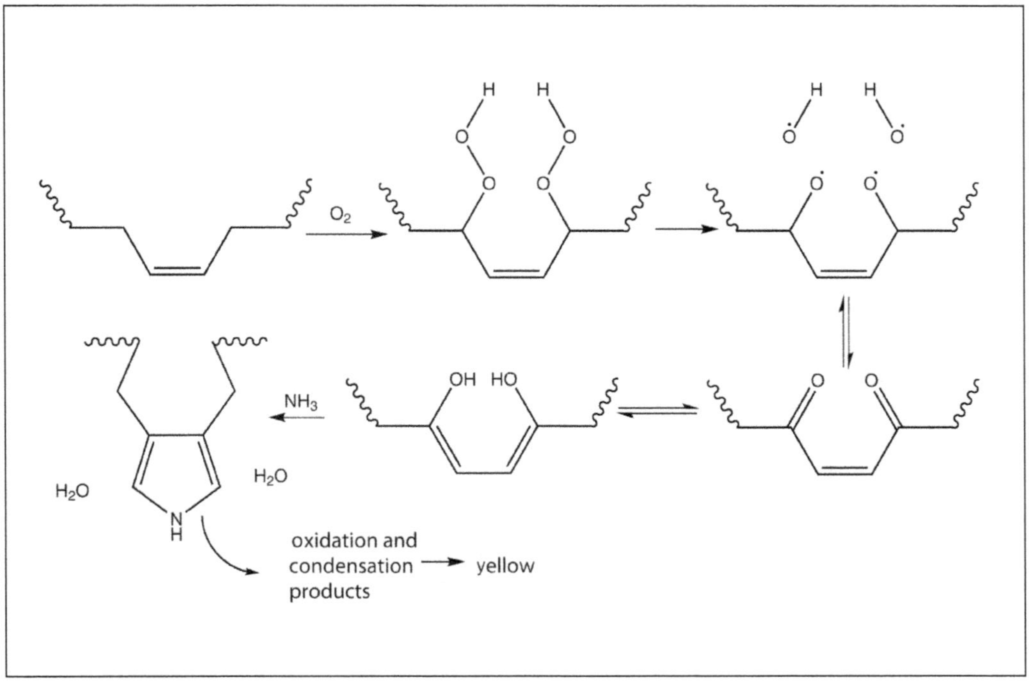

Figure 4.3: Indoor decay of the pendent fatty acids

It is obvious that alkyds are more stable when using polyols that have no α-hydrogen in relation to the carbinol group.

- Polyols used in alkyd synthesis may be di-, tri-, tetra-and hexa-hydroxyl functional. In general the acids used are mono-, di- and sometimes tri-carboxyl functional.
  - difunctional alcohols
  - trifunctional alcohols
  - tetrafunctional alcohols
  - hexafunctional alcohols
- Polyacids employed in alkyd manufacturing are given in the following.

Monofunctional acids are used to limit the molecular weight growth or to improve hardness or other properties. Largely used are: rosin, benzoic acid, t-butyl benzoic acid.

4.2.1.4 Curing of alkyd resins

The fatty acids play an important role in the air drying process, for which the unsaturation degree and the conjugation of the double bonds on the fatty acid determin the drying speed [23].

- oxidative drying of alkyd resins is described in Figure 4.4

The process continues by a radical polymerisation mechanism employing the autoxidation of formed radicals and the double bonds of the fatty acids.

- curing of alkyds by condensation

Other drying mechanisms for alkyd resins involve the alkyd resin functionality, hydroxyl or carboxyl. The mostly employed cross-linking of alkyd resins, if not air drying systems, uses the hydroxyl functionality in reaction with amino resins, or isocyanates.

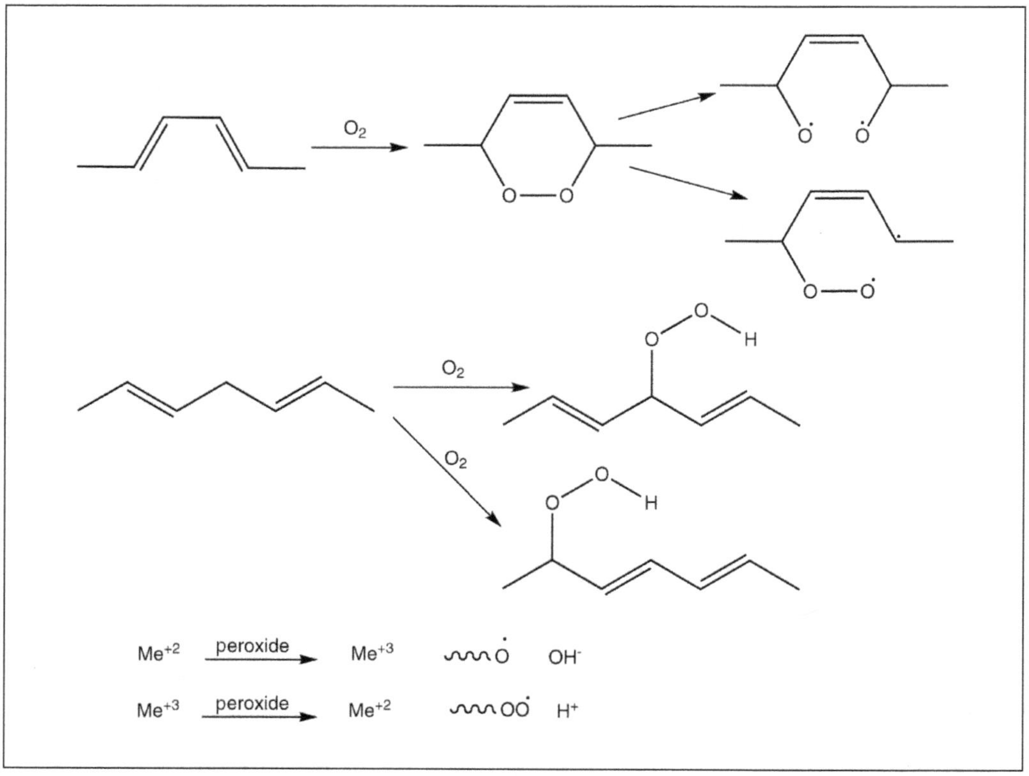

Figure 4.4 Oxidative drying of alkyd resins

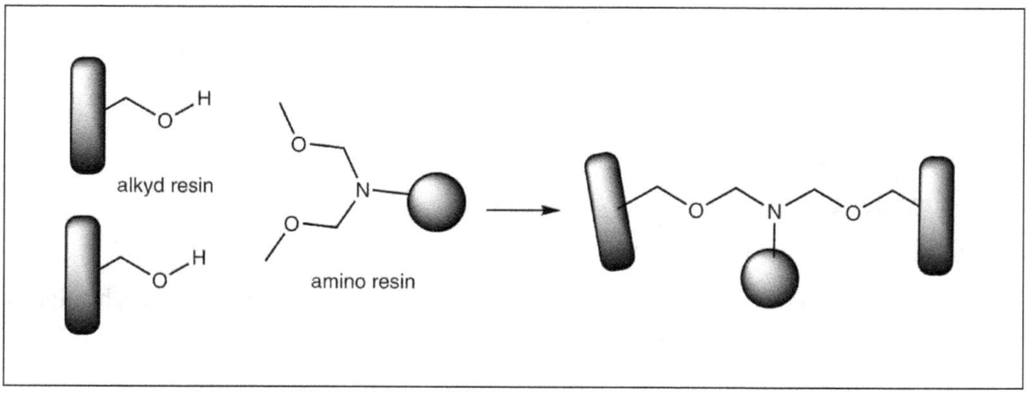

Figure 4.5: Thermosetting curing of alkyd resins

4.2.2 Chlorinated rubber

Nature is capable of the synthesis of cyclic terpenes. Some plants also join isoprene in a 1,4 polymerisation forming polyisoprene.

Hevea Brasiliensis produces 1,4-cis-polyisoprene which is harvested from the notches cut in the three. Other natural rubbers in which 1,4-trans-polyisoprene is present, known as gutta-percha or balata have achieved less technical importance [24-26].

In coatings natural rubber is submitted to chemical modifications to achieve the properties required for a film forming polymer. The chemical modification of rubber aims to improve

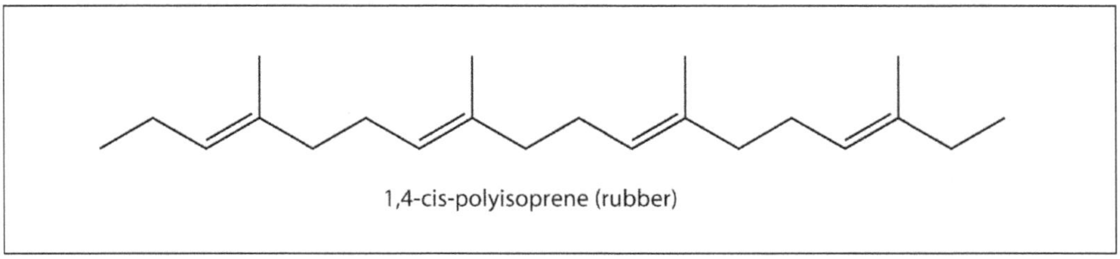

Figure 4.6: Natural rubber

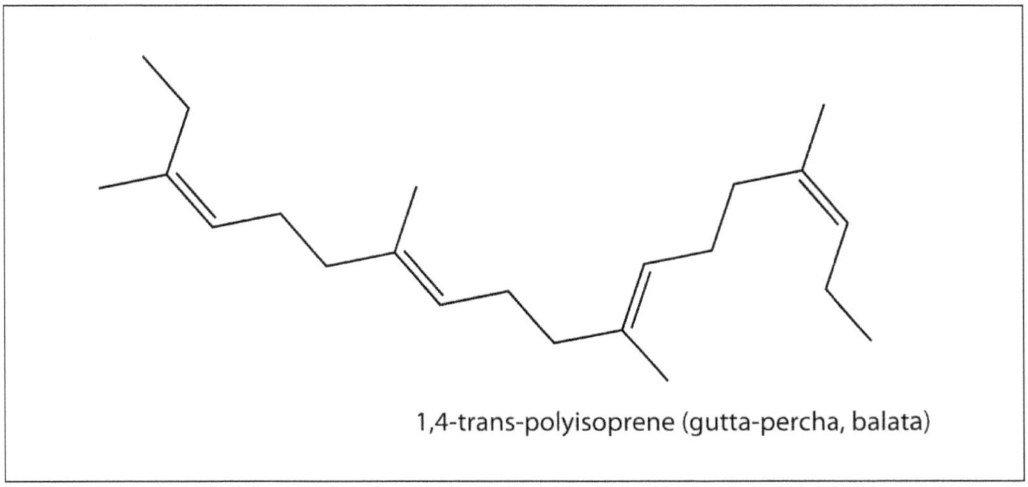

Figure 4.7: Gutta-percha

certain properties such as resistance to oxygen and light. The approach to improve the oxidation and light stability of the hydrocarbon backbone of the rubber, otherwise hydrolytically stable, is to consume the double bonds by addition of halogens or intramolecular cyclisation, thus conferring enhanced anticorrosion properties of the polymer which by said modifications offers less attack points in the backbone.

In the presence of catalysts the process is intended to reduce the number of double bonds by almost 80 %. The process yields the so-called cyclorubber. Cyclorubber has improved solvent solubility, a reduced number of double bonds and a reduced molecular weight, along with an increased melting point of 130 to 140 °C and solubility in aliphatic and aromatic hydrocarbons. The compatibility with other film forming polymers such as alkyds, drying oils, bituminous materials is also improved. The chemical resistance and thermal properties are excellent, as well as a good acceptance for basic pigments due to the absence of acidic groups is obtained.

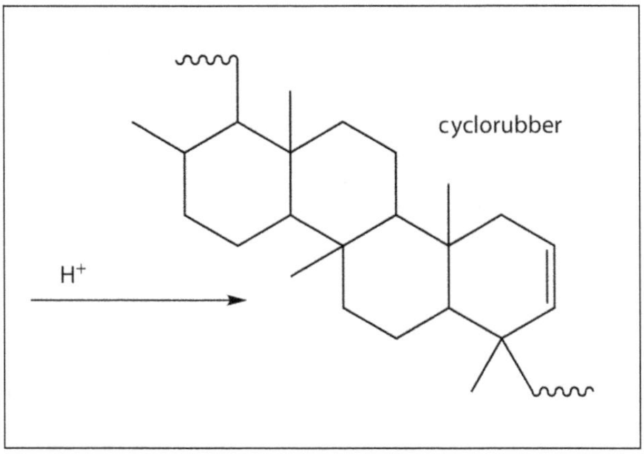

Figure 4.8: Cyclic rubber

Another approach of using rubber in the coating industry is the chlorination of the polymer. In a

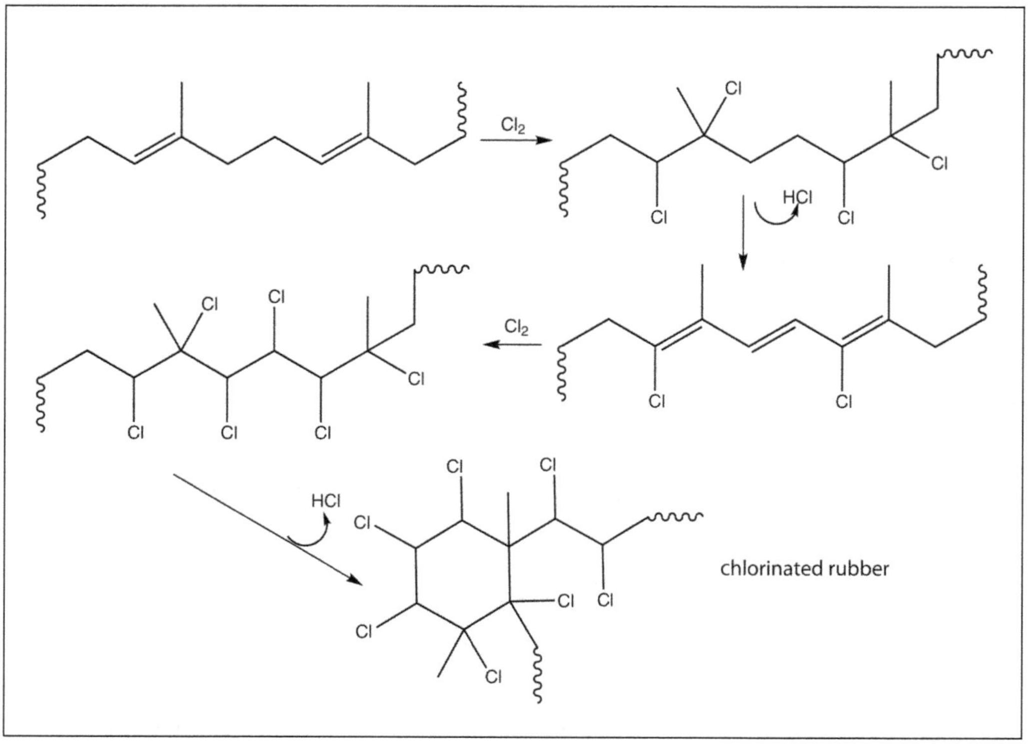

Figure 4.9: Chlorinated rubber

synchronous reaction the chlorine is attached to the double bonds followed by dehydrochlorination and addition of chlorine to the newly formed double bonds, see Figure 4.9.

The reaction is of radical type and may be thermally or UV initiated. However, the chlorinated rubber exhibits limited light and heat resistance, undergoing degradation in which chlorine is released via the same mechanism as explained for polyvinyl chloride.

4.2.3 Polyvinyl chloride

Polyvinyl chloride is the third most polymer produced in the world. The manufacture of polyvinyl chloride largely uses the suspension polymerisation process starting from vinyl chloride. Polyvinyl chloride obtained in this way contains 56.7 % of chlorine and has very poor solubility [24-26].

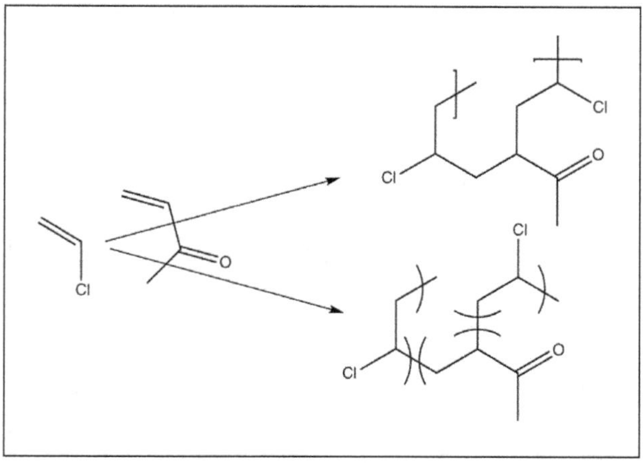

Figure 4.10: Copolymer poly(vinyl chloride-co-vinyl acetate)

Improved solubility is achieved by copolymerisation with other monomers such as vinyl acetate, but the most significant employment of polyvinyl chloride in coatings is as a post chlorinated material. The process is performed by radical chlorination in thermal or UV conditions to yield a polymer containing up to 65 % chlorine. Also, polymers containing up to 74 % chlorine are reported, but the

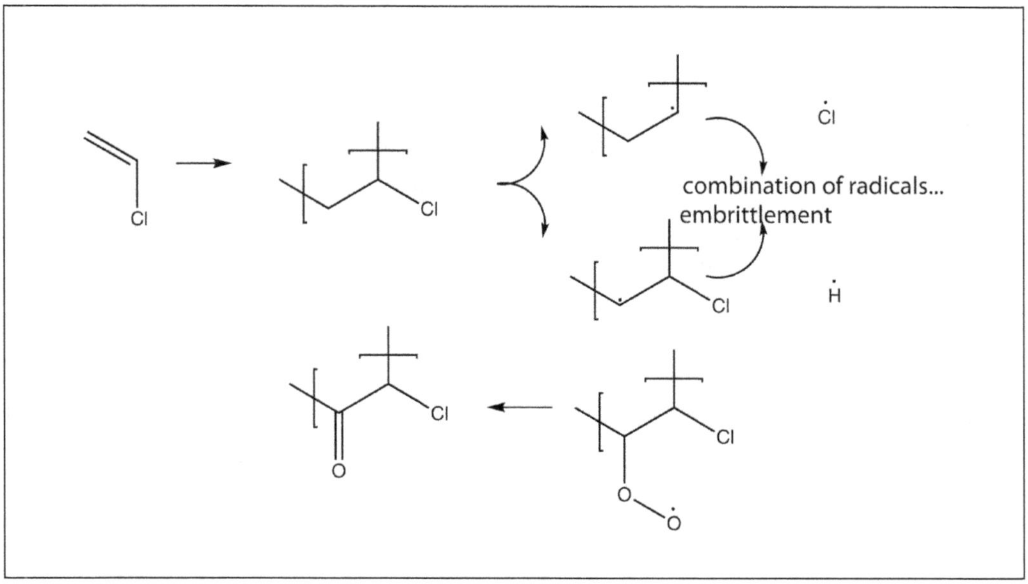

Figure 4.11: Degradation of poly(vinyl chloride)

polymer becomes unstable at a content of about 70 %. The coating industry generally uses polymers with a chlorine content of 65 % to yield coatings by physical drying for heavy duty applications exhibiting good water, solvent and fuel, and chemical resistance.

Polyvinyl chloride exhibits good anticorrosion properties; good hydrolytical and chemical stability, but the main drawback of the polymer is the poor light and heat resistance. Therefore the formulations always contain a range of stabilisers to overcome the problem.

Polyvinyl copolymers are employed in formulations targeting high adhesion coatings on hot galvanised substrates and aluminium, road marking paint, primers and high build anticorrosive paints.

4.2.4 Epoxy resins

Epoxy resins are generally defined as containing a three-membered functional group named epoxy, epoxide, or oxirane [27-29].

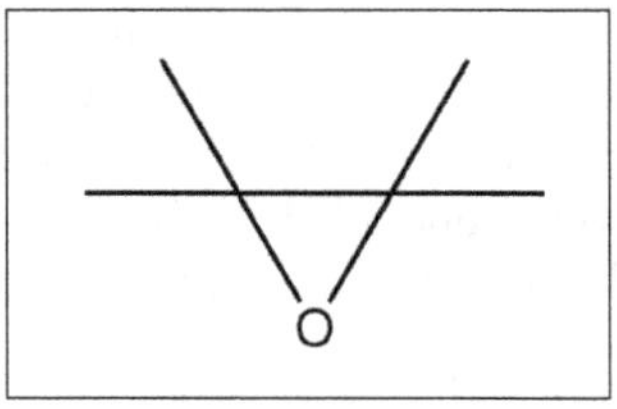

Figure 4.12: Epoxy group

Epoxy polymers present extremely good barrier properties, good substrate adhesion and very good mechanical properties. Several factors may be used to achieve different properties, basically derived from the functionality of the epoxy

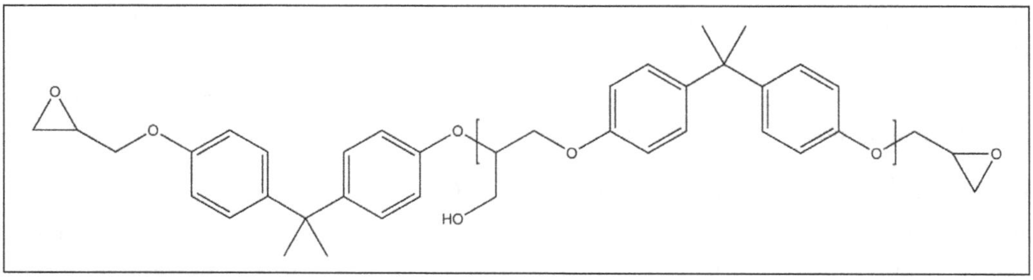

Figure 4.13: Epoxy resin

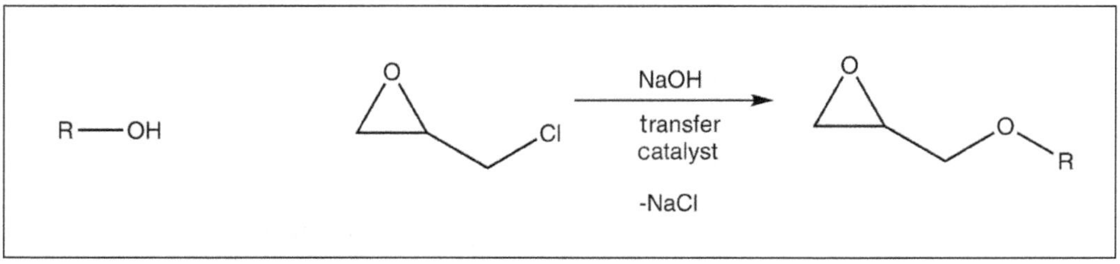

Figure 4.14 Williamson etherification reaction

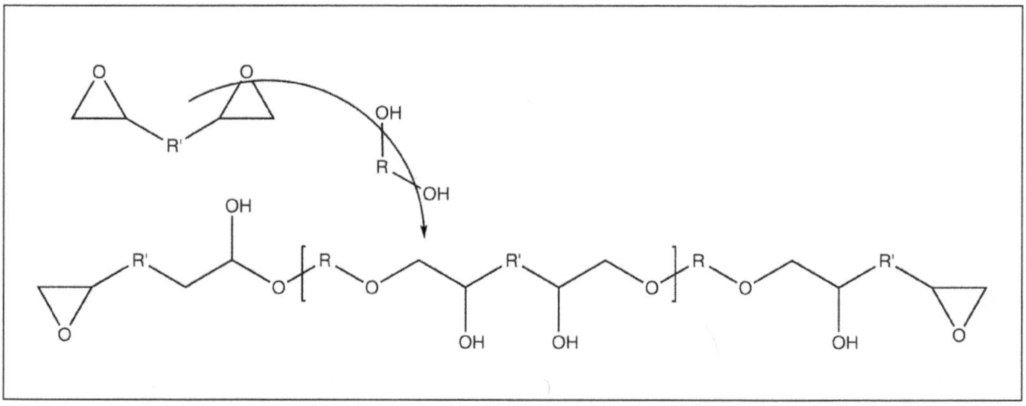

Figure 4.15: Advancement process reaction

binder. The molecular weight of the epoxy resin controls the cross-linking density, hardness and flexibility and resistance to chemicals. Although epoxy resins present a weakness by their poor light stability, generally exhibiting strong chalking, discolouration, loss of gloss and colour, these degradations mainly take place in the surface of the film, leaving the protective properties of the coatings unaffected.

4.2.4.1 Raw materials for epoxy resins

Commercial epoxy resins contain aromatic, aliphatic or cycloaliphatic backbones derived from OH functional substrates. The OH functional substrate is submitted to an etherification reaction with epichlorohydrine in the presence of stoichiometric quantities of sodium hydroxide. Among the aromatic glycidylated substrates, the most interesting are 4,4-methylenediphenol, 4,4'-(propane-2,2-diyl) diphenol, 4,4'-sulfonyldiphenol, 4,4'-(propane-2,2-diyl) bis-(2,6-dibromophenol), phenol and resorcinol.

Some of the aliphatic substrates used for glycidylation are butan-1-ol, 2,2-dimethylpropane-1,3-diol, (tetrahydrofuran-2,5-diyl)dimethanol and 3,3'-(1,1,3,3-tetramethyldisiloxane-1,3-diyl) bis-(propan-1-ol).

The wide majority of the epoxy resins are diglycidyl ether of bisphenol A (DGEBA) also known as bisphenol A diglycidyl ether (BADGE).

The choice of the backbone and the condensation degree in the backbone offer a wide latitude of formulations resulting in a wide range of reactivity and applications.

4.2.4.2 Manufacturing process for epoxy resins

Epoxy resins are produced by direct condensation reaction between components, epichlorohydrine and hydroxyl functional substrate in the presence of basic catalyst at different equivalent ratios epichlorohydrine/hydroxyl (Williamson reaction).

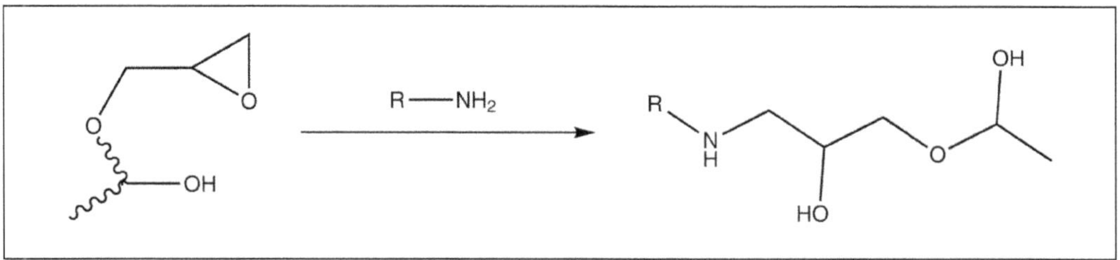

Figure 4.16: Reaction of epoxy group with amines

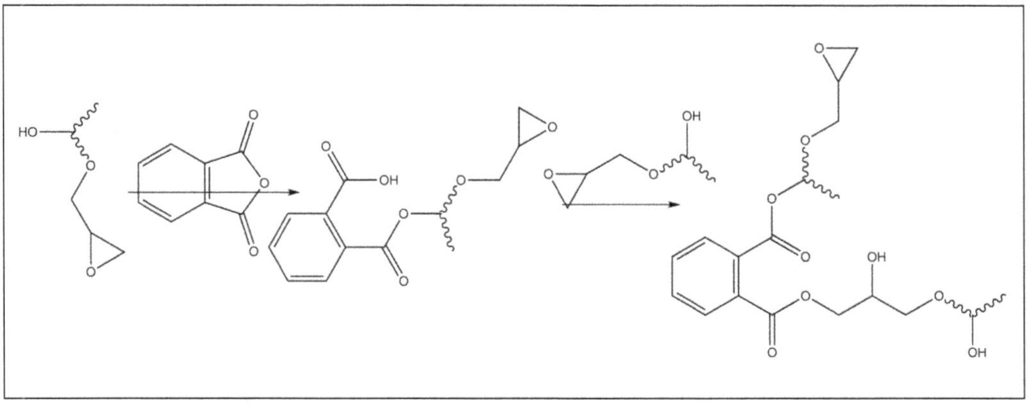

Figure 4.17: Reaction of epoxy group with carboxylic groups

Modern processes (advancement processes) for higher molecular weight epoxy resins involve as starter a low molecular epoxy resin further reacted with a di-hydroxyl functional substrate.

The relation between functionality and molecular weight for bisphenol A diglycidyl ethers is given in Table 4.3.

The amount of bisphenol A necessary to target a certain epoxy index E_f for 1000 parts by weight epoxy resin of initial epoxy index E_i is calculated with the formula:

$$\textit{Bisphenol A} = \frac{1000\ (E_i - E_f)}{8.7 + E_f}$$

E_i refers to the epoxy value of the liquid resin expressed in equivalents per 1000 g using a charge of 1000 g of resin; and E_f refers to the final epoxy value of the desired solid resin product.

Table 4.3: Epoxy resins commercial range (BADGE)

Softening point (orientative values)	Average n value	Molecular weight	Epoxy index (equivalent epoxy/100 g)
9 °C	0.10	400	0.50 to 0.52
64 to 75 °C	2.20	850 to 1000	0.18 to 0.22
90 to 105 °C	5.50	1800 to 2000	0.10 to 0.12
124 to 135 °C	12 to 14	3600 to 5000	0.05 to 0.55
135 to 145 °C	15 to 16	5000 to 8000	0.04 to 0.45

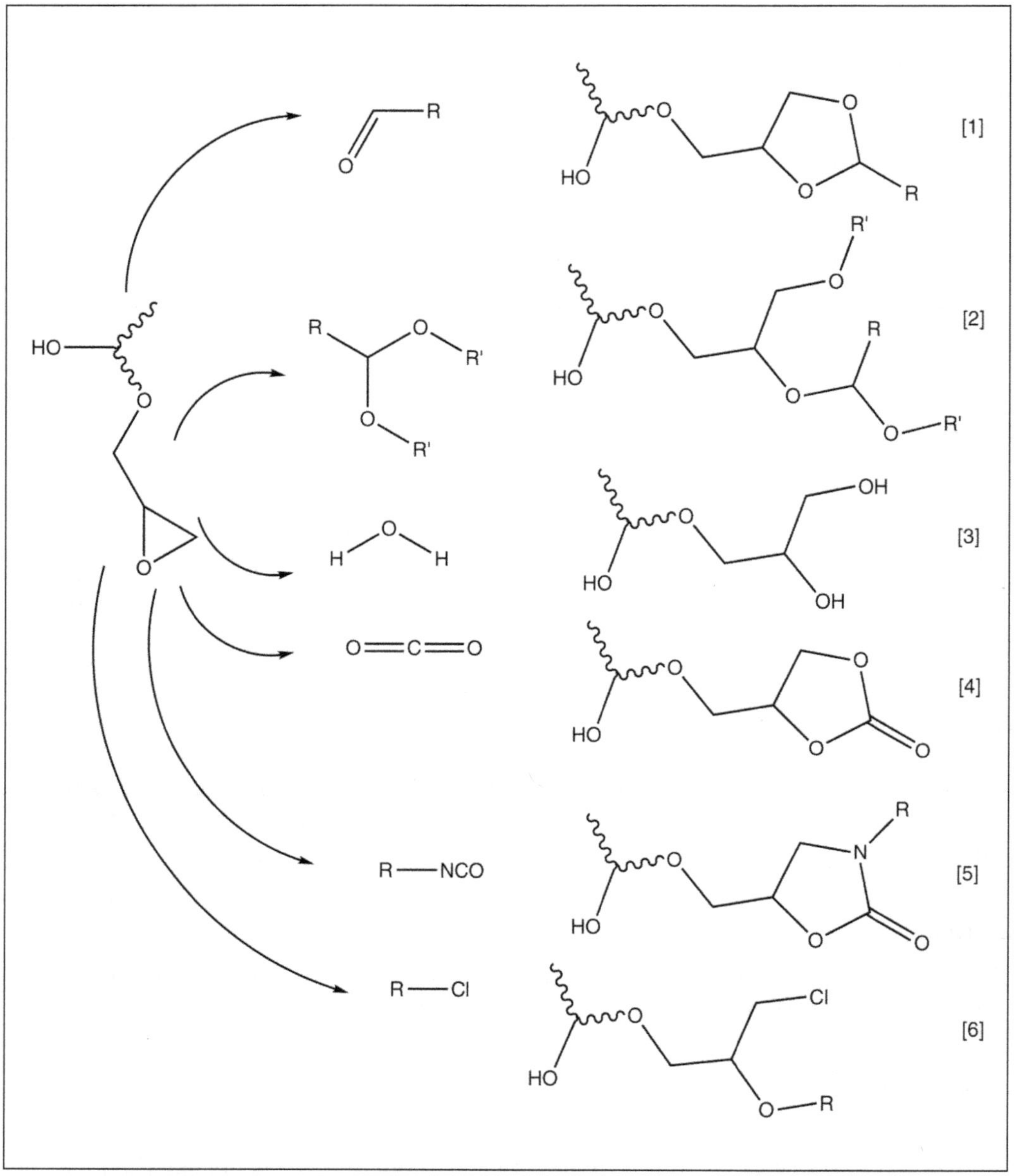

Figure 4.18: Reaction of epoxy group with carbonyl derivatives (1), acetals (2), water (3), carbon dioxide (4), isocyanates (5), halogenated derivatives (6)

Gel permeation chromatography shows that the condensation process from epichlorohydrine and bisphenol A in the presence of stoichiometric quantities of sodium hydroxide yields a number of repeating moieties of 0, 1, 2, 3, etc., while the advancement process yields only even numbers of repeating moieties.

4.2.4.3 Cross-linking of the epoxy resins

The epoxy group has a very high reactivity being capable to react with almost any kind of functional group [30-35]. The epoxy group is capable of self polymerisation in acid or basic catalysis. However the most important cross-linking reactions are the reaction with amines and the reactions with anhydrides and acids.

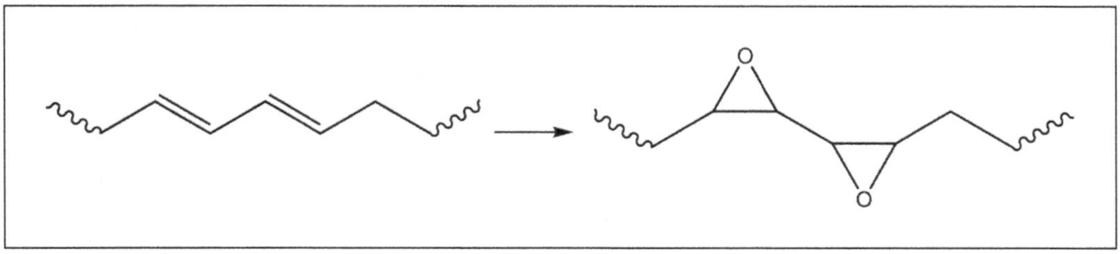

Figure 4.19: Epoxidised oil

Other reactions of the glycidyl group:

An important group of epoxy binders are constituted by the epoxidated oils. Vegetable oils containing double bonds are submitted to partial oxidation to yield epoxy groups. The epoxidated oils find a broad range of applications from covering plasticisers, halogen scavengers, or cross-linkers in reaction with carboxyl functional esters in high solids applications.

4.2.5 Epoxy esters

Epoxy resins like bisphenol A diglycidyl ether present the advantage that the backbone as such is an aromatic polyether having pendent hydroxyl groups [4, 5]. The backbone may be used as a polyol in formulation of ester resins accordingly named epoxy esters. Thus, the epoxy esters show a similarity to alkyd resins using a polyol of moderate molecular weight (800 to 1500) represented by an aromatic polyether. Due to the polyepoxy backbone, epoxy esters exhibit better chemical resistance and mechanical properties than alkyds. In terms of properties, epoxy esters may be assimilated to medium oil length alkyds, but exhibiting higher anticorrosion properties. Due to the polyepoxy backbone they present the same draw backs as epoxy resins, i.e. chalking and yellowing, but as primers in a coating system they exhibit outstanding properties.

This group of resins imparts a range of properties from both backbone and the modification with fatty acids. They exhibit good toughness, flexibility, chemical resistance, durability and adhesion. The resins are easy to formulate, have good air drying or oven drying features and find a wide range of applications in automotive industry, heavy-duty, hardware and metal decorative and UV curing. The coatings using these resins may be applied by all methods and are very versatile in curing conditions.

The content of fatty acids in the binder formulation allows a similar classification as in the case of alkyd resins:

Considering 1 equivalent of epoxy resin as in the Table 4.4, the epoxy esters may be:

- **short oil** using 0.3 to 0.5 equivalents of fatty acids (30 to 40 % oil length)
- **medium oil** using 0.5 to 0.7 equivalents of fatty acids (40 to 50 % oil length)
- **long oil** using 0.7 to 0.9 equivalents of fatty acids (50 to 60 % oil length)

Table 4.4: Esterification equivalent for commercial epoxy resins (BADGE)

Epoxy equivalent weight, [equivalent/g]	Theoretical esterification equivalent weight, [equivalent/g]	Softening point, [°C]
186 to 192	91	liquid
475 to 700	160 to 175	65 to 78
730 to 975	188 to 198	88 to 105
1600 to 2000	233	115 to 130

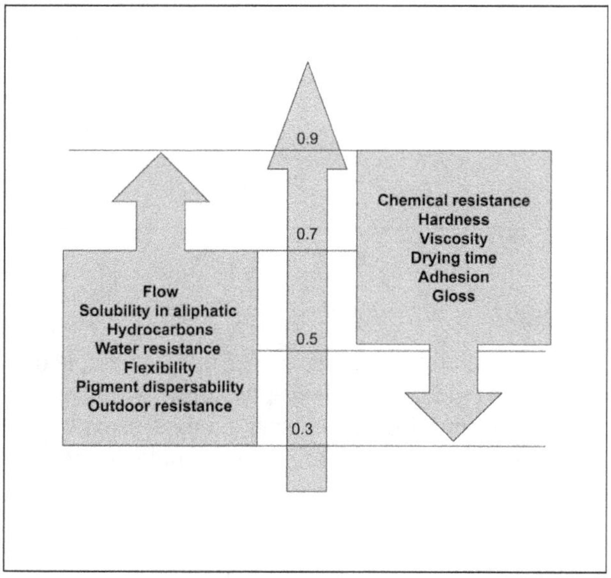

Figure 4.20: Impact of esterification degree on properties

Predicted properties for epoxy esters are as follows depending on the ratio of epoxy to fatty acids:

The choice of the fatty acid is important for the application area:

- linseed oil fatty acids for fast air drying, flexibility and durability
- dehydrated castor oil fatty acids for colour retention and chemical resistance
- safflower oil fatty acids for colour retention and fast drying
- soybean oil fatty acids for flexibility, chemical resistance
- tall oil fatty acids for fair properties and low cost
- coconut oil fatty acids for backing and good colour and chemical resistance

Modification of epoxy esters has the purpose to improve certain parameters of the binder and finally the coating. Dimeric fatty acids improve flexibility and water resistance. Rosin improves hardness and pigment dispersion, benzoic acid improves hardness and mar resistance, adipic acid improves flexibility.

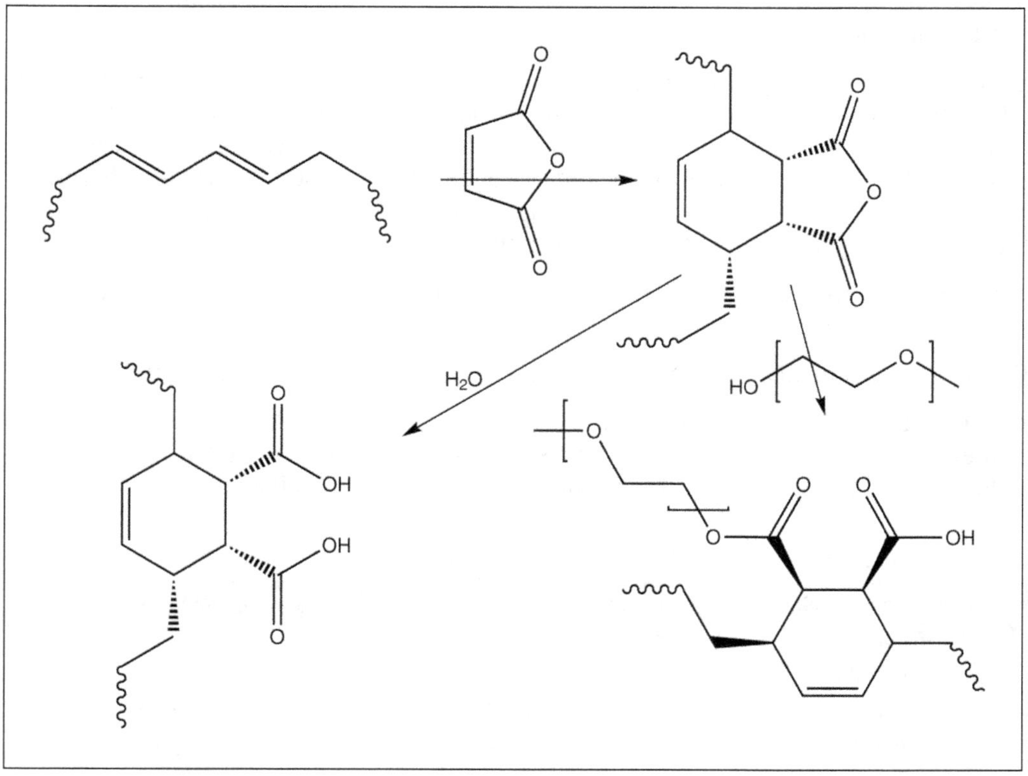

Figure 4.21: Modification of epoxy esters for water transfer

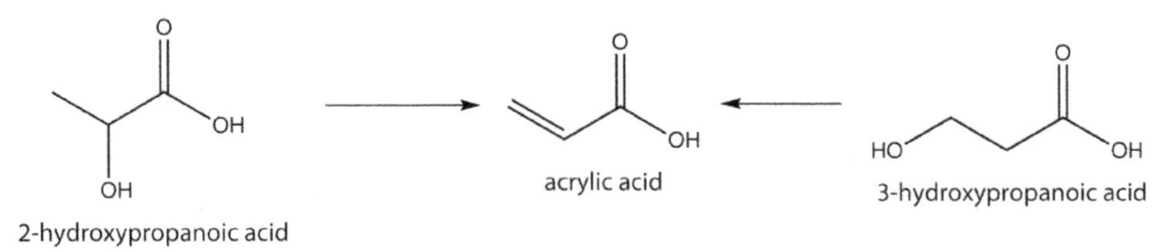

Figure 4.22: Biosynthesis of acrylic acid

Further modification of epoxy esters may be by grafting the fatty acids with vinyl monomers such as styrene, acrylates or methacrylate by the same procedure as in the case of vinylated alkyds.

In most cases the manufacturing process involves catalysts such as sodium bicarbonate, zirconium octoate, or other salts of zinc, calcium or lithium. The process sets in contact the epoxy resin with the fatty acids at increasing temperature up to 240 to 260 °C. The epoxy resin is loaded in small portions or as solution in an appropriate solvent which is later removed, aiming to avoid lumps which deteriorate the quality of the resin and as well make the filtration process very difficult.

In some cases the epoxy resins may be produced in-situ, prior to the esterification process. Bisphenol A glycidyl ether of higher molecular weight is produced by the advancement process starting with a monomeric epoxy resin of equivalent weight of about 0.5 equivalents/100 g to which bisphenol A is added in the presence of a basic catalyst.

The drying process of epoxy esters is similar to that presented for the alkyds. Oven curing with amino resins improves certain properties and applies to short oil epoxy esters while the long oil type cures in the presence of metallic drying agents like alkyd type binders.

An important application of epoxy esters relates to electrodeposition. The anodic electrodeposition is presently used in the appliances coating process. The epoxy esters produced for electrodeposition contain carboxylic groups which are neutralised by amines.

The carboxyl functionalisation of an epoxy ester is generally performed by a Diels-Alder reaction between the pendent fatty acids and maleic anhydride. Further modification is performed by opening the ring of the anhydride.

4.2.6 Acrylic resins

Acrylic resins are polymers emerging from acrylic acid or methacrylic acid derivatives. Generally water clear and exhibiting a great ageing stability, acrylic resins present a great latitude of formulation due to the versatility of acrylic blocks and their capability of copolymerising with other monomers, such as vinyl acetate, styrene, etc [36-44].

Acrylic acid has been first synthesised in 1843. Presently the production of acrylic resins is estimated to exceed 1 million tonnes per year. Today along with the petrochemical path to acrylic monomers, bioacrylic monomers are reported from lactic acid and 3-hydroxy propanoic acid

The nature of the alcohol generating the acrylic ester, as well as the molecular weight of the polymer strongly influence the physical properties of the acrylic resins. Generally, the properties only slightly improve for molecular weights above 100,000 or 200,000.

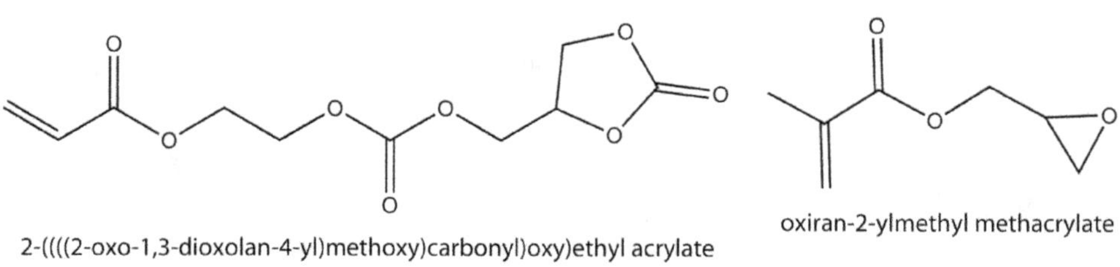

Figure 4.23: Functional acrylates

An important parameter for acrylic resins is the glass transiotion temperature (T_g). Bellow the T_g, the polymer chains have a fixed position and configuration, while above the T_g, the polymer backbone has enough energy for rotational motion and torsion oscillations. All properties such as chemical reactivity, mechanical properties, refractive index, etc. differ markedly below and above the glass transition region. The T_g of a copolymer may be estimated by the following equation:

$$\frac{1}{T_g} = \sum_{i=1}^{n} \frac{W_i}{T_{g_i}}$$

where W_i is the weight fraction of component i having the respective T_{gi}.

Elasticity increase of acrylic polymers may be done by increasing the ratio of ethyl, 2-ethylhexyl or dodecyl acrylate or methacrylate, while the abrasion resistance may be improved by a small ratio of (<1.5 %) of itaconic acid. The use of methacrylic acid esters helps to adjust the melting point of the resin as well as viscosity, gloss, and hardness.

4.2.6.1 Manufacturing of acrylic resins

The manufacturing of acrylic resins is performed by radical polymerisation in solution or emulsion. Sometimes a preferred supply form may be a non-aqueous dispersion. The catalysts for the polymerisation process are generally peroxides, selected according to the synthesis temperature, generally not higher than 150 °C.

For solution polymerisation, the choice of solvent is of importance as it may contribute to chain transfer reactions, reducing the molecular weight of the final binder. The control of molecular weight is generally performed by the choice of process temperature, amount of initiator, use of selected solvent and chain transfer agents such as mercaptanes.

In case the acrylic resin is introduced as an emulsion, the polymerisation process is performed in water in presence of emulsifying agents which secure monomer droplet formation and polymer stabilisation. Additional initiators are present like red-ox systems or water soluble peroxides such as potassium persulphate.

4.2.6.2 Thermoplastic and thermosetting acrylic resins

Acrylic resins present very good outdoor stability, especially when the quantity of aromatic monomer is zero or very low. They present very good gloss retention and colour retention.

Acrylic resins are build on acrylates and methacrylates from saturated alcohols, styrene and vinyl toluene. In this case acrylic resins are thermoplastic. The film formation process is based on solvent release at room temperature or forced drying. They may be used as

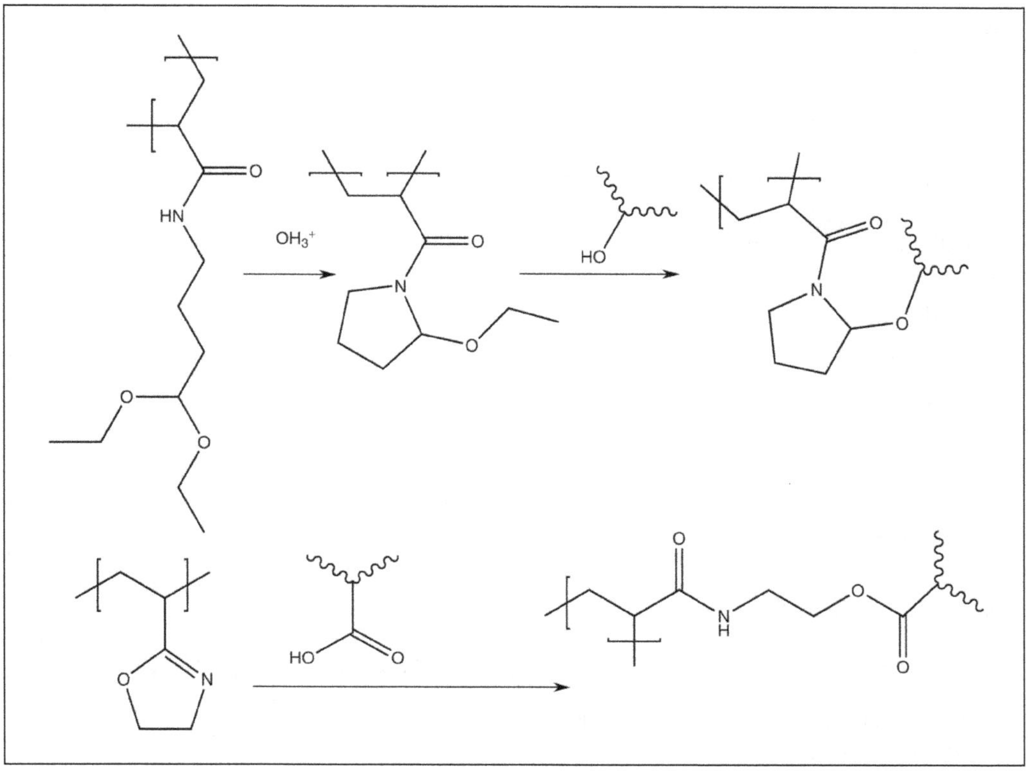

Figure 4.24: Curing of acetal and oxazoline functionalised acrylic polymers

such or in combination with other resins such as alkyds, nitrocellulose, etc. Thermoplastic acrylic resins are presented as organic solvent solutions, non-aqueous dispersions or as water borne emulsions.

Thermosetting acrylic resins are build on the same range of acrylic monomers as thermoplastic resins, which control solvent choice, compatibility with other polymers, T_g, viscosity, etc., but to a certain extent, some acrylic monomers are functional or contain a precursor bearing a functionality. These functional groups may cover a wide range of cross-linking reactions providing the coating with the necessary chemical resistance [45, 46].

Hydroxyl functional acrylic resins may cross-link with other functional binders containing reactive groups prone to react with hydroxyl groups: etherified methylol groups, such as amino resins, glycidyl groups, isocyanates, etc.

Other functional acrylic resins may contain reactive groups as precursors such as cyclic carbonate (Figure 4.33), acetal groups or oxazoline groups.

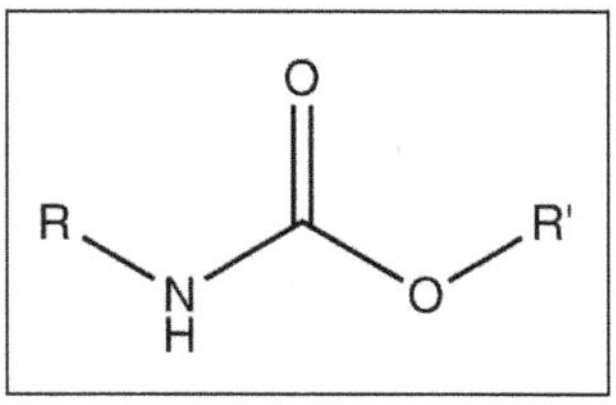

Figure 4.25: Urethane bond

4.2.7 Polyurethanes

Synthetic resins containing a repeatable group as shown in Figure 4.25 are named polyurethanes. Polyurethanes are reaction products of isocyanate derivatives which are homologues from the isocyanic acid, in which different hosts (R) are connected to the functional group [47, 48].

The key benefits presented by polyurethanes making them strongly recommendable for the paint and coating industry are:

- good abrasion resistance
- good chemical resistance
- good adhesion
- good solvent resistance
- good mechanical properties
- good low-temperature impact strength
- good blocking resistance
- excellent film forming properties
- low migration of plasticiser
- high-gloss “wet look”

H—N=C=O (isocyanic acid) R—N=C=O (isocyanate)

$$R-N^{-}-C^{+}=O \longleftrightarrow R-N=C=O \longleftrightarrow R-N=C^{+}-O^{-}$$

Figure 4.26: Isocyanate resonance

The special properties of the isocyanate reaction may be summarised as follows:

- one polymer forming chemical reaction
- randomly reacting functional groups
- monomers disappear early in the polymerisation chain
- molecular weight rises slowly and is high at high conversion degree
- molecular weight distribution is very broad until the very end of the reaction when all oligomers have reacted
- precise stoichiometric ratio is required
- high-purity monomers and solvents are required

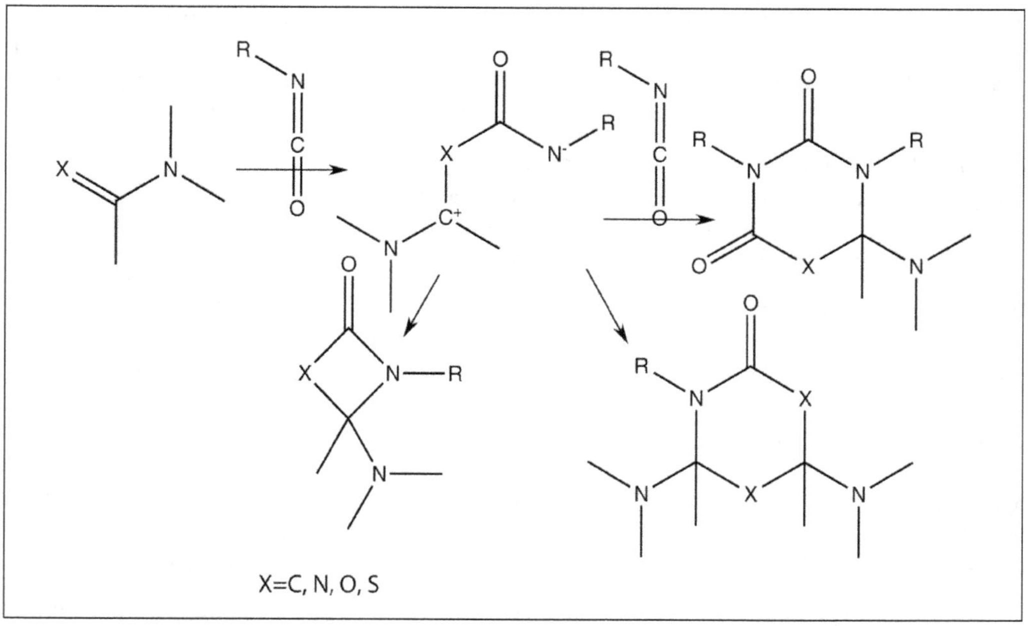

Figure 4.27: Isocyanate cyclo-addition reaction

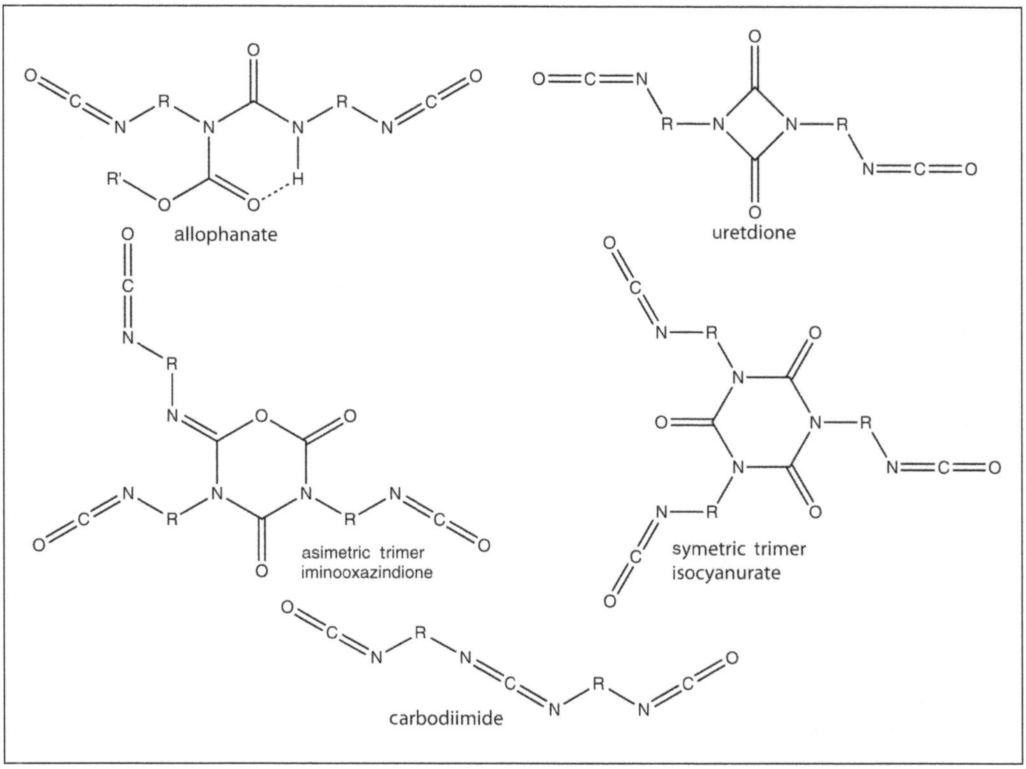

Figure 4.28: Isocyanate dimmers

4.2.7.1 Reactivity of isocyanate group

Isocyanates are highly reactive and undergo addition reactions with a wide range of organic compounds across the double bond of C=N. The cycloaddition of the C=N group to C=O group is also a possible reaction leading to new product development. The driver of the addition reaction is the low electron density at the C atom in the functional group. The reactivity of the isocyanate group is strongly influenced by the host and its electron withdrawing or donating capability [48].

Isocyanates give insertion reactions

- with an acidic C-H group from a 1,3-dicarbonyl derivative generally yielding a carboxyamide at room temperature. Higher temperature favours cyclic compounds.
- with electron rich C=C and C=N double bonds, when the isocyanate group attacks the double bond.
- with a C-O group:
  Isocyanates give cycloaddition reactions. These reactions have been deeply explored. They lead to heterocycles, offering a wide industrial application. Substrates having double bonds such as C=C, C=O, C=S, C=N yield a variety of compounds, see Figure 4.27.

The isocyanate group may polymerise to dimers and trimers which are the forms mostly used in the coating industry as cross-linkers.

Catalysts generally involved in the isocyanate group polymerisation are: 2,4,6-tris-([dimethylamino]methyl)phenyl or 1,3,5-tris-(2-[dimethylamino]ethyl) pyrimidine-2,4,6-(1H,3H,5H)-trione.

Isocyanates react with different functionalities in cross-linking reactions yielding different polymeric networks or new reactive substrates:

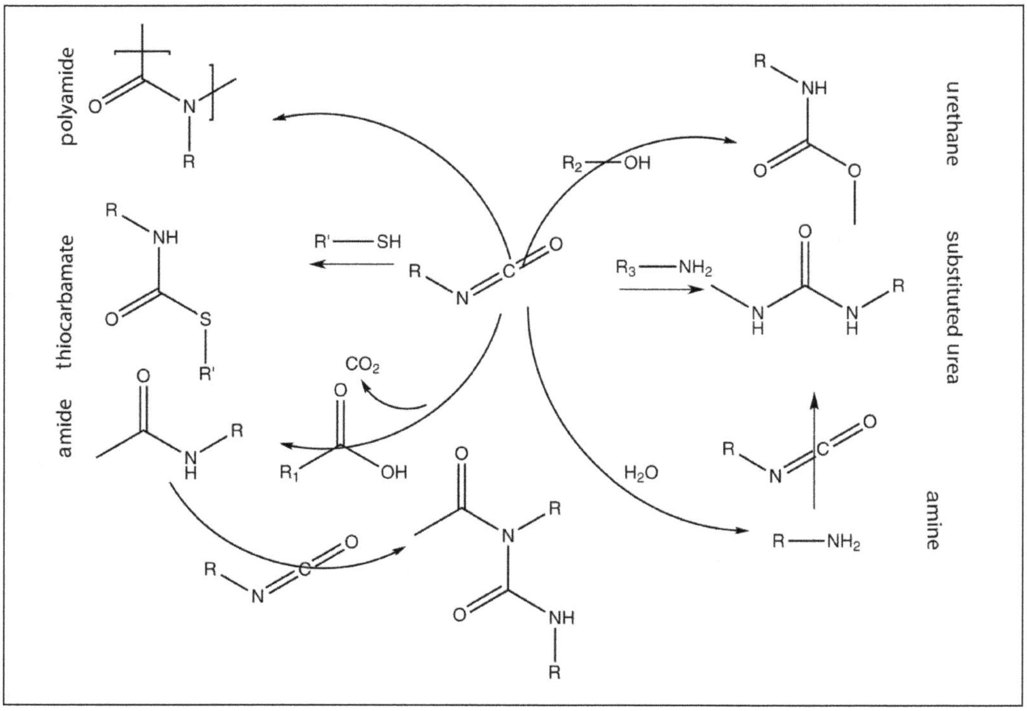

Figure 4.29: Isocyanate cross-linking reactions

Macro-polyols are extensively used, that have a different nature depending on the main backbone. The mostly used types are polyether, polyester and polyacrylic. Depending on the targeted application area, these polyols may cross-link with aliphatic or aromatic polyisocyanates. For higher temperature, blocked isocyanates are used and the coating is one-component. The cross-linking temperature is higher than the deblocking temperature. In any case catalysts are used to ensure the complete cross-linking. The isocyanate reaction with macropolyols is a stoichiometric reaction and therefore the ratio between components is chosen to be 1:1, or with a little excess of the isocyanate component.

Further it must be observed that the hydrogen bond between the polyurethane chains and the very important urea groups that are present to a certain extent-, in but small quantities, give strength to the polyurethane. In this respect the polymeric chain may be considered as constituted from soft and hard segments.

The isocyanate reactions are almost always assisted by catalysts. In general, metal catalysts are used, but recent banning of tin catalysts brought into attention a new range of materials. Metal catalysts such as Zn, Zr are used, but amine catalysts are also quoted: N,N-dimethylcyclohexyl amine, 2,2'-oxy-bis-(N,N-dimethylethane amine)

4.2.7.2 Waterborne polyurethanes

The transfer to water of a polyurethane system may be done by using a hydrophilic monomer, anionic, cationic, non-ionic, as a building block in the polymer backbone or as pendent group in a polyisocyanate which thus becomes water dispersible.

Two-component waterborne isocyanate systems comprise a hydroxyl functional polymer which is waterborne and by nature may be polyester polyol, acrylic polyol, polyurethane polyol, and the hardener which is at least a diisocyanate having a hydrophilic group in order to be water dispersible.

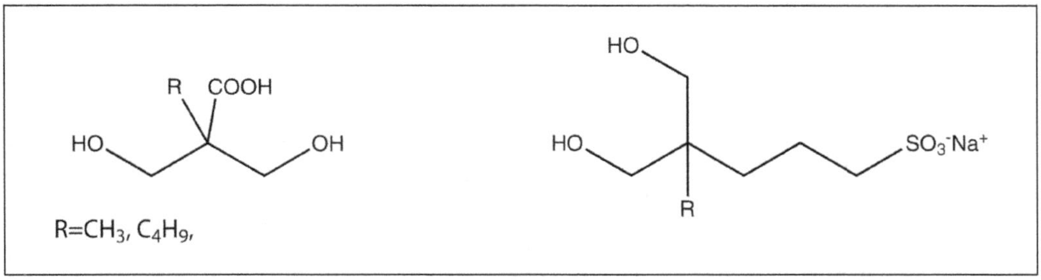

Figure 4.30: Hydrophilic monomer for polyurethane water transfer (anionic)

The new group of polyurethanes presented as dispersions in water is generally known as PUD (polyurethane dispersions) and combines the benefits of polyurethane coatings with a one-pack water borne system, virtually solvent free.

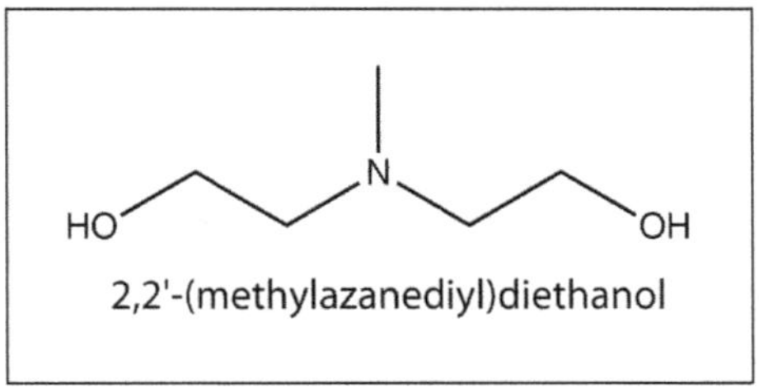

Figure 4.31: Hydrophilic monomer for polyurethane water transfer (cationic)

The hydrophilic moieties of

- anionic,
- cationic,
- non-ionic type

are incorporated as building blocks in the polymer chain in order to enable the dispersion of the polymer in water.

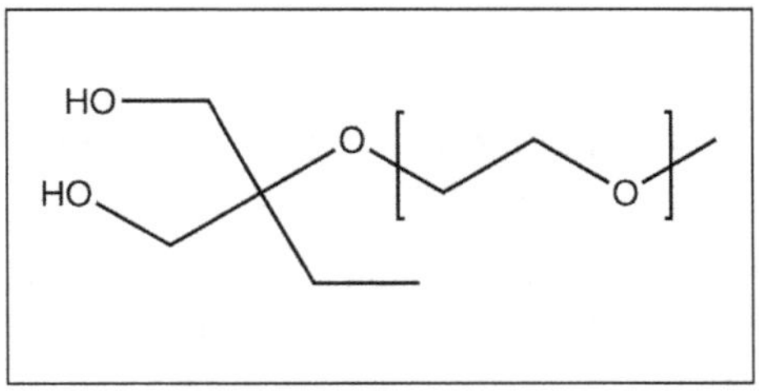

Figure 4.32: Hydrophilic monomer for polyurethane water transfer (non-ionic)

4.2.7.3 *Isocyanate free polyurethanes*

Following the concern of health hazards posed by the contact with polyisocyanates, a market request has been noted as the necessity to produce NCO-free polyurethanes. The so far

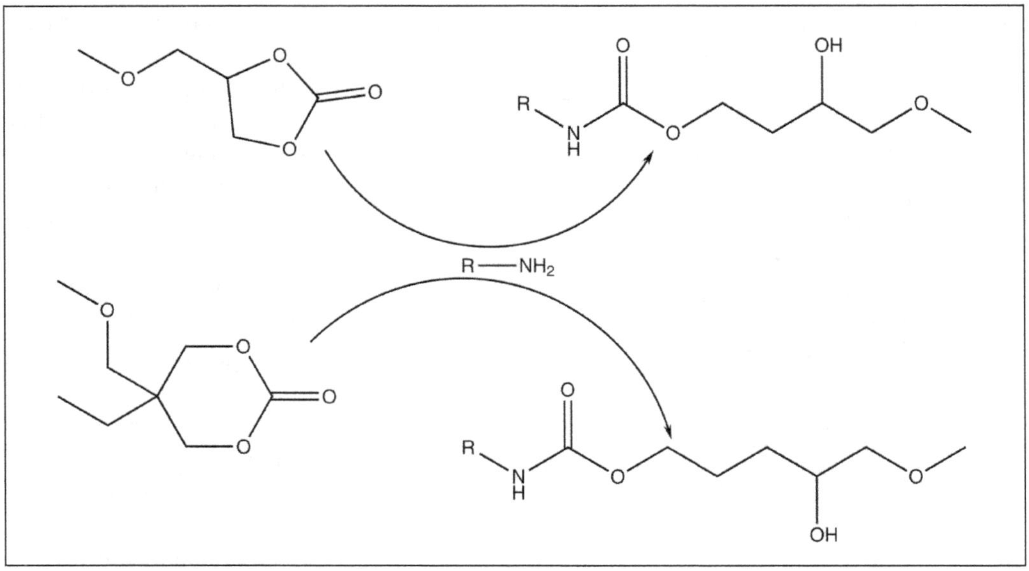

Figure 4.33: Isocyanate-free polyurethane

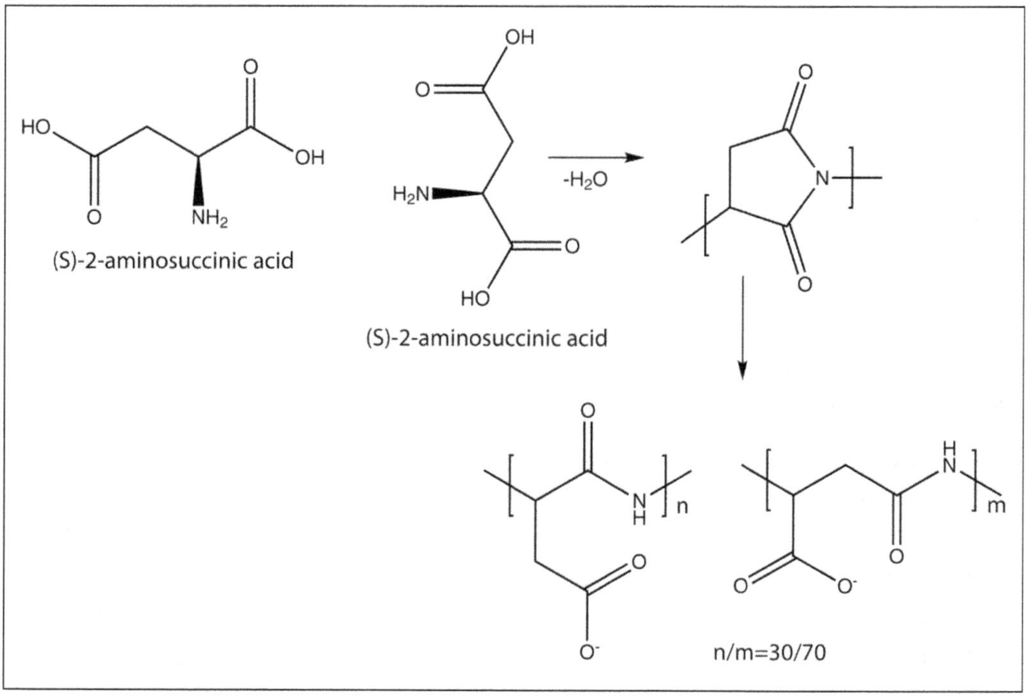

Figure 4.34: Polyaspartic acid

marketed approaches are building on the reaction between a polymer backbone which is functionalised with

- cyclic carbonate groups and a polymer backbone which is amino functional
- oxaozoline groups and a polymer backbone which is carboxyl functional (Figure 4.24).

Six membered cyclocarbonates are by far more reactive but also more difficult to produce [49, 51].

4.2.8 Polyaspartics

Polyaspartic acid offers a new technology with a broad application range. Polyaspartic acid as a biodegradable alternative to polyacrylates has been employed successfully as dispersant, antiscalant and superabsorber. The advantage of polyaspartic acid is the synthesis from L-aspartic acid, a natural amino acid, and as well the biodegradation into benign products such as carbon dioxide and water. The course of synthesis is shown in Figure 4.34.

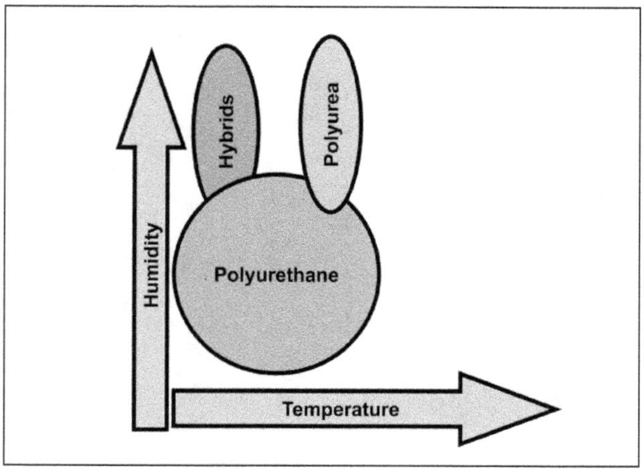

Figure 4.35: Positioning polyurethanes, polyurea and polyaspartics

However in the coating industry, the significance of polyaspartics is different. The concept relates to the reaction of a polyisocyanate with a polyaspartic ester which is an aliphatic diamine [53, 54, 55]. The technology has been primarily employed in solventborne

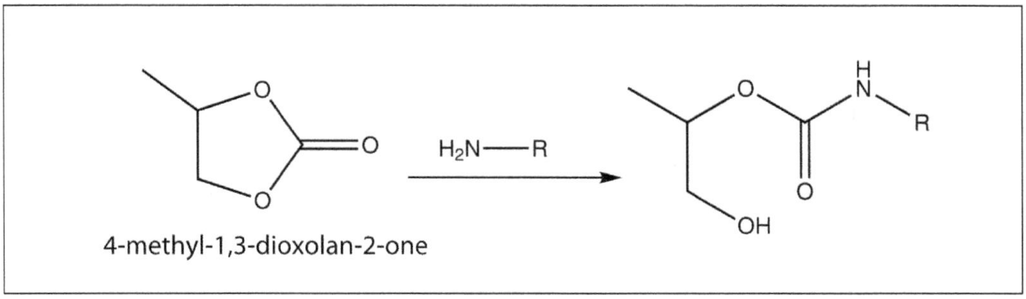

Figure 4.36: Reaction of the reactive solvent in polyureas

coatings, two-component polyurethanes, where the polyaspartic ester was employed as a reactive diluent to yield high solids polyurethane coatings. Recent developments are able to yield formulations with virtually zero VOC, in which the polyaspartic ester is the main component.

The polyaspartic coatings are characterised by

- increased speed of cure
- increased film build
- good weatherability
- excellent corrosion and chemical resistance

For a better understanding the isocyanate reaction may be summarised as follows:

- Polyurethane: the result of the reaction between an isocyanate and a hydroxyl functional component
- Polyurea: the result of the reaction between an isocyanate and a polyether amine
- Polyaspartics: the result of the reaction between an isocyanate and an aspartic ester

The chemistries may be combined to match a range of application conditions and requirements. In the polyurethane landscape the name of hybrid generally refers to the combination between a hydroxyl functional prepolymer and an aspartic ester cured with an isocyanate.

4.2.8.1 Polyurea systems

Polyurea systems combine isocyanate, generally prepolymers from MDI (bis[4-isocyanatophenyl]methane) with an isocyanate content of 10 to 20 %, with propylene glycol carbonate.

The polyether amine (from ethylene oxide or propylene oxide) usually has a molecular weight in the range of 200 to 5000. Further a reactive diluent is employed. In most cases

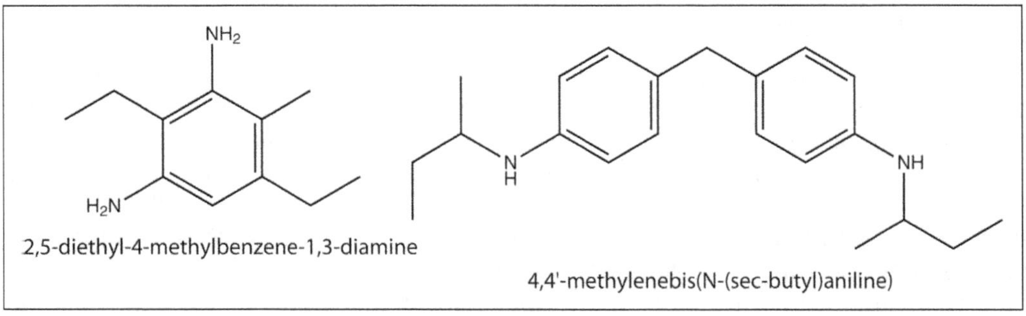

Figure 4.37: Chain extenders in polyureas

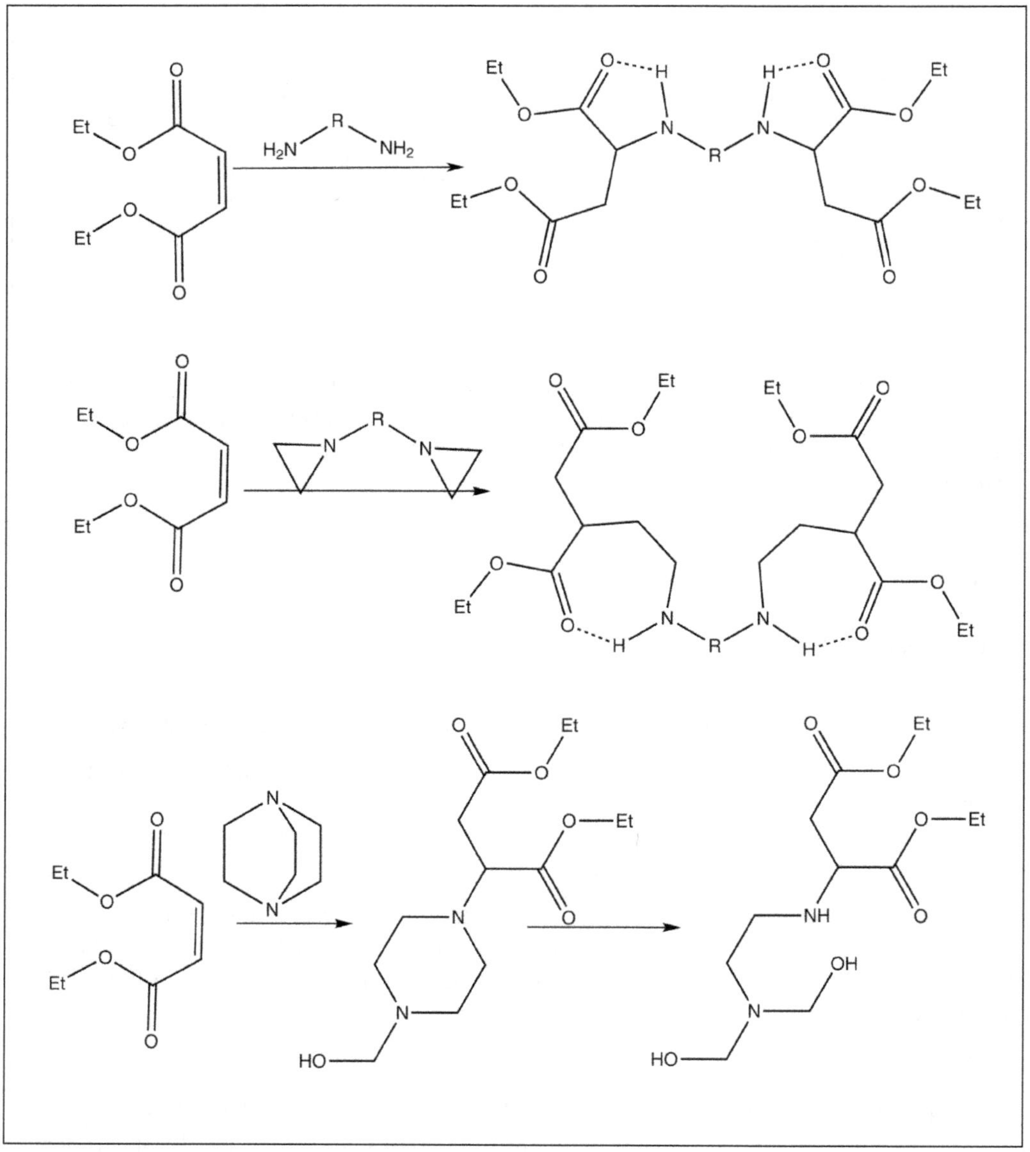

Figure 4.38: Synthesis of aspartic esters

propylene glycol carbonate is used for its high boiling point and easiness of reaction with an amine to yield a carbamate.

4.2.8.2 Polyaspartic systems

Polyaspartic systems employ esters of aspartic acid with different substituents which allow a very good control of the reactivity and gel time. They may be regarded as sterically hindered amines which react with isocyanate to form high build weather-resistant coatings. The intra-molecular hydrogen bond formed between the carbonyl group and the amino group keeps the viscosity low and pot-life in place. Humidity will disturb this intra-molecular hydrogen bond and release free hydrogen atoms, capable to react with isocyanate groups. Polyaspartic esters are different from other amine functional polymers in their reaction with isocyanate groups, the internal hydrogen bond offering longer pot life in absence of humidity.

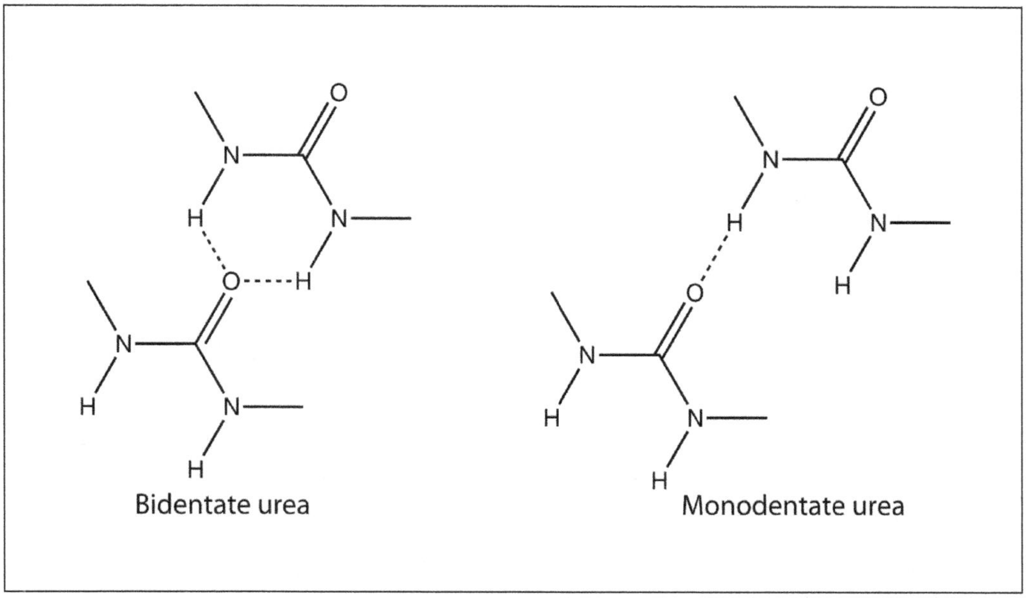

Figure 4.39: Supramolecular monodentate and bidentate urea build up in amino isocyanate curing

Polyaspartic systems have been developed in the beginning of the 1990s and have been mainly used as reactive thinners for the hydroxyl functional binder further used in combination with an isocyanate. Recently the use of aspartic esters has moved to the use as main binders in combination with isocyanates as a broadening of two-component polyurethane coatings. The characteristic of the coatings using aspartic esters present a tolerance for excessive film thickness and lower sensitivity to climatic conditions, combined with a rapid cure. The coatings applied in one layer achieve the same performance as in the case of using primer and topcoat. The synthesis of polyaspartic components is given in Figure 4.38.

The use of amine functional binders is an important approach to corrosion resistant coatings. One of the most interesting results is the fact that the molecules build supramolecular systems by H-bonding between urea moieties and therefore the strength of the coating system is so high and so performing [60-62]. The supramolecular system self assembled by hydrogen bonding is shown in Figure 4.39, and the strength of the coating is higher with the higher content of bidentate moieties.

4.2.9 Alkyl silicate

The general formula $M_2O_mSiO_{2n}H_2O$ where M is an alkali metal and m and n are the respective numbers of moles of SiO_2 and H_2O, per mole of M_2O. For the most common form of soluble silicate m has the value of 3.3, the composition of commercial alkali silicates is typically described by the weight ratio of SiO_2 to M_2O.

Sodium silicates [1344-09-8] are the most common; potassium silicate [1312-76-1] and lithium silicate [12627-14-4] are manufactured to a limited extent for special applications. These materials are usually manufactured as glasses that dissolve in water to form viscous, alkaline solutions.

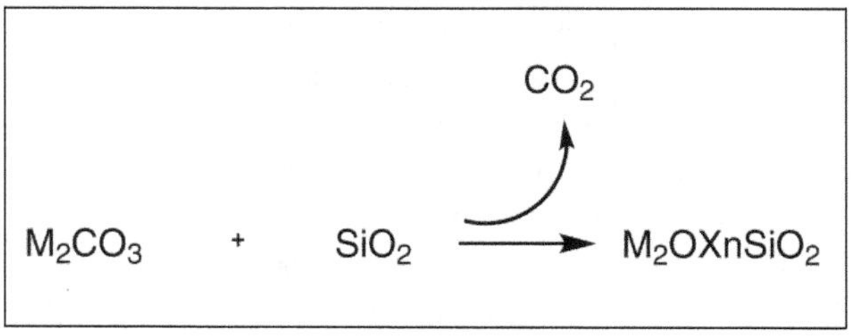

Figure 4.40: Synthesis of a silicate

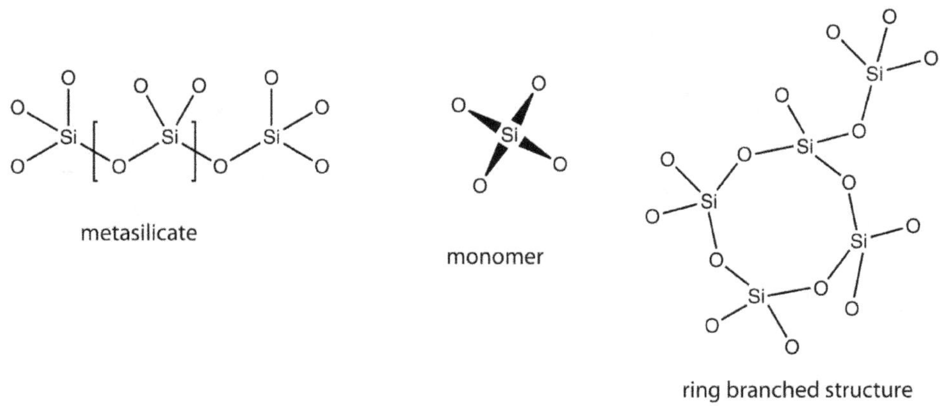

Figure 4.41: Silicate structures

Waterborne alkali silicates in coatings are believed to be harmless to the environment on the basis that:

- cured systems are silicate based, which is effectively very close to what already exists in nature
- are non-hazardous substances
- exhibit no significant health risk during application
- have no allergenic potential from silicate binders
- are zero VOC products
- can provide a good economic lifetime of the coating (leading to a lower requirement to reapply the coating) due to long term stability.

4.2.9.1 Nomenclature of the silicon chemistry

The nomenclature of silicon chemistry is summarised as such in Table 4.5 [57].

Soluble silicates are primarily used as sources of reactive silica, in detergency, in pulp and paper production, adhesives and binders, and in other applications. The understanding of silicate chemistry by modern analytical techniques contributed to the development of new markets. Thus, the sodium silicates constitute a versatile, stable, and growing commodity and are ranked among the top 50 commodity chemicals.

Table 4.5: Nomenclature of silicon chemistry

Nomenclature	Designation
silicon	Si the element
silane	substituted silicon compound
oxysilane	silicon based compound with at least on substituent selected from alkoxide, hydroxide or arylxoide
silicate	metal salt of a silicon-oxygen anion
silica	silicon oxygen compound (sand)
siloxane	silicon compound with two or four oxygen atoms bonded to a silicon atom
polysiloxane	polymeric compound having a silicon-oxygen backbone
silicone	polysiloxane with one or typically two organic substituents bonded to each silicon atom in the backbone

Dissolution of sodium silicate glass undergoes a two-step mechanism that involves ion exchange and network breakdown.

Silicon esters are silicon compounds that contain an oxygen bridge from silicon to an organic

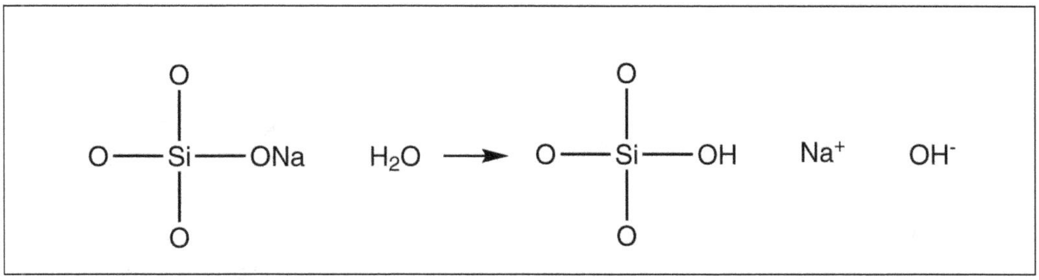

Figure 4.42: Dissolution by ion exchange

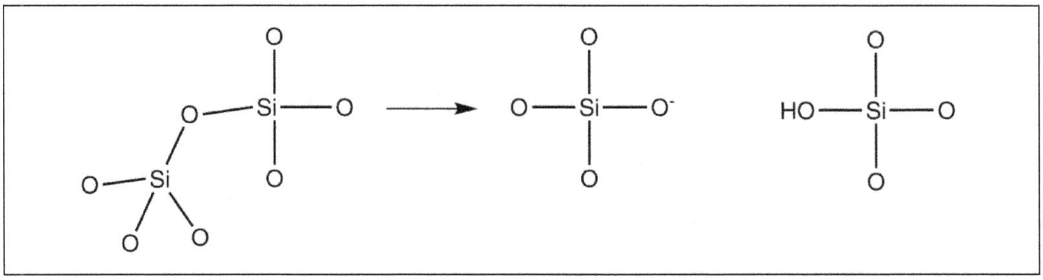

Figure 4.43: Hydrolytically network breakdown

group, such as Si-OR. The earliest reported organic silicon compounds contain four oxygen bridges and are often named as derivatives of orthosilicic acid, $Si(OH)_4$.

The best known material is tetraethyl orthosilicate $Si(OC_2H_5)_4$, (CAS 78-10-4). The use of organosilanes that contain silicon-carbon bonds, Si-C, initiated an organic nomenclature by which compounds are named as alkoxy derivatives. For example tetraethyl orthosilicate $Si(OC_2H_5)_4$ becomes tetraethoxy silane. Acyloxy silanes, such as tetraacetoxy silane, $Si(OOCCH_3)_4$, are also members of this class. The chemistry and applications of acyloxy silanes are significantly different from those of the alkoxy silanes.

The tetraalkoxy silanes exhibit excellent thermal stability and liquid behaviour over a broad temperature range that widens with length and branching of the substituents. The physical properties of the silane esters, particularly the polymeric esters containing siloxane

Table 4.6: Silanes

Homologue	CAS number	Formula
tetramethoxysilane	[681-84-5]	$Si(OCH_3)_4$
tetraethoxysilane	[78-10-4]	$Si(OC_2H_5)_4$
tetrapropoxysilane	[682-01-9]	$Si(O\text{-}n\text{-} C_3H_7)_4$
tetraisopropoxysilane	[1992-48-9]	$Si(O\text{-}i\text{-}C_3H_7)_4$
tetrabutoxysilane	[4766-57-8]	$Si(O\text{-}n\text{-} C_4H_9)_4$
tetrakis(s-butoxy)silane	[5089-76-9]	$Si(O\text{-}sec\text{-}C_4H_9)_4$
tetrakis(2-ethyl-butoxy)silane	[78-13-7]	$Si(OCH_2CH(C_2H_5)_2)_4$
tetrakis(2-ethyl-hexoxy)silane	[115-82-2]	$Si(OCH_2CH(C_2H_5)(C_4H_9))_4$
tetrakis(2-methoxy-ethoxy)silane	[2157-45-1]	$Si(OCH_2CH_2OCH_3)_4$
tetraphenoxysilane	[1174-72-7]	$Si(OC_6H_5)_4$
hexaethoxydisiloxane	[2157-42-8]	$(C_2H_5O)_3SiOSi(OC_2H_5)_3$
ethylsilicate 40	[18954-71-7]	$(OSi(OC_2H_5)_2)_{4-5}$

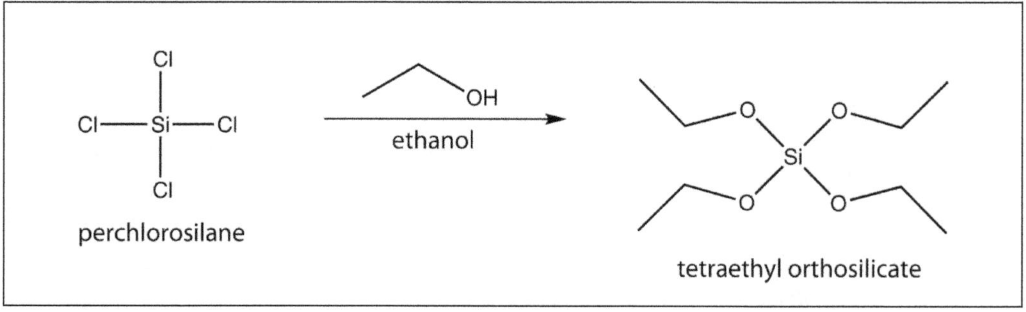

Figure 4.44: Preparation of ethyl silicate

bonds Si-O-Si, are often likened to the silicone oils. These have low pour points and similar temperature-viscosity relationships. The alkoxy silanes generally have fruity odours that become less apparent as molecular weight increases. The alkoxy silanes generally exhibit very low levels of toxicity, with the exception of tetramethoxy silane, trimethoxy silane, triethoxysilane, and a few other closely related compounds that can be absorbed into the corneal tissue, causing eye damage.

4.2.9.2 Manufacturing of alkyl silicates

The principal method of silicon ester production is described by *Von Ebelman's* 1846 synthesis

Continuous processes involve pumping the alcohol and chlorosilane together in a mixing section, introducing the chlorosilane vopour counter-current to liquid alcohol, or introducing chlorosilane vopour in a two-column reaction distillation scheme in which substoichiometric alcohol is introduced center-column to a chlorosilane. Removing hydrogen chloride (HCl) and unreacted alcohol in the overhead is followed by a slight excess of the alcohol beeing introduced center-column, to recover the excess overhead and remove product from the bottom of the column. All processes provide a method for removal of the by-product HCI. The energy of activation for the reaction of ethanol with silicon tetrachloride in the vopour phase is 64.9 kJ/mol (15.5 kcal/mol). The initial stages of the esterification processes are endothermic because the heat of evopouration of HCl cools the reaction mixture. In the last stages of esterification, the mixtures are usually heated during the final addition of alcohol.

In batch processes, the alcohol is always added to the chlorosilane. In the batch production of tetraethoxysilane from silicon tetrachloride, the initial reaction product contains at least 90 wt % tetraethoxysilane with SiO_2 content. By distillation the alcohol and high boiling impu-

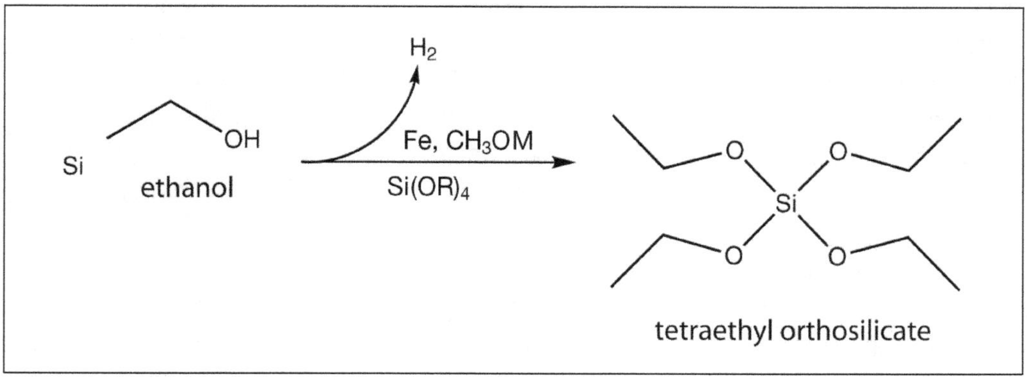

Figure 4.45: Ethyl silicate synthesis from Si and ethanol

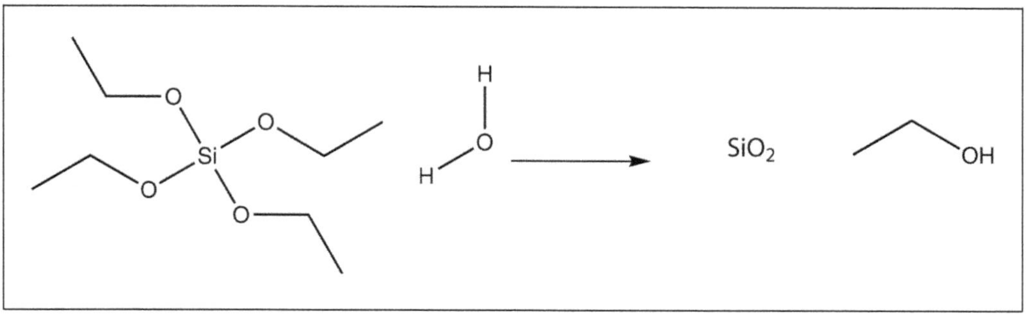

Figure 4.46: Hydrolysis of ethyl silicate

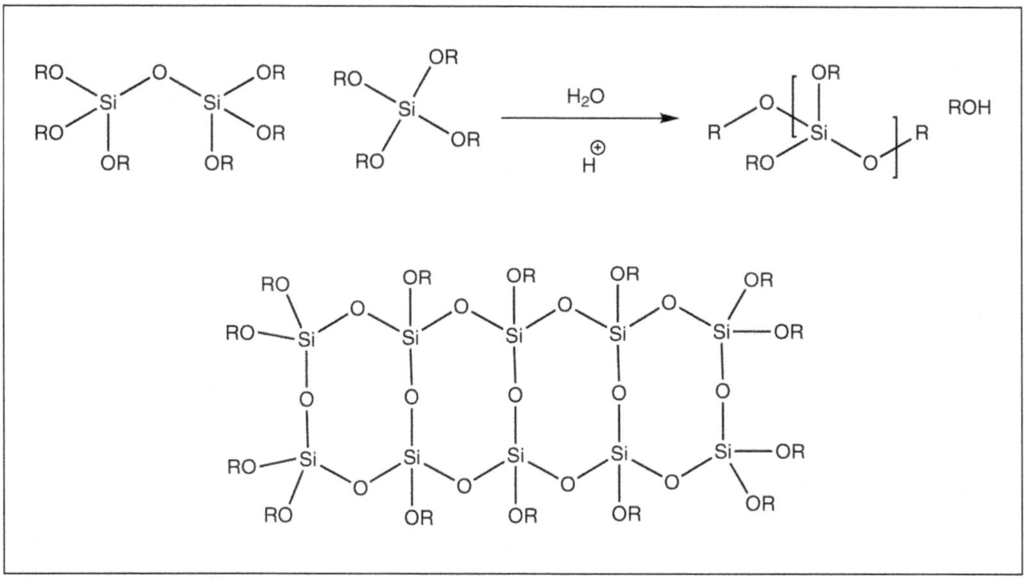

Figure 4.47: Curing of alkyl silicates

rities are removed, and the distilled product, containing at least 98 % tetraethoxy silane, is called pure ethyl silicate. Partially hydrolysed or polymeric versions, where substantial portions have an average of 4 to 5 silicon atoms and 40 % SiO_2 or have an average of 5 to 8 silicon atoms and a 50 % SiO_2 content, are respectively named ethyl silicate 40 and ethyl silicate 50.

Direct catalysed reactions of alcohols using silicon metal have become an important commercial technology for the production of lower esters. Patents have reported the reactions of methanol and ethanol with silicon in high boiling solvents, or in contained reaction products to give high yields of trialkoxy silanes and tetraalkoxy silanes. In the presence of a methoxy compound, where M is a metal, and under moderate pressure, substantial improvements in yield can be achieved.

4.2.9.3 Reactions of alkyl silicates

Ethoxy silanes are used in high-temperature, zinc-rich coatings. Methyl- and phenyltrialkoxy silanes are used to prepare abrasion-resistant coatings for plastics, especially polycarbonate, and dielectric coatings and seals for high voltage electrical components, including television tubes. Tetraethyl orthosilicate easily converts into silicon dioxide. This reaction occurs upon the addition of water:

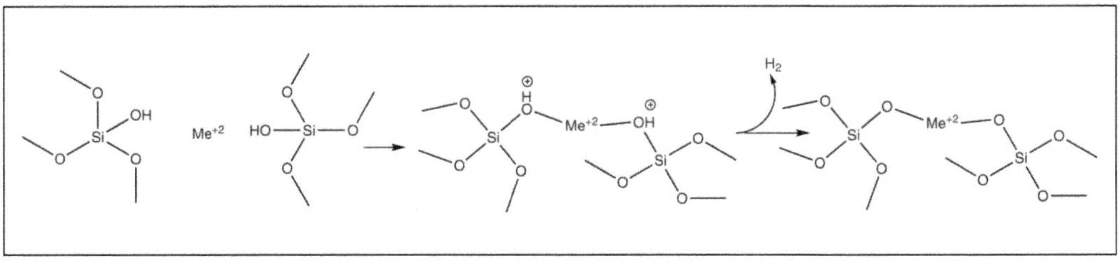

Figure 4.48: Curing of alkyl silicate in the presence of metal (Zn)

This hydrolysis reaction is an example of a sol-gel process with the side product ethanol. The reaction proceeds via a series of condensation reactions that convert the tetraethyl orthosilicate molecule into a mineral-like solid, via the formation of Si-O-Si linkages. The rates of conversion are sensitive to the presence of acids and bases, both having a catalytic activity. At temperatures over 600 °C, tetraethyl orthosilicate converts to silicon dioxide and diethylether.

In general, the curing of alkyl silicates goes through a range of reactions in presence of water and an acid catalyst.

The quantity of water added to the alkyl silicate should not be excessive as this leads to premature gelling. Water and elevated pH leads to silica precipitation and consequently reduces the capacity of formation of inorganic polymer with high molecular weight. Acids may also have negative impact as in excess they slow down the reaction by the repulsion of protonated hydroxyl groups. The target is the formation of a polymer of metallic silicates when metal is used in the formulation, such as zinc:

4.2.10 Polysiloxanes

Generically polysiloxanes relate to a group of polymers based on inorganic siloxane and organic-inorganic siloxane hybrids [56]. Silicon based polymers are inherently temperature resistant. Resistance against UV irradiation and oxidation also are important properties that have motivated attempts to use the chemistry for property improvement of both organic and inorganic systems, including good resistance to certain acids and solvents, high nonvolatile content and very good weatherability.

The use of silicon based polymers covers a history of more than 50 years. Silicate primers and waterborne silicate primers cured respectively by heat or by acid catalysis were commercial in the 1940s and 1950s. A significant progress has been recorded in 1978 by improving the curing chemistry at ambient temperature through the introduction of trialkyl silane. Later on in 1981 the development of binders based on interpenetrating networks from a polysiloxane network and an epoxy-polyamine network was an important progress yielding coating systems with good adhesion, flexibility, acid and solvent resistance.

The name siloxane is pointing out the chemical bond Si-O-Si. The reactions of these compounds may be listed as follows [58]. Siloxanes, considered as hybrids between glass and organic resins, exhibit an inertness beneficially compared to other organic polymers and may be used to upgrade other binders or as such as they may react with all materials containing hydroxyl groups. They exhibit an excellent weatherability, good colour and gloss retention and less chalking.

4.2.10.1 Reactions of siloxanes

Reactions of monofunctional siloxanes yield hexaorganosiloxanes.

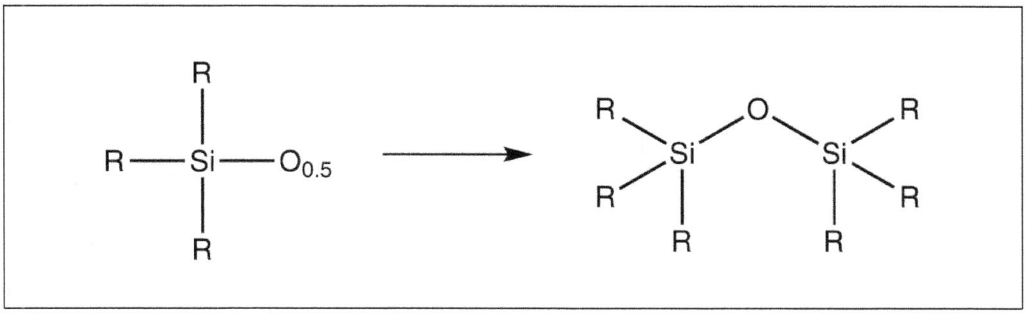

Figure 4.49: Reaction of monofunctional siloxanes

Difunctional siloxanes yield privileged rings.

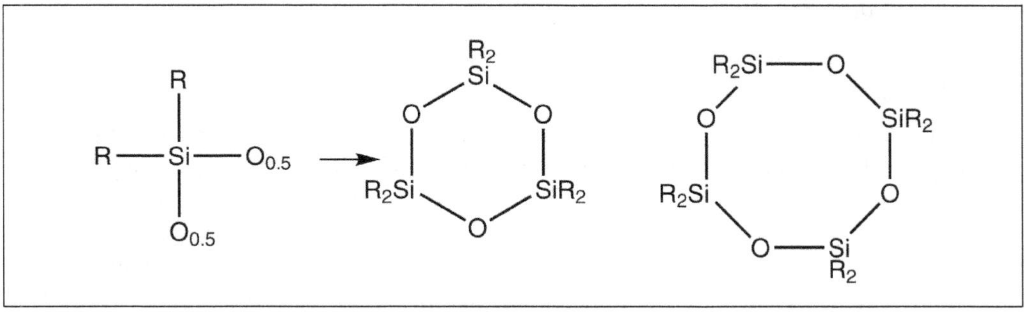

Figure 4.50: Reaction of difunctional siloxanes

Trifunctional siloxanes yield randomly cross-linked macromolecules.

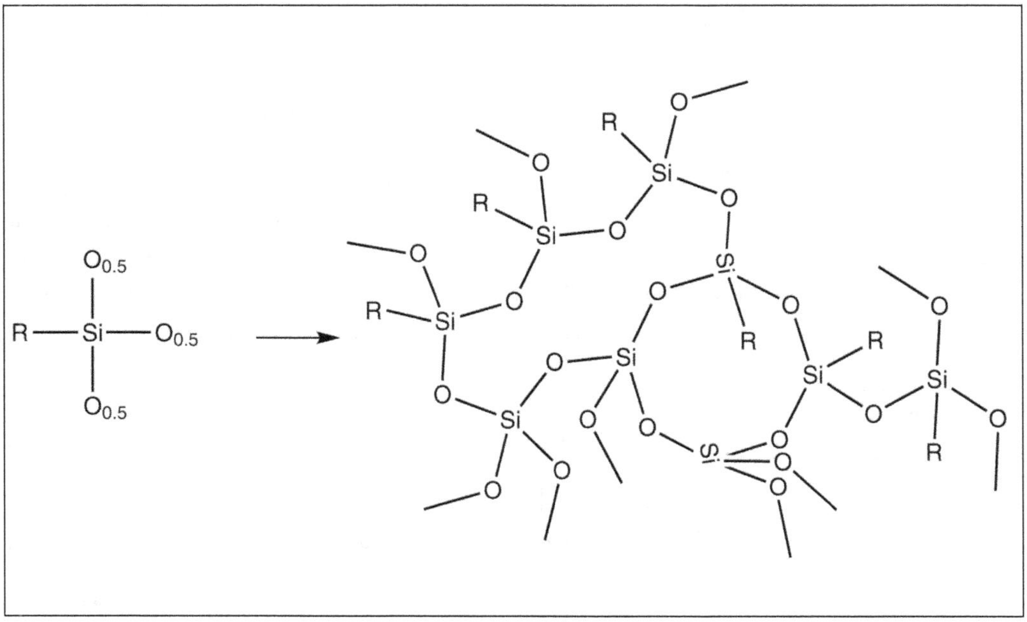

Figure 4.51: Reaction of trifunctional siloxanes

Di- and tetra functional siloxanes yield tetrahedral arrangements.

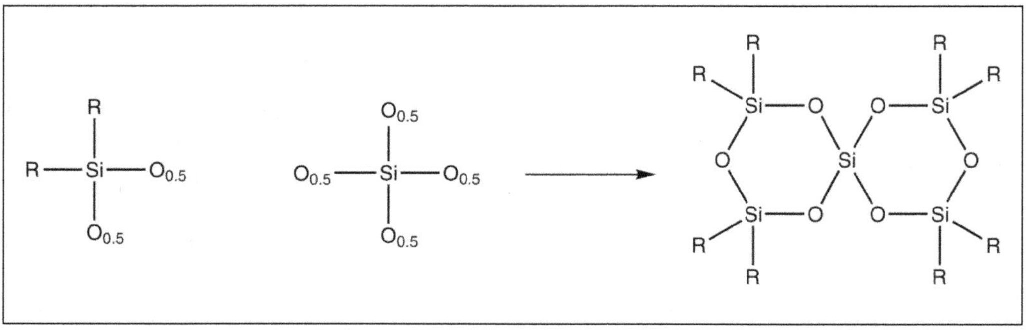

Figure 4.52: Reaction of di- and tetrafunctional siloxanes

For almost all industrial products, R is methyl or phenyl.

4.2.10.2 Manufacturing of siloxanes

The process of employing the siloxane in binder manufacturing follows the scheme as shown in Figure 4.53:

Polymethyl siloxanes have the lowest carbon content and therefore exhibit a long-term heat resistance at 180 to 200 °C. As well, these resins exhibit good water repellency and solvent resistance. At temperatures over 400 °C, the methyl groups are oxidised and the skeleton of the newly formed binder is extremely stable in combination with lamellar pigments.

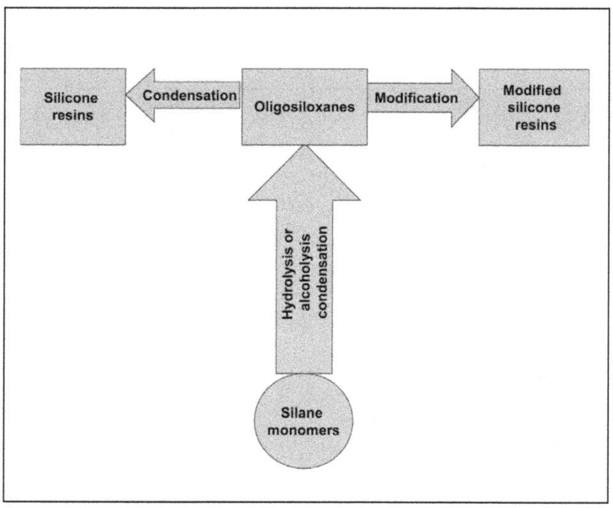

Figure 4.53: Manufacturing scheme for siloxanes

Polyphenylsiloxanes contain more than 20 % phenyl groups. The heat resistance is generally in the range of 200 to 250 °C. The polyphenyl siloxanes exhibit better pigmentability and very good solubility in aromatic solvents.

Polyesters, medium and long oil alkyds, and acrylic binders are reacted with siloxanes in a ratio of generally 70/30 to obtain resins

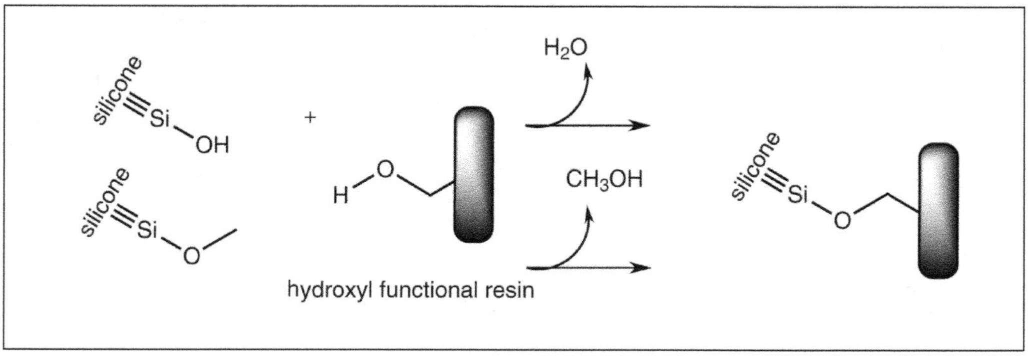

Figure 4.54: Siloxane modification of a hydroxyl functional polymer

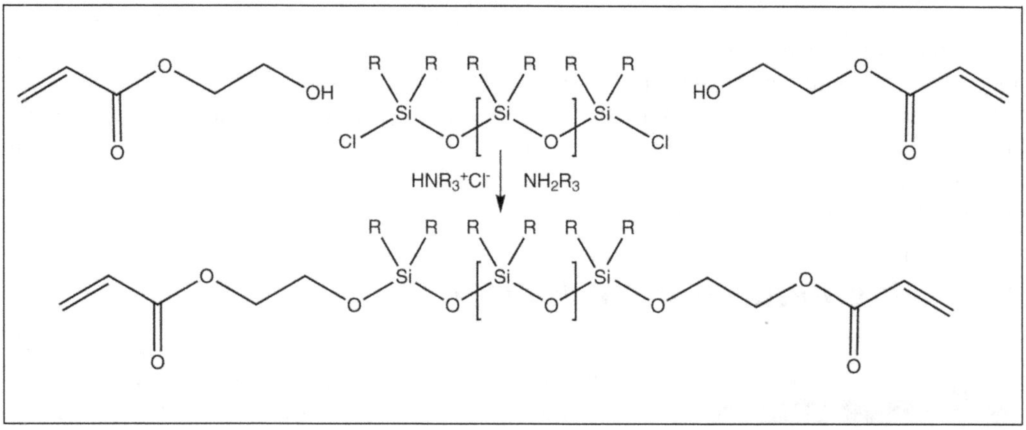

Figure 4.55: Acryl modified polysiloxane

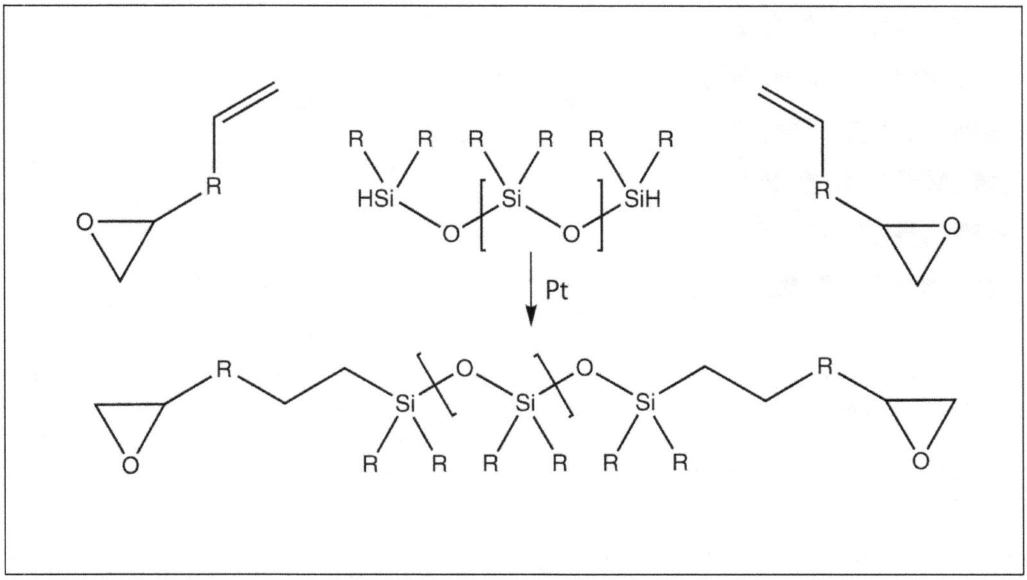

Figure 4.56: Glycidyl modification of a siloxane

with improved properties. This reaction is a transesterification reaction proceeding in solvent at a temperature of about 140 to 180 °C.

A UV curing system comprising ethylenically modified siloxane backbones is extremely interesting.

Glycidyl functional siloxanes are produced by the following reaction:

Further, the glycidyl group may be functionalised to any other functional group.

A particular case of the siloxane chemistry is room temperature vulcanising (RTV). The cross-linking of the resins may be performed by:

- addition between the vinyl groups and the Si-H groups
- vinyl polymerisation in the presence of peroxides
- condensation with α,-ω-dihydroxy-polydimethyl siloxanes and silica esters.

The products obtained by this method are castable, brushable, kneadable and exhibit permanent flexibility, good electrical insulating properties, good chemical resistance, etc.

Waterborne polysiloxanes are prepared by polymerisation of low molecular weight siloxanes in the presence of intermediates of tri or tetra alcoxysilanes and the aid of an emulsifier. Secondary emulsions are prepared from organic solvent solutions transferred to water followed by the solvent recovery by distillation. However an elegant method is to employ low molecular weight amino modified silicone resins neutralised by acetic acid to yield microemulsions. These microemulsions may be produced shortly before use as their curing process starts in the frame of few hours.

4.3 Pigments - introduction

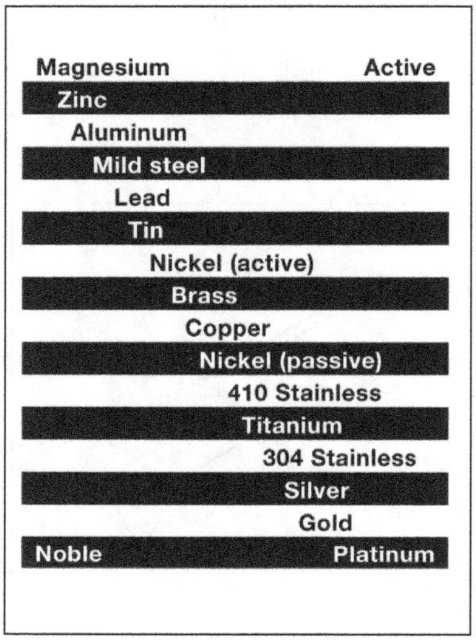

Figure 4.57: Electromotive series of metals in sea water

It is well-known that the incorporation of pigments with appropriate characteristics has a distinctive influence on the performance properties of a protective coating for metal substrates. If the resin is the most important part of a coatings formulation, the pigment and its ratio to the whole matrix is second [63].

Only a few metals occur naturally, namely e.g. the noble gold, silver and platinum. Other metals must be produced from their ores by smelting which requires enormous energy. Following the rules of Lenz the same energy will be released when the metal is reconverted back to its ore [63].

As a result, corrosion is the natural tendency of metals to move towards their thermodynamically most stable form, for example oxides, sulfides or carbonates. The more energy is required for isolating the metal by smelting, the higher the readiness of the produced metal to return to its lower energy state, which means corrosion. In the electromotive series of metals, the most active metals are those with ready release of energy, which means a higher affinity to corrode.

Corrosion is often described to be an electrochemical process of oxidation and reduction reactions. At the anode the metal dissolves as metal ions with release and flow of electrons. The electrons migrate through the internal phase to the cathode where the primary reaction is the consumption of the electrons by reaction with the environment (water/oxygen). The so-formed hydroxyl ions then react with ferrous ions to produce rust [64].

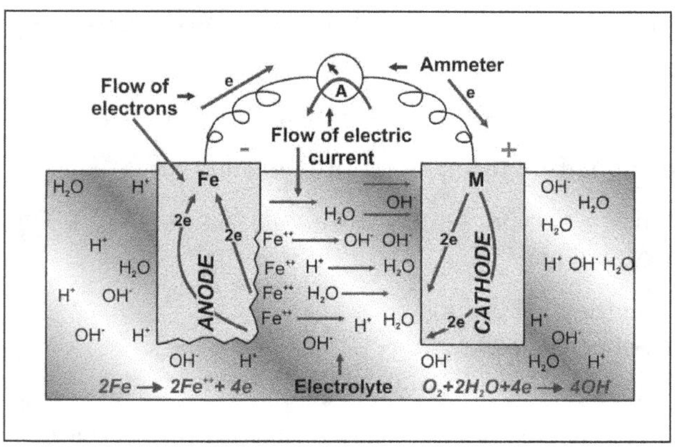

Figure 4.58: Electrochemical corrosion cell [63]

Every given metal shows heterogeneities in its composition, deriving from surface contamination or grain structures which have an impact on the readiness to corrode. Once being in contact with an electrolyte like moisture, those areas have a higher or lower tendency to corrode and will act as local anodes or cathodes in a galvanic cell.

Electrolytes might be made up of rain, snow and dew and contain water-soluble corrosion stimulators like chlorides and sulphate salts. These inorganic salts dissociate into their respective ions which provide an excellent pathway for the transmission of electric currents generated in the corrosion process. Corrosion control by means of protective coatings has to interfere with the above described electrochemical mechanism [63].

4.3.1 Corrosion protection pigments

According to accepted standards, the modern meaning of "pigment" is associated with a substance consisting of small particles that is nearly insoluble in the applied medium and is used for colouring, protective or magnetic properties [65]. Corrosion protection pigments need to have certain solubility in order to display a chemical or electrochemical activity. Nevertheless, if the solubility behaviour of the anticorrosive pigment is not adjusted properly, it could lead to osmotic effects and resulting blister formation with deterioration of the anticorrosive properties of the paint film.

Anticorrosive pigments have a strong influence on the performance of protective coatings in several ways, for example [66]:

- Increase of the durability of the coating system
- Prevention of undercutting in damaged areas and underfilm corrosion
- Decrease of the corrosion velocity
- Cathodic and/or anodic passivation of the metal surface

Inorganic anticorrosive pigments may be classified by the way they work as

- Active pigments (chemically/electrochemically)
- Barrier pigments (physically) or
- Sacrificial pigments (electrochemically/physically)

Often, optimal protection is achieved through a combination of differently active pigments.

Active pigments show a certain controlled solubility which results in either chemical and/or electrochemical activity. Passivation of the metal substrate can be achieved with pigments which show a strong oxidising potential such as chromates or nitrites. These are termed to be active in the cathodic area by consumption of the free electrons which are released when the metal at the anode passes into solution. This means a limitation of the anodic process. The pigments that inhibit the corrosion process by stimulating the growth of the adherent passive layer on the metal surface are said to be active in the anodic area. Moreover, neutralisation of acidic or corrosion stimulating substances like chlorides or sulphates is another pathway which can be chosen by utilisation of active anticorrosive pigments [67].

Barrier pigments like chemically inert aluminium, glass flakes and micaceous iron oxide are also said to be able to reduce the water permeability of paint films due to their lamellar or platelet structure. These pigments are packed in layers within the paint film and thus increase the diffusion path for water, oxygen and corrosion stimulating ions to the substrate. They can also improve adhesion between the metal surface and the coating [68].

Sacrificial pigments: An alternative way of protecting ferrous substrates is the utilisation of metallic pigments like zinc. Zinc-rich primers contain high levels of powdered zinc, in general more than 90 % on total formula weight is not unusual in order to ensure optimum contact of the zinc particles with the steel surface. The zinc itself acts as a sacrificial anode if its electrochemical activity in the electromotive series is higher than that of the metal which has to be protected. In this case zinc becomes the sacrificial anode and passes into solution whereas the substrate acts as the cathode. The whole process is therefore termed cathodic protection mechanism. In practice only zinc dust or flakes are of economic importance as sacrificial pigments. The electrochemical efficiency of zinc-rich coatings does not last forever due to corrosion products formed at the particle interfaces. Nevertheless, this reaction of zinc with the atmosphere might be beneficial in terms of filling pores or pinholes of the paint film with insoluble zinc compounds (cementation) [69].

Besides possible classification of anticorrosive pigments discussed above, there is one more class often described as having film-reinforcement and adhesion promoting properties. For example, micaceous iron oxid (MIO) is said to be able to reinforce the binder matrix. The ratio of length to thickness and the alignment of the MIO flakes in a coating toughen and strengthen the film. Coatings formulated with MIO can show greatly improved resistance to blistering and increased substrate adhesion. For example, MIO is used to promote adhesion in coatings formulated for galvanised steel [70].

Extender pigments, especially calcium silicate (wollastonite), have also achieved increasing importance as reinforcers for paint films. Due to its ability to maintain high pH values and decreased permeability via improved pigment/binder interfaces, wollastonite has found increasing use in protective coatings.

4.3.1.1 Lead and chromate pigments

4.3.1.1.1 Lead pigments

Lead-based paints have disappeared from consumer sales for residential use because of toxicity concerns. However, the use of lead tetroxide, or red lead, can be traced back to ancient Roman times where it was used simply as a red pigment. The Latin name **minium** derives from the Minius River in northwest Spain where it was first mined. Chemically, red lead is lead tetroxide or triplumbic tetroxide Pb_3O_4. It is a bright red or orange pigment, virtually insoluble in water. By contrast, its solubility in hydrochloric acid present in the stomach is responsible for its toxicity [71].

Lead tetroxide is usually produced on industrial scale by calcination of lead (II) oxide in a stream of air at 450 to 480 °C.

$$6\ PbO + O_2 \rightarrow 2\ Pb_3O_4$$

Its electrochemical effectiveness derives from the oxidation potential of red lead (reduction of Pb(IV) to Pb(II)). It is able to form insoluble plumbate complexes with iron and iron oxides. The ability to form lead soaps with fatty acids present e.g. in linseed oils is said to be responsible for the anticorrosive properties in paint films [72]. Moreover it has been suggested that the basic content of lead pigments promotes the reinforcement of the paint film itself with regard to mechanical sustainability, water resistance and adhesion, even on poorly prepared and rusty surfaces [73].

Among all lead based anticorrosive pigments, red lead gained the most industrial importance in the past, followed by lead silico chromate, a so-called core pigment type of anticorrosive. It is a composite pigment in which the active lead chromate is precipitated on a core of silica particles. Due to the silica core, this pigment shows considerably lower density (4.1 g/cm^3) compared with red lead (8.8 g/cm^3) [73-75]. However environmental and human health awareness referring to the lead and also chromate content prevented this pigment from having an economic breakthrough.

4.3.1.1.2 Chromate pigments

There is almost no other group of anticorrosive pigments that has been the focus of so many detailed investigations than chromates. The wide success of hexavalent chromium (VI)

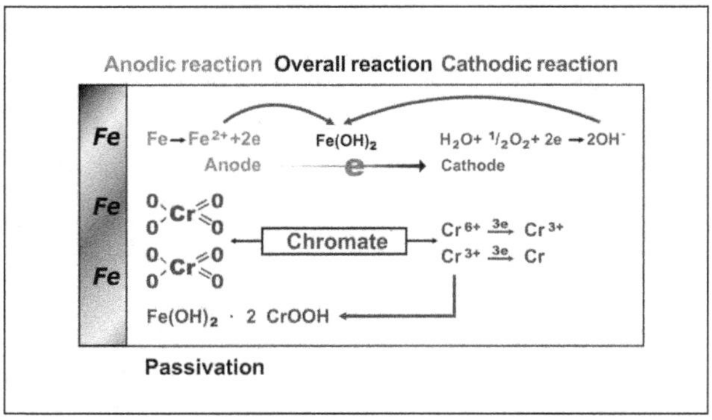

Figure 4.59: Corrosion of iron and inhibitive effects of chromate pigments [73]

pigments in the past is based on their ability to be effective in both, the anodic and also in the cathodic areas of ferrous substrates. The anticorrosive properties are based on chemical and electrochemical reactions. Protective paints should be accurately formulated at controlled ratios of pigment volume concentration and critical pigment volume concentration (PVC/CPVC), to ensure a controlled rate of moisture permeability within the paint film. When this rate is adjusted to the solubility of the chromate pigment, chromate ions will migrate to the metal surface to form passive films [76].

The high amount of water soluble chromate which can be reduced at the cathodic areas is responsible for the passivation of metal. For steel, a layer of mixed oxide, Fe(III) and Cr(III), was found which is constantly "repaired" due to the high oxidising potential of hexavalent chromium compounds. Nevertheless, the layers do breakdown in the presence of depassivating or corrosion stimulating ions like chlorides or sulphates. As these ions originate from outside the paint film, a properly formulated primer paint film which should be more permeable to water than to salt solutionshould restrict the access of corrosion stimulators to the metal interface.

Electrochemical passivation and chemical reaction are illustrated in Figure 4.59.

However, if the PVC/CPVC ratio of the formulated primer is chosen so high that the film is also permeable to ionogenic solutions (e.g. sodium chloride), the passivation process of the substrate may be prevented [63].

The effectiveness of chromate based pigments is therefore related to the:

- Amount of water-soluble chromate ions
- Ratio of water-soluble chromate to amount of corrosion stimulating ions
- PVC of the primer formulation
- Particle size and distribution
- Dispersibility
- pH stabilising effect of $Zn(OH)_2$ present in zinc containing chromate pigments

As described before, besides the solubility of the chromate within the paint film, the $Zn(OH)_2$ content in zinc containing pigments plays an important role. The chemical effectiveness is

Table 4.7: Solubility of chromates in water at 25 °C [77]

Pigment	Formula	Grams of CrO_3/ 1000 ml of H_2O
Calcium chromate	$CaCrO_4$	17
Zinc potassium chromate	$4\ ZnO \cdot K_2O \cdot 4\ CrO_3 \cdot 3\ H_2O$	1.1
Zinc tetrahydroxychromate	$ZnCrO_4 \cdot 4\ Zn(OH)_2$	0.6
Strontium chromate	$SrCrO_4$	0.6
Barium chromate	$BaCrO_4$	0.02

due to the neutralising and pH stabilising properties towards acidic decomposition products of the resin itself.

Zinc potassium chromate or zinc yellow

$K_2CrO_4 \cdot 3\ ZnCrO_4 \cdot Zn(OH)_2 \cdot 2H_2O$

Zinc chromate based pigments are now classified as either confirmed or at least suspected human carcinogens. Zinc potassium chromate gained major economic importance primarily due to the broad application latitude and compatibility with resins and extender pigments in industrial primer paints [69].

Zinc tetrahydroxy chromate $ZnCrO_4 \cdot 4\ Zn(OH)_2$

Zinc tetrahydroxychromate has been widely used for wash and shop primer applications based on polyvinyl butyral (PVB) resin systems. The phosphoric acid which is present in PVB wash primers reacts with zinc hydroxide and chromate to form adhesive complexes on the metal substrate [69].

Strontium chromate $SrCrO_4$

One of the key areas of application for strontium chromates is in primers for aluminium. It still maintains commercial importance as it is considered to be one of the most effective anticorrosive pigments for aircraft and coil coating primers [69].

To summarise the benefits of chromate based anticorrosive pigments [78]:

- Chromates can be stored as conversion coatings and as pigments in paints.
- Soluble chromates are able to refresh scratched areas of a coating. The released chromate ions are in equilibrium with the chromate in the coating.
- Chromate is soluble and mobile in solution, so it migrates to exposed areas on the metal surface. Chromate absorbs on the active site of the surface to form a protective monolayer of a Cr(III) species.
- Inhibition of oxygen reduction at the cathodic area is an important part of the overall corrosion inhibition mechanism.

Chromate based pigments are toxic substances. Government regulations, concerning both protection of the environment and protection of human health have already prompted paint manufacturers and end-users to favour the use of non-chromium pigments [79].

4.3.1.2 Phosphate based pigments

4.3.1.2.1. Zinc phosphate

The use of lead and chromate based pigments has been restricted by increasing awareness for human health protection. Thus, a large amount of detail had been placed upon zinc phosphate and its protective properties. Due to the very low solubility compared to chromate based pigments it can be used with a large variety of protective binder systems which explains the ongoing economic importance in the protective coatings world· [68, 69]. Still, zinc phosphate is one of the most commonly used phosphate anticorrosive pigments in the market.

Zinc phosphate is produced on industrial scale by reacting zinc oxide with phosphoric acid in a wet precipitation process. ISO 6745 - Zinc phosphate for paints - Specifications and methods of test - defines a zinc phosphate to be either a dihydrate type ($Zn_3(PO_4)_2 \cdot 2\ H_2O$), a tetrahydrate (($Zn_3(PO_4)_2 \cdot 4\ H_2O$) or a mixture of di- and tetrahydrate. According to ISO 6745 there may be differences in the protective behaviour depending on the type of zinc phosphate which is used [80].

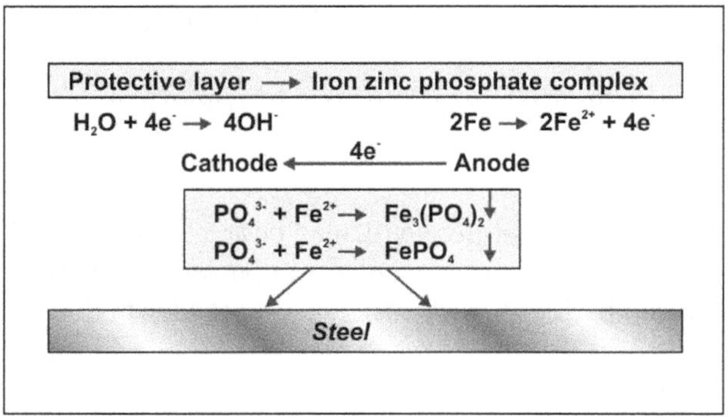

Figure 4.60: Performance of zinc phosphate

Table 4.8 summarises the main requirements on zinc phosphate pigments.

Pure zinc phosphate is extremely insoluble in the neutral pH range. Nevertheless, the secondary phosphate ions formed due to the very low solubility are said to be responsible for the formation of adherent phosphate inhibitor complexes on the metal substrate. Another theory for the mechanism of protection, claims the formation of tribasic iron phosphate films upon the metal surfaces.

It is well-known that the protective behaviour of zinc phosphate correlates strongly with the pigment structure, particle size distribution and dispersibility of the pigment. Micronisation improves the anticorrosive properties [76, 81]. Furthermore, investigations have shown that the inhibitive action predominantly takes place in slightly acidic environments of pH 4.

Under moist conditions, the formation of basic complexes by reaction of zinc phosphate with inorganic ions or with carboxylic groups of the resin, which leads to so-called adhesion by reaction with metal ions, cross-linking and inhibitor complexes, is being discussed in the literature [73, 76, 92]. In contrast to chromate based anticorrosives, the formation of inhibitive layers on the metal surface is somewhat delayed. The complexation reaction does not start before the permeability of the coating allows enough moisture to penetrate from outside for the zinc phosphate to hydrolyse. Practical experience has shown that zinc phosphate is an active anticorrosive pigment with a minor chemical or electrochemical effectiveness compared to lead and chromate based pigments. The performance of those pigments can only be achieved in certain protective primer systems [73, 76].

4.3.1.2.2 Modified orthophosphates

As mentioned before, although the low reactivity as well as the low water solubility of zinc phosphate is advantageous in terms of paint stability, it does also limit the rate at which the active ions are released to protect the substrate [68]. Phosphates also require oxygen to be effective. Nevertheless, it is possible to increase the activity by modifying the pigment with other functional groups, different anions, organic surface treatments or by blending with other inorganic anticorrosives (taking advantage of synergistic interactions). Decreased particle size, particle size distribution and particle shape have also been found to have an effect on the performance of anticorrosive pigments. Therefore, the pigment industry has put a major focus on developing phosphate based pigments with improved properties [76, 81].

Table 4.8: Quality requirements of zinc phosphate

Characteristic	Unit	Requirement
Zinc as Zn	%	50.5 to 52.0
Phosphorous as PO4	%	47.0 to 49.0
Conductivity	µS cm^{-1}	max. 154
pH		6.0 to 8.0
Density	g cm^{-3}	3.0 to 3.6
Sieve Residue (45 µm)	%	max. 0.5
Colour		white

With basic zinc phosphates, for example $Zn_2(OH)PO_4$ with an increased hydroxyl anion concentration, it is possible to stabilise the pH value in the coating and also the equilibrium of the cathodic reaction, resulting in an inhibition of the emission of electrons [83, 84].

Zinc aluminium phosphate is produced by wet precipitation of aluminium phosphate with zinc oxide. This yields a pigment with higher phosphate ion content than standard zinc phosphate. Also, the aluminium ions hydrolyse, causing slight acidity which in turn increases the solubility and phosphate concentration. The improved performance is attributed to this higher phosphate content, providing greater interaction with the resin's carboxyl groups (soap formation) and the metal substrate [83-85].

Basic zinc molybdate phosphates have an improved anticorrosive activity due to the molybdate anion as anodic passivator with a passivating capacity which is only slightly less than that of the chromate anion [68].

The electrochemical activity of the surface treatment is responsible for the efficiency of organically modified zinc phosphate. Improved resin/pigment bonding and coating/substrate adhesion are discussed [83, 86].

Phosphate/borate combinations are used due to their improved readiness to hydrolyse. Phosphate/silicate types of pigments are so called mixed phase or core pigments [73, 86, 82, 92, 93].

All of the mentioned improvements in performance properties were possible by taking advantage of synergies. Using synergistic values is not a new concept in the coatings industry. The most important argument for taking the synergy pathway might be the increased water soluble content resulting in a better supply of inhibitive ions combined with other synergistic properties [67].

Table 4.9: Orthophosphate based anticorrosive pigments

Pigment	Modification
Zinc aluminium phosphates	with aluminium phosphate [83-85]
Basic zinc phosphates	contains basic contents (e.g. zinc hydroxide) [83, 84]
	with different organic compounds treated [83, 86]
	with zinc molybdate and/or calcium molybdate [73, 86, 88]
	with basic zinc borate [73, 86]
	with iron phosphate [89]
	with calcium phosphate [88]
	with potassium phosphate [84]
	with barium phosphate [88]
	with aluminium phosphate and zinc molybdate [90]
	with inorganic compounds treated [86]
Zinc-free phosphates	altering the cations e.g. with calcium phosphate and/or magnesium phosphate [91, 92]
Zinc phosphate silicates	with calcium silicate and strontium phosphate [90, 91, 93]
	with barium phosphate [94]
	with calcium carbonate [95]
	with different organic compounds treated [91, 94]
Zinc-free phosphate silicates	variation of the cations, e.g. with calcium phosphate silicate [94]
	with barium phosphate [94]
	with strontium phosphate [94]

4.3.1.2.3 Modified polyphosphates

Though the improvement in the performance of standard zinc phosphate by chemical modifications is remarkable, there are still some special applications, e.g. coil coatings and aircraft primers, where the effectiveness of chromates could only be approached to a limited extent. Intensive research in the field of pigments for those special high-performance applications has led to a new family of anticorrosive pigments based on modified polyphosphates [75].

While orthophosphates are produced by reaction of orthophosphoric acid with basic or amphoteric compounds, polyphosphates are manufactured by condensation of acidic orthophosphates at high temperatures [67].

- Orthophosphates
- $3\ ZnO + 2\ H_3PO_4 \rightarrow Zn_3(PO_4)_2 + 3\ H_2O$
- Polyphosphates
- $Al(H_2PO_4)_3 \rightarrow AlH_2P_3O_{10} + 2\ H_2O$

Recently, new applications have been opened up again by modifying the chemical structure, resulting in different electrochemical effectiveness. Most of the modified polyphosphates are conversion products of acidic aluminium tripolyphosphate with compounds based upon zinc, strontium and calcium [96]. One key property of the polyphosphate grades is the significantly increased phosphate content. The corrosion protection is attributed to the ability of the tripolyphosphate ion to chelate metal ions (passivation) and to depolymerise into orthophosphate ions, providing higher phosphate levels than orthophosphate pigments [97]. Aluminium tripolyphosphate has limited solubility in water and is modified with zinc, calcium or strontium to control both solubility and reactivity.

Also, the development of modified polyphosphates as active pigments was possible by taking advantage of synergistic inorganic-inorganic interactions within one pigment [86].

4.3.1.3 Inorganic/organic synergies

Synergy is defined as a mutually advantageous conjunction of distinct elements. Intense research work in the field of new anticorrosive pigment concepts has led to the conclusion that there is also the possibility to combine the advantages of modified phosphate pigments with the advantages of organic corrosion inhibitors. This so called "2-way synergy" is advantageous in terms of providing improved substrate protection in the initial phase of exposure and thus resulting in improved long-term protection of the application [98].

Table 4.10: Polyphosphate based anticorrosive pigments

Pigment	Modification
Zinc aluminium phosphates	with aluminium phosphate [86, 96]
	partly with calcium silicate [86]
	partly with strontium chromate [86]
	with silicon dioxide [73]
	with different organic compounds [86, 96]
Zinc-free aluminium phosphates	with strontium phosphate [86]
	with calcium silicate [86]
	with magnesium phosphate [96]
	with barium phosphate [88]
	with inorganic compounds [86]

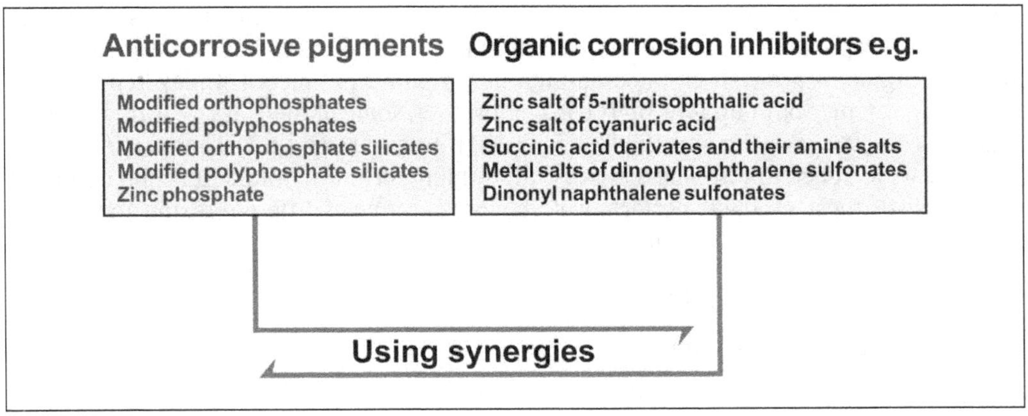

Figure 4.61: Inorganic/organic synergies

Taking advantage of synergies by combining modified phosphate pigments with an organic corrosion inhibitor leads to improvement of early substrate protection resulting in increased long-term substrate protection, a reduced tendency of blistering and enhancement of adhesion, especially wet adhesion [90, 91, 98].

The requirements on organic corrosion inhibitors are as follows:

- High activity at low concentration levels (typ. 0.5 to 2 %)
- Sufficient thermal stability
- High activity in the range of pH 5 to 9 (preferably pH 2 to 14)
- Compatibility with a broad variety of resin systems
- Easiness to add in
- Low water-solubility

Of the variety of potentially suitable corrosion inhibitors, the following compound classes are generally considered to be useful for coating applications [99]:

- Organic acids and their salts
- Organic bases/amines
- Organic compounds with oxidising properties

The efficiency of organic inhibitors is attributed to different mechanisms. On one hand they are said to interact with either the anodic or cathodic corrosion reactions. On the other hand these inhibitors are able to form protective layers on the metal substrate by adsorption and to improve the barrier function of the paint film [73]. A positive influence on the wet adhesion to the substrate is also being discussed. Nevertheless, to provide early substrate protection with long-term performance, it is still essential to combine inorganic anticorrosive pigments with organic inhibitors [86].

4.3.1.4 *Wide spectrum anticorrosives*

Chromate- based pigments have long been utilised as anticorrosives. In general, the utilisation of chromate anticorrosive pigments has shrunk significantly as compared to previous decades. While unrivalled in their performance and flexibility in inhibiting corrosion, they are classified as either confirmed or suspected human carcinogens and as "dangerous for the environment". International legislation, political and ethical pressures have accelerated the search for viable commercial alternatives [100].

Even though one key feature of modified orthophosphate based anticorrosives is their broad application latitude, it is necessary to select those pigments which are recommended by the pigment supplier for any particular application. For a long time the pigment industry has put focus on developing anticorrosives for universal use [86]. The modern wide spectrum anti-corrosive pigments are well-engineered high-performance products with layered systems of active inorganics built up over inert cores. Moreover, some of them are post-treated with organic or organo-metallic sheaths [101]. Table 4.11 summarises typical properties of two modified wide spectrum anticorrosives. Unlike many of the non-toxic phosphate based anticorrosives these products perform well in a broad range of different resins and in both accelerated and natural environments.

4.3.1.5 Phosphites and phosphides

4.3.1.5.1 Zinc hydroxyphosphite

Zinc hydroxyphosphite is produced by reacting zinc hydroxide with phosphorous acid and a complex promoter. It has the theoretical chemical composition [2 $Zn(OH)_2 \cdot ZnHPO_3 \cdot x\ ZnO$]. Zinc hydroxyphosphite is a white pigment with basic characteristics and low water solubility [69, 102]. It is described to form passivating iron phosphite and phosphate layers on steel substrates. The zinc oxide content helps to inhibit corrosion by forming zinc soaps in oleoresinous resin systems [68].

Table 4.11: Typical properties of two assorted pigments for universal application [90]

Specific data	Unit	Zinc aluminium molybdenum phosphate	Zinc calcium strontium aluminium phosphate silicate
Zn as ZnO	%	62	37
Phosphorous as P2O5	%	27	18
Aluminium as Al_2O_3	%	2	3
Molybdenum as MoO_3	%	1	-
Calcium as CaO	%	-	14.5
Silicon as SiO_2	%	-	15.5
Strontium as SrO	%	-	5
Conductivity	µS/cm	max. 300	max. 100
pH		7,.	7.5
Specific gravity	g/cm³	3.7	3.5
Oil absorption value	g/100 g	18	30
Average particle size	microns	3	3
Colour		white	white

Table 4.12: Typical properties of molybdates

Specific data	Unit	Calcium zinc molybdate	Calcium zinc phosphate molybdate	Zinc molybdate	Zinc phosphate molybdate
Core material		calcium carbonate	calcium carbonate	zinc oxide	zinc oxide
Colour		white	white	white	white
Specific Gravity	g/cm³	3.0	3.0	5.1	4.1
Oil absorption value	g/100 g	18	18	11	13.5

4.3.1.5.2 Iron phosphide

Commercially available iron phosphide pigments are grey ferro-alloy pigments which primarily consist of Fe_2P with traces of FeP and SiO_2. It is mainly used for partial replacement of zinc dust in organic and inorganic zinc-rich primer systems. These coatings protect steel by cathodic protection if an electrical path between the pigment particles and the steel substrate is given. Iron phosphide is highly conductive and therefore able to provide an electrochemical path between the zinc particles. It is also said to improve the weldability of a coating if up to 40 % of zinc dust is replaced by iron phosphide [104]. Unlike zinc dust or lamellar zinc flakes, iron phosphide type pigments have not yet gained major commercial importance [64].

4.3.1.6 Borates

4.3.1.6.1 Barium metaborate

Barium metaborate $BaO \cdot B_2O_3 \cdot H_2O$ is the oldest borate and has been available since the 1960s. The alkalinity helps to neutralise acidic ions and decomposition products of the binder. In order to reduce the high solubility of barium metaborate it is often coated with silica [63]. Silica coated barium metaborate is produced by precipitation of barium sulphide and borax in the presence of sodium silicate. Inhibition is derived through the alkalinity and the ability of the metaborate ion to provide anodic passivation. It should be noted that soluble barium is an acute toxicant [103].

4.3.1.6.2 Zinc borate

Usually zinc borate is used as a flame retardant in plastics and cellulosic fibres. It is known that the combination with barium metaborate or zinc phosphate exhibits synergistic effects resulting in improved anticorrosive performance properties [103].

4.3.1.6.3 Calcium borosilicate

Calcium borosilicate is particularly effective in long and medium oil alkyds [103]. It is sometimes termed as a core pigment like phosphate silicates. It takes advantage of the borate and silicate chemistry, both soluble and basic, which helps to neutralise acid ions but also to form borate scales on the anode [74].

4.3.1.7 Molybdates

The basic molybdates introduced in the 1970s are anodic inhibitors and are discussed to form a protective layer of insoluble ferric molybdic oxide, quite similar to that produced by chromates. The high cost of molybdenum has limited the acceptance of molybdate pigments. [68, 105]. Many molybdates nowadays are blended with zinc oxide, zinc phosphate or fillers like calcium carbonate to overcome the price issue. As both zinc oxide and zinc phosphate are corrosion inhibitors, those blends show good anticorrosive performance properties in most solvent-borne resins but also in waterborne acrylic paints [103].

Table 4.13: Typical properties of commercial ion-exchange pigments [86, 108]

Specific data	Unit	1	2	3
Calcium content	%	6	6	6
Oil absorption value	g/100 g	60	80	90
Specific gravity	g/cm³	1.8	1.8	2.2
pH		9	9	7.5
Average particle size	microns	3	5	3

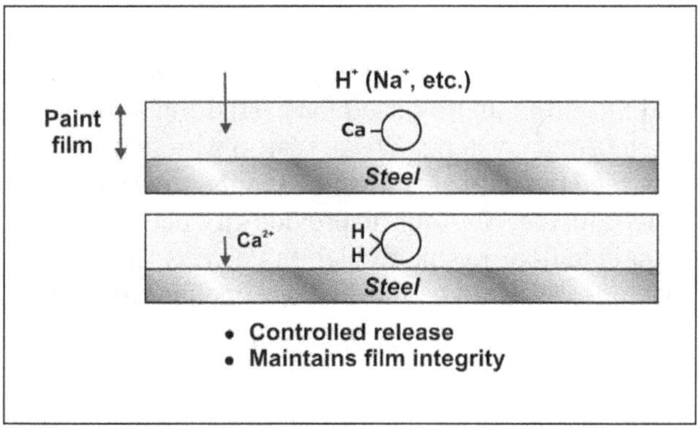

Figure 4.62: Protective behaviour of ion-exchanged pigments

4.3.1.8 Ion-exchange pigments

Ion-exchange pigments are the results of calcium ion exchanging on a microporous silica gel [106]. The surface area is quite high compared to conventional anticorrosive pigments. The surface is covered with calcium ions which are responsible for the alkaline nature of these pigments.

Their anticorrosive performance is attributed to their ability to adsorb aggressive ions on the silica surface. At the same time, the release and subsequent migration of the calcium ions lead to the formation of a protective layer on the surface of the metal substrate [108]. The advantage of this mechanism is that the anticorrosive species is used only where and when required. The pigment morphology does not change during the weathering which prevents the paint film from breakdown and formation of cracks. Due to the very low specific gravity of the silica based pigment there is a high number of active particles available for a given weight [86, 108]. Therefore these pigments offer a different type of chemical and physical constitution and protective behaviour than most other anticorrosive pigments [64].

4.3.1.9 Zinc cyanamide

Zinc cyanamide, $Zn(CN)_2$ is a white to beige powder and is used as non-toxic replacement for lead cyanamide in mirror back coatings, electrodeposition coatings and primer applications [109]. It is produced from industrial grade calcium cyanamide and zinc oxide. Until now this pigment has not reached the economic importance of lead cyanamide.

4.3.1.10 Hybrid anticorrosive inhibitors

Hybrid corrosion inhibitors are described as composite pigments providing the best of both inorganic and organic worlds in terms of anticorrosion synergy. Due to increasing regulation of anticorrosive pigments, combined with the requirement for chromate-like performance properties, there has been a steady growth in the value of non-toxic additives that are synergistic or multifunctional [110]. It is proven that synergy can produce positive effects with various corrosion inhibitors [77]. There are several pigments available on the market referred to as hybrid corrosion inhibitors. One example is an anticorrosive based on an oxyamino salt of magnesium or calcium. It is said that compositions of oxyaminophosphate salts were found to form stable layers, composed of magnesia (MgO) on the steel surface to be protected. The layers act as an effective barrier against corrosion on aluminium (including alloys), and other metals in a wide range of applications. [111].

US 6,139,610 patent describes a solid phase inorganic/organic hybrid with a selection of metal cations and different anions like phosphates, polyphosphates, phosphites, molybdates or silicates. The organic part includes ammonium salts of mercapto- and thio compounds and corresponding derivatives which form a stable unitary hybrid with the inorganic portion [119]. There is another pathway of corrosion prevention by utilisation of hybrid inhibitors.

Early research into the use of electroactive conducting polymers (ECPs) as corrosion inhibitors showed that the ECPs can provide corrosion inhibition to various substrates. Unfortunately, the ECPs formed in this manner often display poor film quality. In order to facilitate industrial application, polypyrrole (PPy) coated aluminium flakes were developed, acting as anticorrosive pigments in a binder system, synthesised through a new aqueous chemical oxidation process [112].

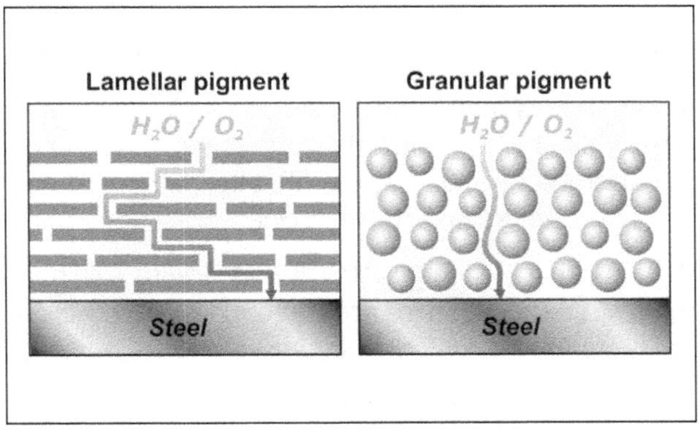

Figure 4.63: Barrier effect of lamellar pigments

4.3.2 *Barrier pigments*

Lamellar pigments such as mica and micaceous iron oxide form a barrier of flat and thin particles which limits the permeability of the paint film. The degree of reducing the permeability depends on:

- Film thickness
- Type of binder
- Cross-linking ratio of the binder
- Pigment volume concentrations
- Type of particle and particle shape

The main difference between barrier pigments and active anticorrosive pigments is that barrier pigments need to be absolutely insoluble and inert to provide their best possible performance [97].

4.3.2.1 *Micaceous iron oxide*

Micaceous iron oxide (MIO) is a crystalline form of iron oxide (α-Fe_2O_3, haematite) that differs from the more familiar red, yellow, and black forms of iron oxide pigments. The name derives from mica. Its crystals are very easily fractured into thin flakes giving it a physical form similar to mica. MIO is normally dark grey to black in colour.

MIO flakes, when applied properly, align themselves parallel to the substrate surface, producing a shield or barrier of overlapping plates. This alignment enhances protection by providing barrier protection, ultraviolet (UV) absorption, film reinforcement and increased intercoat adhesion. Because the flakes are impermeable, a physical barrier is formed that prevents the access of water, oxygen and corrosion stimulating ions, thus preventing corrosion of the steel and degradation of the binder [115]. MIO reinforces the binder matrix. The aspect ratio (ratio of length to thicknewss) and the alignment of flakes in a coating toughen and strengthen the film. Coatings formulated with MIO can show greatly improved resistance to blistering and increased substrate adhesion [116]. The incorporation of MIO might produce, after weathering, a surface with an excellent physical profile for subsequent coats. Alternatively, coatings may be formulated with higher levels of MIO to boost the intercoat adhesion of freshly applied systems [117].

For effective barrier properties PVC of 25 % to 45 % should be used with material of at least 80 % purity [97].

The most popular example of the use of MIO pigments is probably the Eiffel Tower in Paris. Since its erection in 1889 MIO has been used in maintenance paints on this tower.

Newer studies in protective coatings have shown no major differences in the performance properties of MIOs with different shapes i.e. granular and more lamellar [118]. Therefore, recent studies focus on possible alternative protective mechanisms besides the barrier effect of lamellar shaped iron oxide.

MIOs are specified in ISO 10601 – Micaceous iron oxide pigments for paints – Specifications and test methods. There are three different grades according to ISO 10601 with different lamellar particle content [121]:

- Grade 1: thin-flake content >50 %
- Grade 2: thin-flake content 10 to 50 %
- Grade 3: thin-flake content <10 %

4.3.2.2 Aluminium flakes

There are two different types of aluminium pigments available: Leafing and non-leafing. Leafing aluminium pigments show a tendency to float to the surface of the coating and to align themselves parallel to the surface of the coating film. It is necessary to overcoat leafing aluminium coatings with a top coat in order to protect the leafing metallic finish from unwanted corrosion processes and pigment abrasion [120].

Non-leafing pigments have a more random orientation in the coating and are very effective in barrier coatings. Systems containing non-leafing aluminium pigments perform significantly better under mechanical strain, with less abrasion of the metallic pigments and notably higher chemical resistance against corrosive chemicals compared to coatings based on leafing aluminium pigments [120].

4.3.2.3 Zinc flakes

Zinc flakes should not be mixed up with zinc dust which is used in zinc-rich primer systems. The platelet pigments consisting of metallic zinc might offer cathodic corrosion protection like zinc dust. Nevertheless, this should be difficult in practice since cathodic protection requires a gap-free electrical contact of the zinc flakes with each other and the metallic surface. Due to their lamellar structure they provide an additional barrier effect in the coating film, resulting in a significantly reduced permeability to corrosive agents especially in powder coatings [120]. Zinc flakes have been successfully proved in combination with other lamellar pigments (MIOs or aluminium flakes), spherical active particles like zinc phosphate or even zinc dust.

4.3.3 Sacrificial pigments

4.3.3.1 Zinc dust

In so-called zinc-rich coatings a high level of zinc dust is incorporated as a pigment. Particles of the less noble zinc in the paint film protect the ferrous substrate by acting as a sacrificial anode which corrodes in favour to the ferrous cathode. The sacrificial anode protects the ferrous cathode as long as zinc is able to pass into solution. Cathodic protection in this way is only possible if the zinc particles are in electrical contact with each other and the metal substrate. The basic requirement for sacrificial protection is for the sacrificial metal to be more electrochemically reactive than the metal substrate under the conditions of use. A guide to choosing a metal for sacrificial protection of another is their relative locations in the electrochemical series in seawater (see Chapter 4.3).

The most favourable loading level of zinc dust is still under discussion but normally primers have a content of more than 80 % in the dried paint film [122, 123].

Zinc is also able to react with atmospheric components like oxygen, moisture or carbon dioxide to form voluminous insoluble salts that might fill voids in the paint film and decrease the overall permeability. This process is often termed as cementation [123].

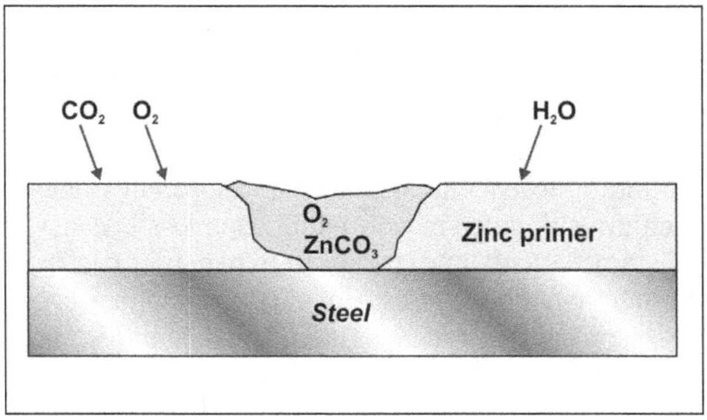

Figure 4.64: Cementation with zinc dust

Today, zinc dust coatings are widely used in organic and inorganic matrices for the protection of structural steel, marine coatings and underwater steel constructions [123]. Zinc dust is a blue-gray powder of spherical particles, which is produced from zinc metal by evopouration and condensation under controlled conditions. Usual particle sizes obtained by different condensation and sieving processes vary from 2 to 8 μm [122].

4.3.3.2 Magnesium

The need for chromium-free methods of protecting aluminium and its alloys from corrosion is very urgent, especially with respect to high-strength aerospace alloys [124]. Recently, a new concept of a magnesium-rich primer which contains magnesium metal powder for cathodic corrosion protection of aluminium alloys was developed which is quite analogous to a conventional zinc-rich primer for protection of steel (see Chapter 4.3.3.1). Magnesium is less noble and consequently more reactive than aluminium in the galvanic series of metals in seawater. Therefore the pigment should corrode as sacrificial anode preferentially to the substrate when there is an electrical contact between the magnesium particles and aluminium. At the time being, there are no aluminium skinned military aircraft which have authorisation to use completely chromate-free coating systems. Due to the cathodic protection mechanism, good performance requires no electrically insulating surface treatment layer be to present when magnesium is used. This effectively eliminates the need for a chromate- or other heavy metal containing conversion coating surface treatment. Since magnesium is a very active metal, concerns over the flammability of the pigment powder are alleviated through the presence of a very thin oxide layer present on the surface of the particles. [125].

4.3.4 Colouring agents

4.3.4.1 White pigment, titanium dioxide

Today, there is only one white prime pigment of economic importance, titanium dioxide. In protective coatings only the rutile form with a compact crystal structure is used. Nevertheless, even rutile has to be treated with metallic oxides to prevent chalking processes and provide exterior durability. Rutile is produced by either the sulphate or the chloride process [126, 127].

It is inert, has moderate oil absorption and specific gravity and is easy to disperse. Since 1999, there has been a new type of anticorrosive pigment on the market based on an inhibiting phosphate modified titanium dioxide [128, 129].

4.3.4.2 *Red pigments*

There are two general types of reds, the inorganic red oxides and high performance organic reds such as quinacridones. The best performing pigments used for coatings in aggressive environments are definitely synthetic iron red oxides with good chemical resistance, opacity and hiding levels and also low oil absorption values [130]. Cadmium reds made of cadmium selenium sulphide are now being used less because of toxicity concerns. They are also not resistant to acids and are quite expensive. The inorganic oxides have a lower degree of chroma than the organics. The high performance organic quinacridone pigments offer excellent chemical resistance, light and heat stability. Like all organic pigments they suffer from being relatively difficult to disperse, having high oil absorption values and low opacity [77].

4.3.4.3 *Yellow pigments*

Like the reds, there are two groups of yellows, inorganic and organic. For low-cost rather impure yellow ferrites are used (hydrated iron oxide). At high temperatures they turn red due to the loss of water of crystallisation [77].

Chrome yellows (lead chromate and lead sulphate) have a high chroma ranging from pale orange to greenish, with increasing lead chromate content [131]. They are sensitive to alkaline vehicles. Cadmium yellows offer excellent light stability but show poor acid resistance.

Bismuth vanadates are excellent chrome-free alternatives with high lightfastness, opacity and low oil absorption values.

Titanium yellows are very inert pigments with good opacity and low oil absorption. Despite their high price they are commonly used as inert high temperature pigments in high performance coatings [132].

Like the reds, the organic yellows suffer from low opacity and high oil absorption. High performance yellows are benzimidazolones with improved heat, weather and chemical resistance [133].

4.3.4.4 *Green pigments*

Chromium oxide green is a trivalent chrome product with good opacity which is inert and suitable for most protective applications.

The most important organic greens are phthalocyanines which offer excellent tinting strength and chemical resistance but low opacity in mass-tone [134].

4.3.4.5 *Blue pigments*

The only pigment of importance for the application in high-performance coatings is phthalocyanine blue. It is highly valued for its superior properties such as light fastness, tinting strength, covering power and chemical resistance to alkalis and acids [77].

4.3.4.6 *Black pigments*

Virtually all black pigments are inorganic, with carbon blacks and iron oxide blacks being the most common. Carbon blacks are fine pigments which are difficult to disperse. They offer good hiding power and relatively high oil absorption. Black iron oxide is not as strong a black as carbon black. All blacks are excellent UV absorbers and therefore impart excellent durability to the paint film [77].

4.4 Extender pigments

Traditionally, the accurate choice of extender pigments has not received much attention in the protective coatings world. Today, formulators use them as engineering tools to optimise and adjust the function of modern coatings.

Extender particles may be described by three different particle shapes: blocky, platy or lamellar, and acicular. Nevertheless, these pigments should better be classified by their chemistries i.e. carbonates, sulphates, silicates or oxides [77]. They do not only contribute to the aesthetics of a coating system but may also have a distinct influence on the overall performance.

4.4.1 Carbonates

Today, the only extender of economic importance is calcium carbonate. It is commercially available in both natural and synthetic forms. Natural calcium carbonate is a low-cost pigment with low oil absorption and specific gravity, which enables cost savings at high loading levels. Calcium carbonate pigments are designed to improve the rheological and mechanical properties of a coating. Their alkalinity might be advantageous for in-can corrosion protection and metal primers in non-acidic environments. On the other hand the alkalinity limits the utilisation due to the implied sensitivity to acids or acid rain for exterior use [135].

4.4.2 Sulphates

Like calcium carbonate for carbonates, in the case of sulphate extender pigments, barium sulphate is the only product of economic importance. It is chemically inert, insoluble in water, which makes this pigment attractive for use in high-performance chemically resistant coatings. Like calcium carbonate, it is available as a natural pigment (barytes) or synthetic as Blanc Fixe. It is easy to disperse and is extremely resistant to water, acids, alkalis and corrosive gases. Its low oil absorption combined with a dense structure enables good film build and packing density also in combination with other suitable extender pigments [77].

4.4.3 Silicas

The only oxide type of extender pigment with a refractive index low enough to be classified as extender is silica. All silicas are chemically inert and have low specific gravities. There are natural and also synthetic silicas available. Their oil absorption varies, depending on the amorphous, crystalline or synthetic types. Since the crystalline silicas are limited for use in coatings due to their hardness, the amorphous and softer types have found their way into the pigmentation of protective coatings. In primer applications, these pigments promote intercoat adhesion and might be used as auxiliary fillers in high solid coatings. The major use of the fluffier, diatomaceous types is as flatting agents. They are also said to reduce the blistering tendency because of their ability to control the vapour permeability of the paint film [136].

4.4.4 Silicates

Silicates represent the largest class of extender pigments. They are all inert materials and should be divided into talcs, clays, micas and miscellaneous materials.

4.4.4.1 Talc

Talc products are perhaps the most broadly functional of the extender pigments. Magnesium silicate or talc is the most popular extender in coatings of all types. It is available in different particle shapes such as fibrous, platy, acicular or nodular. Moreover it is inert and low in cost. The platy talcs are suitable for protective coatings as their particles overlap and increase the diffusion path for corrosion stimulators to the metal substrate from outside. Nodular talcs are said to improve adhesion properties of the paint film. Acicular types might improve the film strength like the platy talcs. They are all easy to disperse and show good anti-settling properties in the paint [77].

4.4.4.2 Kaolin

Kaolin or clay is a hydrous alumino-silicate and is one of the most popular extender pigments in today's coatings industry. It has a platy structure and is said to reinforce the paint film. Because of its hydrophilic nature it is not widely used as extenders for anticorrosive coatings. Its popularity derives more from the contribution to optical properties in architectural finishes [137].

4.4.4.3 Wollastonite

Wollastonite or calcium silicate is available in grades ranging from 10 to 100 micrometers in particle size. It is an acicular pigment and acts as a reinforcer of the paint film. It shows low oil absorption and moderate specific gravity. Wollastonite is slightly alkaline and water soluble and was originally used in latex paints to maintain high pH levels. In the surface treated form, the extender is widely used in anticorrosive primers, where it is said to contribute to the corrosion resistance and reduction of blistering [138].

4.4.4.4 Mica

Mica has gained some importance in the formulation of anticorrosive coatings. Only the Muscovite mica is in use, which is a platy pigment with a thickness no more than 1/25th of the flake diameter. The wet-ground form is preferred for utilisation in protective coating primers. The particle structure decreases moisture ingress and reduces blistering problems. Like talc, the platy particles overlap and increase the diffusion path for corrosion stimulators. In zinc-rich primers mica has sometimes been used for partial replacement of zinc. The high oil absorption as well as the high specific gravity have a distinct influence on the critical pigment volume concentration (CPVC) of the coating [77].

4.5 Other additives

Apart from solvent carrier, binder and pigmentation to obtain colouring and corrosion protection, organic coating chemicals contain a variety of additives [138, 139]. Any of these additives are chosen for their specific purposes and their amount delicately tuned in a low percentage, in order to achieve the desired effect without sacrificing the overall performance of the coating.

Cross-linking catalysts may be added, like antimony and high-molecular tin compounds as starter for the cross-linking of blocked isocyanates, in particular in cathodic electro-dip paints, and boron trifluoride, BF_3, to catalyse oxirane ring opening in epoxy polyaddition.

Extender pigments like silicates, carbonates, talc etc. are often added to give extra volume, however they may also exhibit additional benefits. For instance, the doting of calcium carbonate with magnesium (dolomite, $[Ca,Mg]CO_3$) leads to improved acid resistance, kaolin (aluminium magnesium silicate) adds to the mechanical resistance, baryte and synthetic "blanc fixe" barium sulphate, $BaSO_4$, improve resistance vs. acids and other chemicals. The scratch resistance of a coating is supported by addition of nanoparticulate compounds, e.g. alumina or silica.

Anti-microbial properties are achieved by addition of general-purpose antibiotics like isothiazolinone derivatives or formaldehyde generators, or compounds that interfere specifically with the metabolism of bacteria, like dithiocarbaminates or silver doted zeolites. Chlorinated phenols, tin and mercury compounds have been banned due to their eco-toxic behaviour.

Corrosion and weathering resistance can be furthered by addition of polyolefinic waxes, stearates and polysiloxanes that bring about a controlled hydrophobicity at the substrate/coating interface, thereby increasing the insensitivity vs. moisture and water uptake. Micaceous iron oxide is used as diffusion path extender (cf. Chapters 2.3 and 4.3.2.1). In metallic paints, aluminium flakes are used that are passivated to improve the water resistance of a film. The weathering resistance is also supported by UV absorbers (e.g. the colourless and stable additives o-hydroxy benzophenone, oxalic dianilide, or 2-[2'-hydroxyphenyl] benzotriazole, that undergo mesomeric rearrangement under UV radiation) and radical quenchers, i.e. so called "hindered amine light stabilisers" (HALS) like 2,2,6,6-tetramethyl piperidine N-ethers that form nitroxide radicals under UV radiation in the presence of oxygen or peroxides as depicted in Equation 4.1 [140].

Equation 4.1:

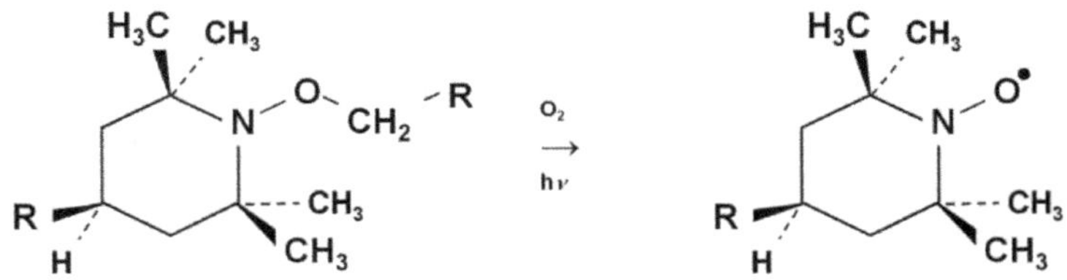

Lubrication is brought on by solids or paraffinic wax additives. Solid lubricants comprise graphite, molybdenum disulphide (MoS2) or polytetrafluoro ethylene (PTFE) and other fluorinated hydrocarbons. The latter are unsuited when over-paintability is required; however, they are preferred as anti-adhesive and filler (extender) in single-coat paints.

Further auxiliary additives are needed for rheology control (bentonites, polyacrylic acid, modified siloxanes for surface slip control), as matting agents (silica and polyamide waxes), or plasticisers (phthalic, phosphate, or adipic esters) that reduce the glass transition temperature (T_g, cf. Chapters 2.2.1 and 5.2.1). Long primary alcohols, e.g. 1-dodecanol, help to prevent boiling blisters, and various surfactants act as wetting agents and spreaders, dispersants, defoamers, and degassing agents.

Fire resistant, intumescent coatings extend under increased temperature, thereby forming effective heat insulation. Typically, they contain thermoplastic resins, e.g. copolymers of vinyl acetate with ethylene or acrylate, ammonium polyphosphate or melamine phosphate that release ammonia and water at temperatures of 300 and 550 °C, respectively, or liberate nitrogen, carbon dioxide and water, as extender gases, in the presence of oxygen. Furthermore, polyalcohols, titanium dioxide (TiO_2, rutil), china clay (kaolin) and inorganic fibres may be incorporated.

Flame retardants may contain alkali silicate water glass, antimony trioxide, Sb_2O_3, or brominated cycloaliphatic hydrocarbons. Chloroparaffins are being replaced by less toxic materials, as they release hydrogen chloride, HCl, upon excessive heating.

Wash primers are particularly suitable as primer for repair areas. They contain a high-molecular polyvinyl butyral resin, phenolic resin, phosphoric acid, and zinc chromate or phosphate [141, 142].

4.6 Solvents

Solvents are important constituents, with a lot of parameters depending on them, like film integrity, appearance, application, and even the service life of the coating system. Solvents in general are transient components in coatings and polymer solutions. Other solvents are functional (reactive solvents, plasticisers, etc.) and are not transient, however. They confer particular properties to the liquid as well to the dry coating and the finished system in service. Two parameters related to solvents must be observed in order to make a proper solvent choice [9]:

- The solvency (or solvating power) of a solvent must allow to properly dissolve a polymer and adjust viscosity in different stages of the manufacturing and application process. The major principle of choice is generally known as "like dissolves like" (Latin: similia similibus solvuntur).
- The evopouration rate of solvents, when used as a transient component, controls the drying process and the viscosity of the film throughout its formation.

However the generally accepted definition of solvents emphasises the fact that they are volatile in specific drying conditions and can dissolve film forming agents without any chemical reaction as well giving the guarantee of a processability of coatings from the liquid phase [10–15].

Beyond coil expansion of the molecules of the film forming agent, the solvents generally exhibit a lower surface tension promoting substrate wetting. As well the solvents assist the application process by spray, dipping, rolling, etc. through the rheological properties they confer to the coatings.

In general, solvents are considered as environmentally incompatible and also hazardous. Recent legislation limits the level of solvent content and recommends the replacement of hazardous solvents by more water which is environmentally compatible.

Organic solvents may be categorised as follows [4, 5], shown also in Table 4.14:

- Aliphatic hydrocarbons are solvents with good solvent power for non-polar polymers. They may consist in mixtures of similar individual components having different boiling points. Flash point temperatures below 21 °C require stringent manufacturing and application conditions while solvents with flash points higher than 21 °C are good sol-

Table 4.14: Organic solvents for coatings

Class	Example	Features
Aliphatic hydrocarbons	solvent naphtha, cyclohexane	solvate binders of low polarity, for polyvinyl ethers, neoprene; not suited for epoxies, polyesters, or polyurethanes
Aromatic hydrocarbons	xylene, toluene	particularly suitable for polyesters, polyacrylates, polyvinyl ethers, vinyl acetate mixed polymers
Alcohols	n-butanol	for aminophenolic resins, and polyvinyl acetates form hydrogen bonds
Ketones	isophorone	low ketones: for polar resins high ketones: for unpolar resins
Esters	butyl glycol acetate	suited for most resin types, good spreading, avoid bubbles
Glycol ethers	butyl diglycol propyl ether	for polar resins, particularly alkyds, improve wetting and spreading, co-solvents for aqueous paints

vents for poor polarity polymers such as chlorinated rubber, long oil alkyds, and some copolymers.

- Cycloaliphatic hydrocarbons are solvents presenting low solvency and generally a low evopouration rate. They are less used in the coating industry but may significantly influence film forming properties at low and very low concentrations.
- Terpene hydrocarbons: are mainly used in combination with mineral spirit in air drying organic film formers.
- Aromatic hydrocarbons are the "working horses" in the coating industry. They cover a large range of boiling points and volatilities as well as they present a very good solvency power. Benzene is effectively forbidden, but substituted derivatives are largely used in coating formulations.
- Chlorinated hydrocarbons present very good solvency but for environmentally and physiological reasons are limited in use. However they are still used at relatively large extent for surface cleaning of substrates primarily in North America and Asia.
- Alcohols are solvents with high solvency and hydroxyl functionalisation group. They are utilised in combination with high polarity polymers. The carbon chain length of alcohols makes them preferably suitable either as a diluent or as a storage stabiliser in combination with other solvents.
- Esters are very good solvents for all polar polymers. The solvency and volatility of ester type solvents are mainly related to the length of the alcohol used, considering the fact that the main acids employed for these type of solvents are acetic acid and propionic acid. A class of esters that must be mentioned here as making inroads in the coating market comprises di-basic esters and cyclic carbonates from ethylene and propylene glycol.
- Ketones are strong solvents for polar polymers due to the capability of forming hydrogen bridge bonds with polar groups of the polymer.
- Glycol ethers present high capability in hydrogen bonding and are significantly used in water-borne coatings or high-temperature drying coatings.
- Nitrogen containing solvents such as dimethyl formamide and N-methyl pyrrolidone have limited use in the coatings market but are of significant importance in relation to polyurethane polymers.

4.7 *Raw materials for powder coatings*

Powder paints are considered to be particularly environmentally friendly, as they allow the coating process to be performed without solvents and provide an efficient application. They are applied onto cleaned and pretreated metal surfaces, by electrostatic methods (e.g. gun spraying), by thermal fusion in fluidised bed processes or as slurries in water [145]. When weathering resistance is required, it is common to use preprimed stock, e.g. in automotive parts (electro-dip paint) or coil and sheet metal for architectural and appliance uses (conventional coil coating primers).

Most powder paints are single-package, reactive systems. Curing in conventional, infrared, near-infrared (wavelength 0.8 to 1.5 μm) or induction ovens comprises two steps: First fluidising and coalescing occur in the temperature range between 80 and 150 °C. The subsequent cross-linking is carried out at typical curing temperatures and times of 150 to 200°C and 20 to 30 min [146]. Coating gauges usually are in the order of 50 to 100 μm, however, also higher thicknesses can be obtained [147]. By induction preheating up to 260 °C, also substantially shorter curing times in the order of 1 min are possible, enabling the technology to be used in compact coil and sheet lines [148].

Powder primers and so called hybrid topcoats are usually based on epoxies and blends of epoxy/acidic polyester, respectively. Admixing of polyester resins improves the resistance

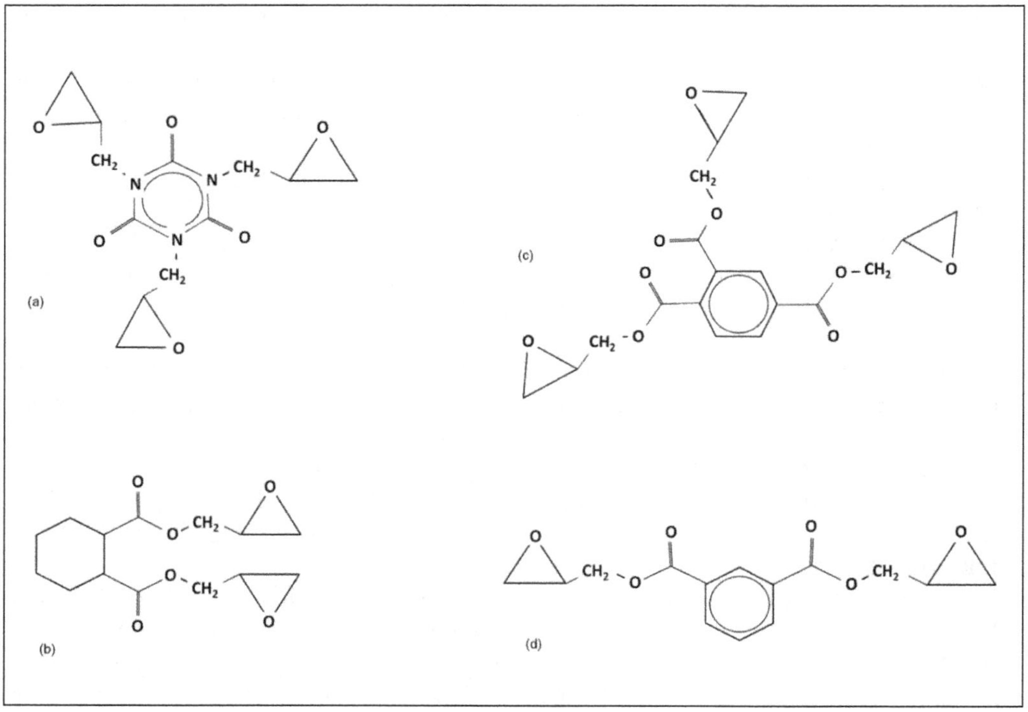

Figure 4.65: Cross-linking agents for epoxy powder paints

against chalking and yellowing when exposed to elevated temperatures [149]. For increased weathering resistance, polyester or polyurethane powders are used. In particular for clear-coats in the automotive industry, polyacrylate based powders have become common that are mixed from glycidyl functional compounds and dicarboxylic acid cross-linkers. They are also commercialised as finely dispersed aqueous slurries that cure at particularly low temperatures of about 140 °C. Thermoplastic powders (polyamide, PVC or PVDF based) are only of minor importance [150].

Epoxies are commonly cured in the presence of di or trifunctional cross-linkers with epoxy (glycidyl) groups. The use of triglycidyl isocyanurate (TGIC) has been largely discontinued due to the mutagenicity of this compound. Alternatives include glycidyl esters of hexahydro phthalic, isophthalic or trimellitic acid (cf. Figure 4.7.1) or -hydroxy alkyl adipic amides and blocked isocyanates or uretdiones in the case of urethane cross-linking [151-153].

a) tri-glycidyl isocyanurate (TGIC)
b) di-glycidyl hexahydrophthalate
c) tri-glycidyl trimellitate
d) di-glycidyl isophthalate

Further common additives for powder paints are alumina or silica to control pourability while handling, applying and recirculating the overspray, and benzoin derivatives to support degassing during the melting and gelling phase.

4.8 *Literature*

[1] EN 971-1
[2] Linak, E., Kishi, A., Paint and coating industry overview, CEH Industry overview, 2005
[3] Linak, E., Kishi, A., Urethane surface coatings, CEH Marketing research report, 2004
[4] Blaga, A., Robu, C.,Lacuri si vopsele: chimismul reactilor, ISBN 973-31-0427-2
[5] Motoiu, I., Motoiu, M., Rasini Sintetice pentru lacuri, vopsele si cerneluri poligrafice, Edit. Tech. 1972
[6] Contribution of Working Group I to the forth Assessment Report of the Intergovernmental Panel on Climate Change
[7] Determination of volatile matter content, water content, density, volume solids, and weight solids of surface coatings-Method 24
[8] EPA website www.epa.gob/ttn/atw/coat/coat.html
[9] J. Brandrup and E. H. Immergut, eds., Polymer Handbook, 3rd ed., Wiley-Interscience, New York, 1989
[10] Provder T., ed., Chromatography of Polymers: Characterization by SEC and FFF, ACS Symposium Series No. 521, American Chemical Society, Washington, D.C., 1993
[11] Rosen S. L., Fundamental Principles of Polymeric Materials, 2nd ed., Wiley-Interscience, New York, 1993
[12] Prane, J.W., Introduction to polymers and resins, Federation of Societies for coatings technology, Philadelphia, PA, June 1986
[13] U.S. Pat. 4,413,110
[14] Elis, W.H., Solvents, Federation of Societies for coatings technology, June 1986, Philadelphia, PA.
[15] Norris, T., Powder Coater's handbook
[16] D. H. Solomon and J. J. Hopwood, J. Appl. Polym. Sci. 10, 1893-1914 (1966)
[17] E. T. Turpin, Hydrolysis of water dispersible resins, J. Paint Tech. 47(602), 40-46 (1975)
[18] P. Oldring and G. Hayward, eds., Resins for Surface Coatings, Vol. 1; Vol. 2, SITA Technology, London, 1987
[19] Yang H., et al, J. Polym. Sci. Polym. Chem. Ed. 31, 1261, 2015 (1993)
[20] US4369283
[21] Patton, T.C., Alkyd resin technology, 1962, John Wiley & Sons, Inc.
[22] Manczyk, K., et al., Highly branched high solids alkyd resins, Progress in Organic Coatings, 44, (2002), pp 99-109
[23] Holmberg, K. High solids alkyd resins, ISBN 0-8247-7778-6
[24] L. V. McAdams and J. A. Gannon, in J. I. Kroschwitz, ed., Encyclopedia of Polymer Science and Engineering, 2nd ed., Vol. 6, Wiley-Interscience, New York, 1986
[25] Dreyfuss P., et al, in J. I. Kroschwitz, ed., Encyclopedia of Polymer Science and Engineering, 2nd ed., Vol. 16, Wiley-Interscience, New York, 1989
[26] Kirk-Othmer Encyclopedia of Chemical Technology, Dialog on Disk 2001
[27] W. G. Potter, Epoxide Resins, Springer-Verlag, New York, 1970
[28] The Epoxy Resin Formulators Training Manual, The Society of Plastics Industry, Inc., New York, 1984
[29] C. A. May, ed., Epoxy Resins Chemistry and Technology, 2nd ed., Marcel Dekker, Inc., New York, 1988
[30] Charlesworth J., J. Polym. Sci. 18, 621 (1988)
[31] Billmeyer F. W., Jr., Textbook of Polymer Chemistry, Interscience Publishers, a division of John Wiley & Sons, Inc., New York, 1957
[32] Emulsion Polymerization of Acrylic Monomers, CM-104, Rohm and Haas Co., Philadelphia, Pa.
[33] Uibel J. J., Dinon F. J., J. Polym. Sci. Polym. Chem. Ed. 21, 1773 (1983)
[34] Bovey F. A., High Resolution NMR of Macromolecules, Academic Press, New York, 1972
[35] Matheson M. S., Aner E. E., Bevilacqua E. B., Hart E. J., J. Am. Chem. Soc. 73, 5395 (1951)
[36] Maduga L., An. Quim. 65, 993 (1969)
[37] Griffiths P. G., Rizzardo E., Solomon D. H., J. Macromol. Sci. Chem. 17, 45 (1982)
[38] Kloosterboer J.G., Bressers H.L., Polym. Bull. 2, 205 (1982)
[39] The Manufacture of Acrylic Polymers, CM-107, Rohm & Haas Co., Philadelphia, Pa.
[40] O. Bayer, Angew. Chem. A59, 257 (1947)
[41] Spindler R., Frechet J. M. J., Macrom. 26, 4809 (1993)
[42] Sperling L. H., Klempner D., Utracki L. A., Interpenetrating Polymer Networks, American Chemical Society, Washington, D.C., 1994
[43] Schnabel W. J., Kober E., J. Org. Chem. 34, 1162 (1969)
[44] Laas H. J., Halpaap R., Pedain J., J. Prakt. Chem. 336, 185 (1994)
[45] Aksoy, B., et al., Effect of zirconium insolubilizers on silica coating rheology, Western Michigan University, Kalamazoo MI 49008
[46] Moles, P.J., Zirconium based crosslinkers for water borne coatings, Symposium proceedings Northwest Region, Harcos chemicals (UK), Ltd., Manchester 1996
[47] Schollenberger C. S., Stewart F. D., Advances in Urethane Science and Technology, Vol. 1, Technomics, Stamford, Conn., 1971

[48] Mircea Manea, High solid binders, ISBN 3866308434
[49] Webster, D., Cyclic carbonate functional polymers and their applications, Progress in organic coatings, 47(1), 2003, pp77-86
[50] Imaz, A., N-vinylcaprolactam-based microgels: effect of the concentration and type of cross-linker, J.Polym Sci Part A, Polym Chem 46: 2766/2775
[51] Sarkar, S., Non-isocyanate PU-A new generation polyurethane Paint India (2007), vol 57(12), pp99-110
[52] Broekaert, M, Polyurea spray coatings, Huntsman Polyurethane
[53] Girasa, E, De Wispelare, M, Polyaspartate a new alternative for the conditioning of cooling water, 14th International Conference on the Properties of Water and Steam in Kyoto
[54] Narsavage, D., Thermal Polyaspartate as a biodegradable alternative to polyacrylate and other currently used water soluble polymers, Univ. of Scaranton
[55] Roschu, R., Schmitz, J., Melchiors, M., Surface Coatings Australia, Innovative VOC compliant raw materials for the European coatings industry, July 2007, pp20-25
[56] Pereyra, A.M., Giudice, C.A., Ethyl silicates with different hydrolysis degree like non-flamable impregnating material for wood, ISSN 0717-3644
[57] Mowerer, N.R., Polysiloxanes, November 2003
[58] Heilen, W., Silicone resins and their combinations
[59] UV-driven reversible switching of a polystyrene/titania nanocomposite coating between superhydrophobicity and superhydrophylicity, Hou, W., Wang, Q., Langmuir2009, 25(12) 6875-6879
[60] Moisture curing kinetics of isocyanate ended urethane quasi-prepolymers monitored by IR spectroscopy and DSC, Daniel-da-Silva, A.L., Bordado, J.C.M., Martin-Martinez, J.M., Journal of Applied Polymer Science DOI 10.1002/app
[61] Thermoplastic elastomers based on strong and well-defined hydrogen bonding interactions, Söntjens, S.H.M., Renken, R.A.E., van Gemert, G.M.L., Engels, T.A.P., Bosman, A.W., Janssen, H.M., Goevaert, L.E., Baaijens, F.P.T., Macromolecules 2008, 41, 5703-5808
[62] Dielectric Relaxation in Hydrogen Bonded Urea-Based Supramolecular Polymer *N;N0*-di(2,2-dipentylheptyl)urea ,J. Jad_zyn*a*, L. Bouteiller*b*, J.-L. Dejardin*c*, G. Czechowsk, Vol. 110 (2006) ACTA PHYSICA POLONICA A, No. 4
[63] Hare, Clive H., Protective Coatings, Fundamentals of Chemistry and Composition, Technology Publishing Company, Pittsburgh, USA, 1994, p. 331
[64] Buxbaum, G., Industrial Inorganic Pigments, Wiley-VCH, Weinheim, 2005, p. 207
[65] Dören, K., Water-borne coatings: the environmentally friendly alternative, Carl Hanser Verlag, München, 1994, p. 75
[66] Vogelsang, J., Basics about Anticorrosive pigments and Corrosion Inhibitors and Possibilities for their Usage, European Coatings Conference, Berlin, 2000
[67] Krieg, S., Anticorrosive Pigments, European Coatings Conference, Berlin, 2000
[68] Smith, A., Inorganic Primer Pigments, Federation Series on Coating Technology, Philadelphia, PA, 1988, p. 7
[69] Buxbaum, G., Industrial Inorganic Pigments, [64] pp190-210
[70] Fannin, S., New Anticorrosive Pigments for Improved Environmental Performance, European Coatings Conference, Berlin, 2000
[71] Allegheny Environmental Services, Inc., Lead-Based Paint- A Resource Guide on Lead-Cased Paint, Company Information, Randolph, NY, 2005
[72] Appleby, A. J., Mayne, J.E.O., Corrosion Inhibition by the Salts of Long Chain Fatty Acids, JOCCA, 50:897, 1967
[73] Ruf, J., Organischer Metallschutz, Vincentz Verlag, Hannover, 1993, p. 260
[74] Austin, M. J., Inorganic Anticorrosive Pigments, ASTM Manual 17, American Society for Testing and Materials, Philadelphia, USA, 1995, p. 238
[75] Koskinimie, M. S., Conti W. J., Corrosion Inhibition: Past, Present and Future, ABRAFATI Congress Proceedings, Sao Paulo, Brazil, 1995, p. 639-641
[76] Ruf, J., Korrosionsschutz durch Lacke und Pigmente, Verlag W.A. Colomb, Stuttgart, Berlin, Germany, 1972, p. 94
[77] Hare, Clive H., M[63] p. 340
[78] Frankel G., McCreery R. L., The Electronical Society Interface, Winter, 2001, p. 34-37
[79] Bierwagen, G. P., Organic Coatings for Corrosion Control, ACS Symposium Series 698, American Chemical Society, Washinton DS, 1998, p. 356
[80] ISO 1249, Paints and varnishes - Corrosion protection of steel structures by protective paint systems, International Organization for Standardization, Genève, Switzerland, 1998
[81] Schik, J.P., Korrosionsschutz mit wässrigen Lacksystemen, expert Verlag, Renningen-Malmsheim, Germany, 1997, p. 98-117
[82] Brock, Groeteklaes, Mischke, European Coatings Handbook, Vincentz Network, Hannover, Germany, 2010, p. 149-150

[83] Adrian, G., Bittner, A., Grawol, M., FARBE UND LACK, 87 (1981) p. 833-837
[84] SOCIETE NOUVELLE DES COULLEURS ZINCIQUES, Company information, Beauchamp, France, 1990
[85] Andrásné, T., Laszlóné, O., REPORT on the details as laid down within the Research-Development Contract concluded between Holland Colours and Paint Research Ltd., Budapest, Hungary, 1993
[86] Heubach GmbH, Anticorrosives - Time for ultimate Corrosion Protection, Company information, Langelsheim, Germany, 2009
[87] Adrian, G., Bittner, A., European Supplement to Polymers Paint Colour Journal, 1985
[88] KIKUCHI COLOR & CHEMICALS CORPORATION, ANTICORROSIVE PIGMENTS, Company information, Tokyo, Japan, 1995
[89] Leblanc, O., JOCCA, 1991 (8), p. 288-301
[90] Heubach GmbH, Wide Spectrum Anticorrosives, Company information, Langelsheim, Germany, 2003
[91] Krieg, S., Inorganic Anticorrosive Pigments based on Phosphates, European Coatings Conference Proceedings, Anticorrosive Pigments, Berlin, Germany, 2000, p. 93-110
[92] Bittner, A., Ruf, J., Farbe & Lack, 104 (1998) p. 50-55
[93] Austin J., PAINT & INK INTERNATIONAL, September 1993, p. 13-19
[94] Halox, Product Application Guide, Company information, Hammond , USA, 2004
[95] Wayne Pigment Corp., WAYNCOR, Company information, Milwaukee, USA, 1997
[96] TAYCA CORPORATION, K-WHITE ALUMINIUM TRIPHOSPHATE, Company information, Osaka, Japan, 1995
[97] Forsgren, A., Corrosion control through organic coatings, CRC Press Taylor&Francis Group, Boca raton, FL, USA, 2006, p. 36
[98] Kirmaier, L., Waterborne Coatings for Industrial and Protective Applications - Performance Properties of Anticorrosive Pigments and Inhibitors in Waterborne Primer System, European Coatings Conference Nürnberg Germany, 2008
[99] Braig, A., Advances in corrosion protection by organic coatings III, The Electrochemical Society, Pennington NJ, USA, 1998, p. 18
[100] Pianoforte, K., Anti-corrosion coatings: anti-corrosion coatings offer environmentally responsible products that can withstand the harshest conditions, Coatings World, Rodman Publishing, Farmington Hills, MI, USA, February 2008
[101] Hare, C., A Review of Inhibitive Metal Primers, Modern Paint and Coatings. Reprint Courtesy of Mineral Pigments, Beltsville, MD, USA, July 1996
[102] NL Industries, US 4, 386, 059, 1983 (W. Reichert, et. al.)
[103] Hare, C., A Review of Inhibitive Metal Primers, Modern Paint and Coatings. Reprint Courtesy of Mineral Pigments, Beltsville, MD, USA, July 1996
[104] Hare, C., A Review of Inhibitive Metal Primers, Modern Paint and Coatings. Reprint Courtesy of Mineral Pigments, Beltsville, MD, USA, July 1996
[105] Oxy, Ferrophos Pigments, Company Information, Glenn Springs Holding Inc., Columbia, TN, USA, 2008
[106] MOLY-WHITE Pigments Group, Mechanism of action, Company information, Cleveland, USA, 2004
[107] Deflorian, F., Fedrizzi, L., Bonora, P.L., Corrosion Protection of Coil Coating Products with Ion-Exchange Pigments, Advances in corrosion protection by organic coatings III, The Electrochemical Society, Pennington NJ, USA, 1998, p. 45
[108] C.A. Bravo, Non toxic ion-exchange anti-corrosion pigments, ABRAFATI Congress Proceeedings, Sao Paulo, Brasil, 1995, p. 770-787
[109] GRACE Davison, Shieldex AC 3 - Shieldex AC 5, Company information, Worms, Germany, 1995
[110] Evans, E., et al., PPG Industries Inc., US 4 707 405, 1987
[111] Halox, Company information, Hammond , USA, 2009
[112] PIGMENTAN LTD., Pigmentan, Company information, Ramat - Gau, Israel, 2004
[113] Vetter, C. et al., Polypyrrole/Aluminum Flake Pigmented Coatings for Corrosion Inhibition, Department of Coatings and Polymeric Materials, North Dakota State University, Fargo, ND, USA, Triservice Corrosion Conference, 2007
[114] Pokorny, G., Micaceous Iron Oxide in Heavy Duty Coatings, European Coatings Conference Proceedings, Anticorrosive Pigments, Berlin, Germany, 2000, p. 135-143
[115] Koleske, J.V., Paint and coating testing manual: fourteenth edition of the Gardner-Sward handbook, ASTM International, 1995, p. 245
[116] Lamox Industries, Company Profile, Mumbai, India, 2009
[117] Grimme, D., Krüger, J., Korrossionsschutz durch Beschichtungen und Überzüge auf Metallen, Volume 2, Chapter 9/4.6., H.J. Gessner, Barriereprinzip und korrosionsschützende Wirkung von Eisenglimmer, WEKA Fachverlag, Augsburg Germany, 1990 (Chapter 9/4.6., 1999)
[118] Wayne Pigment Corp., US 6, 139, 610, 2000 (J. Sinko)
[119] Eckart Effect Pigments., Leafing aluminium pigments, Company information, Hartenstein, Germany, 2010
[120] ISO 10601, Micaceous iron oxide pigments for Paints- Specifications and test methods, International Organisation for Standardization, Genève, Switzerland, 1993

[121] Kittel, Lehrbuch der Lacke und Beschichtungen, Band 5, S. Hirzel Verlag, Stuttgart · Leipzig, Germany, 2003, p. 175-176

[122] Vogelsang, J., Selbstheilende Beschichtungssysteme - ein Überblick, Schiff und Hafen, DVV Media Group GmbH, Seehafen Verlag, Hamburg, Nr. 8, August 2009, p. 30

[123] Bierwagen, G., Battocchi, D., Simoes, A., Stamness A., Tallmann, D., The use of multiple electrochemical techniques to characterize Mg-rich primers for Al alloys, Progress in Organic Coatings, DVV Media Group GmbH, Seehafen Verlag, Hamburg, Nr. 8, August 2009, p. 30

[124] Johnson, J.A., Magnesium Rich Primer for Chrome Free Protection Of Aluminum Alloys, Air Force Research Laboratory, Wright Patterson AFB, OH, USA, Triservice Corrosion Conference, 2007

[125] Taylor, J., JOCCA, 38, 233, 1955

[126] Kampfer, W.A., Titanium Dioxide, in "Pigment Handbook", Vol.I, Patton, T.C., Wiley & Sons, New York, 1973, p. 1

[127] Dr. T. Rentschler, DEVELOPMENT AND BENEFITS OF AN ANTICORROSIVE PIGMENT BASED UPON TITANIUM DIOXIDE, European Coatings Conference Proceedings, Anticorrosive Pigments, Berlin, Germany, 2000, p. 161-173

[128] Sachtleben, HOMBICOR - The bifunctional anticorrosion pigment with extra hiding power, Company information, Duisburg, Germany, 1999

[129] Austin, M.J., DeVaney, J.G., Extender Pigments for Latex Coatings: Do they make a difference, Company information, Duisburg, Germany, 1999

[130] Bloch, G., Chrome Yellows and Molybdate Oranges, Double Liaison, 161, 51, 1969

[131] Luescher, M., Inorganic Pigments-Other, Surface Coatings Vol. I, Surface Coatings of Australia, Chapman and Hall, London, 1993, p. 449

[132] Mees, P., Benzimidazolone Pigments, Pigment Handbook Vol. I, Patton, T.C., Wiley & Sons, New York, 1973, p. 543

[133] Moser, F.H., Thomas, A.L., Phthalocyanine Compounds, ACS Monograph 157, Reinhold, NY, 1963

[134] Kogel, J.E., Trivedi, N.C., Barker, J.M., Industrial minerals & rocks: commodities, markets, and uses, Society for Mining, Metallurgy, and Exploration (U.S.), 7th ed., 2006, p. 1303

[135] Georgov, W., Chasen, A.A., Synthetic Silicas: Important Pigments to the Paint Industry, Journal of Paint Technology, June 1971, p. 51

[136] Steig, F.B., Effect of Extenders on the Hiding Powder of Titanium Dioxide Pigments, Official Digest, 31, 408, 52, 1959

[137] Hare, C.H., Effect of Extender on Crowding of Titanium Dioxide, Journal of Coatings Technology, November 1989, p. 7

[138] Kittel, H., Streitberger, H.-J. (ed.), Lehrbuch der Lacke und Beschichtungen (Coursebook of Paints and Coatings), Vol. 6, 2nd ed., Hirzel, Stuttgart 2008, pp. 209, 212, 335, 372, 419, 436 ff, 446, 502 ff, 522 ff, 536 f, 538

[139] Goldschmidt, A., Streitberger, H.-J., Basic of Coating Technology, 2nd edition, BASF Coatings AG, Münster 2007, pp. 168 ff, 171 ff, 185, 191 ff, 194

[140] Step, E. N., Turro, N. J., Gande, M. E., Klemchuk, P. P., Mechanism of Polymer Stabilization by Hindered-Amine Light Stabilizers (HALS). Model Investigations of the Interaction of Peroxy Radicals with HALS Amines and Amino Ethers, Macromolecules 27, 1994, pp. 2529 ff

[141] Kittel, Streitberger [138] pp. 354 f

[142] Goldschmidt, Streitberger [139] pp. 84, 118

[143] Goldschmidt, Streitberger [139] pp. 127 ff

[144] Meuthen, B., Jandel, A.-S., Coil Coating, 2nd ed., Vieweg, Wiesbaden 2008, p. 71

[145] Goldschmidt, Streitberger [139] pp. 595 ff

[146] Kittel, Streitberger [138] pp. 429 ff

[147] Meuthen, Jandel [144] pp. 83 f, 139 ff, 157 ff

[148] Ritterbach, A., Schnelle Pulverlack-Aushärtung mit Induktion: Coil-Coating im Kleinformat (Fast Curing of Powder Coats by Induction…), JOT 47, 2007, pp. 12 ff

[149] Goldschmidt, Streitberger [139] pp. 608 ff

[150] Kittel, Streitberger [138] p. 393

[151] Goldschmidt, Streitberger [139] pp. 104 ff

[152] Manea, M., High Solid Binders, Vincentz, Hannover 2008, p. 48

[153] Müller, B., Rath, W. Formulating Adhesives and Sealants, Vincentz, Hannover 2010, p. 137

5 Film formation

Jörg Sander

5.1 Physical drying

Conventional coatings comprise macromolecules being suspended in a solvent, or emulsified and complexed in water. The solvent or water is also used to control viscosity of the liquid. Curing of an organic coating therefore involves evaporation of the carrier solvent before macromolecules can be physically deposited [1].

Polymer binder molecules that are solvated, i.e. disentangled and individually dissolved, in the liquid paint close up and coalesce during this process of physical drying, thereby forming a closed film, cf. Figure 5.3, Chapter 5.2.2.1. A film of suitable hardness and mechanical resistance requires the polymer strands to mechanically entwine and form a spatial structure, to obtain the necessary cohesion. Weak bonding via hydrogen bridge bonds may occur and help in aligning the individual molecules. However, when dissolution is less than ideal, polymer knots will persist throughout the paint manufacture and application and the drying processes, making for failures in the film and cohesion losses. Therefore, the way of dissolving has an effect on the final film forming result, and the mechanical/technological features of coatings may be different, despite an identical composition. Moreover, films formed in this way do not provide chemical fastness, in particular resistance to solvents. Furthermore, they display thermoplastic behaviour, which means that they soften and even may become fluid again with increasing temperature.

The film formation in aqueous dispersions, by contrast, is ruled by the reduction of total surface of particles that were initially dispersed as individuals, resulting in a fused film that ideally has no interstices between particles. This can be interpreted as a gain of free surface energy or as the action of attraction forces between dispersion particles. In effect, the attraction is so strong as to compress the initially spheric particles to a hexagonally stacked layer of rhombic dodecahedrons, as can be shown by electron micrographs. Again, hydrogen bridge bonding does increase the cohesive force. When admixed with a solution of inorganic compounds, the latter can be found along the interfaces of compressed dispersion particles. Reactive inorganics, like chromates, react with the particle surfaces, forming linking bonds that provide a kind of chemical curing. A similar reaction path is known for the interaction of acidic polyester resin with ammonium zirconyl carbonate [2].

This mechanism requires a viscosity that allows dispersed polymer particles to flow easily. Any dispersion, therefore, has a minimum film forming temperature (MFFT), below which the viscosity is too low. When necessary, the MFFT can be decreased by addition of organic solvents that boil above water. Alternatively, the molecular mass of the film builder might be reduced which softens the dispersion, however sacrifices film properties like the abrasion resistance.

The viscosity change during the process of physical drying can be described by Equation 5.1 which illustrates that the logarithm of the viscosity, η, is proportional to the solid content percentage, % SC, and the square root of the average molecular mass, M_n, of a liquid paint. Figure 5.1 depicts the situation for physically drying (a) vs. chemically curing (b) coatings

Jörg Sander et al.: Anticorrosive Coatings

ISBN 978-3-86630-911-1

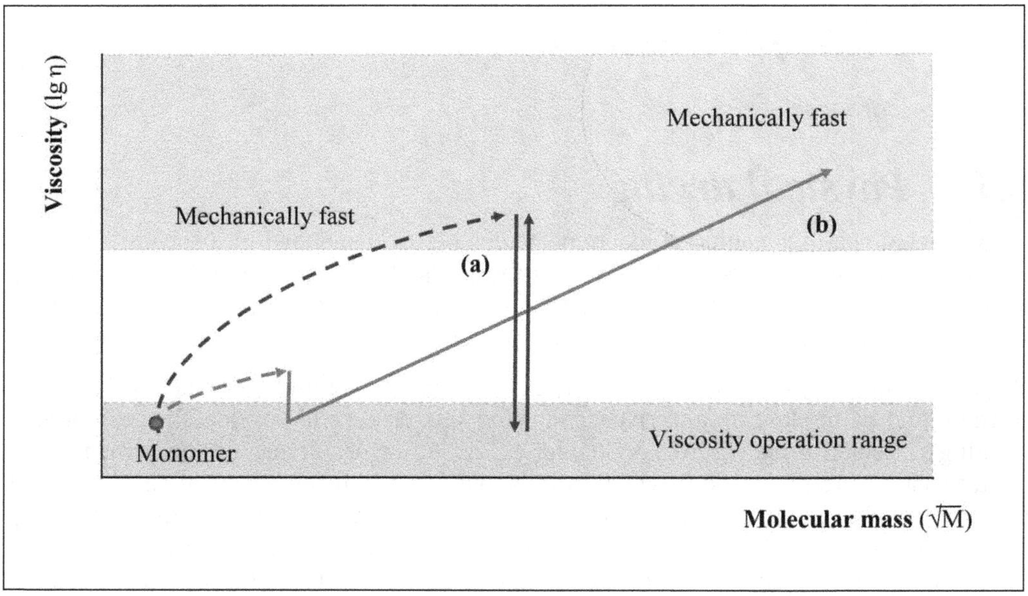

Figure 5.1: Schematic: film formation regimes by physical drying vs. chemical curing [1]

in an $\lg\eta / \sqrt{M_n}$ diagram for a given solids content. Physically drying coatings contain polymer particles with a relatively high molecular mass that have to be derived from monomers during the paint manufacture. These are also relatively hard, and must be softened by addition of solvent. Upon evaporation, the viscosity increases again, forming a film according to the mechanism described above, without a change in molecular weight. This is represented by a vertical and reversible shift of the polymer state in the diagram.

Equation 5.1 $$\lg \eta = \infty \, \% \, SC \cdot \sqrt{M_n}$$

η = viscosity
% SC = solid content percentage
M_n = average molecular mass

5.2 Chemical curing

5.2.1 Thermal cross-linking: chemistry, mechanism, imparted properties

To obtain coatings with chemical and higher mechanical resistance, the formation of a molecular network is necessary. It is brought about by reaction of suitable end or side-chain groups of the polymers that form additional cross-links. The polymer size, obtained by prereaction from the monomer, is relatively low, and only little solvent is added to adjust working viscosity. Curing involves evaporating the solvent plus chemical cross-linking, thereby increasing polymer size (and molecular mass) by magnitudes (cf. Figure 5.1b).

The enlargement of polymers can occur via addition, condensation or polymerisation mechanisms. Polyurethane links form by reaction of isocyanate and hydroxyl (-OH) functions by way of an addition. As isocyanates are reactive species, that undergo conversion with many different partners (amino, thiol, carboxylate, ether) and are also water-sensitive, they are usually blocked to obtain stabilisation. Suitable blocking agents are protic compounds like malonic acid derivatives, lactones or lactams (cyclic esters or carboxylic amides), e.g. ε-caprolactam, that provide thermally reversible blocking. Typical de-blocking temperatures range from 100 to 180 °C [3]. Cross-linking of polyesters and chemically related alkyds (class of polymers that is synthesised from natural or semi-synthetic oils and fatty acids and polyols, cf. Chapter 4.2.1) is obtained by reaction of the -OH groups with co-resins like polyisocyanate, melamines, anhydrides etc. [4], the reaction with the isocyanates yielding polyurethane resins with their typical features of high durability and temperature insensitivity.

When the polymer growth is predominantly linear, the resulting coating will behave as a thermoplastic, i.e. soften and eventually liquefy with increasing temperature. Cross-linking to a three-dimensional network yields duroplastics. These are characterised by the transition between the amorphous, glassy and the crystalline state at the glass transition temperature T_g, where a sudden softening and increase in elasticity occurs, however retaining a residual level of mechanical hardness [5].

Cross-linking paints cure in a temperature/time window that allows relatively low temperatures and a wide temperature tolerance band between under and over-curing at extended curing times, while with shorter times, the required temperature increases as well as the tolerance becomes more sensitive. Under-curing leads to soft, non-resistant coatings, while over-curing usually results in brittle, cracking films with poor adhesion or even thermal degradation indicated by discolouration (yellowing). The required temperature is defined by type of cross-linking, more often also by the temperature where de-blocking occurs. The curing time for batch operations often is in the range of 10 to 30 min [6]. In fast coil coating lines, typical curing temperatures are as high as 240 °C, and reaction times of some 30 s are required to complete solvent or water evaporation, uniform deposition and cross-linking [7, 8].

5.2.2 Radiation curing

5.2.2.1 Chemical principles and intrinsic properties

In recent years, a fair amount of research has been focused around radiation curing systems [9-12]. Ultraviolet (UV) or electron beam (EB) curable formulations have been in use for applications like wood painting and preservation, paper and foil printing, or flexible packaging. The major advantages of this technology are speed of reaction that allows short installations and facilitates immediate onward processing and "cold curing" that is much more energy efficient, i.e. offers up to 90 % energy savings, when compared to conventional thermal curing [13], and enables temperature-sensitive substrates to be coated.

UV coatings are known as solvent or waterborne systems with solid contents up to 85 %. However, also liquid, fully reactable, "100 %" systems are available that utilise "reactive thinners" to control viscosity. Obviously, prior to curing, water or solvents must be thoroughly evaporated from diluted systems by heating (convection, infrared or induction), sacrificing part of energy gain. Upon UV irradiation, the reaction completes within fractions of a second.

The technology of radiation curing involves photo-chemistry, which is utilised to form a cross-linked polymeric film in-situ on the substrate from monomeric or oligomeric precursors. Issues often encountered with radiation cured polymeric films include volume contraction (shrinkage) that occurs during the rapid polymerisation, and brittleness due to over-dense cross-linking. In combination with lack of suitable anchor groups to attach the forming film to a metal surface, adhesion to metallic substrates used to prove difficult to achieve in the past. However, radiation curing also offers undisputable benefits, in particular for the coil coating process, where a flat substrate has to be coated and cured in very

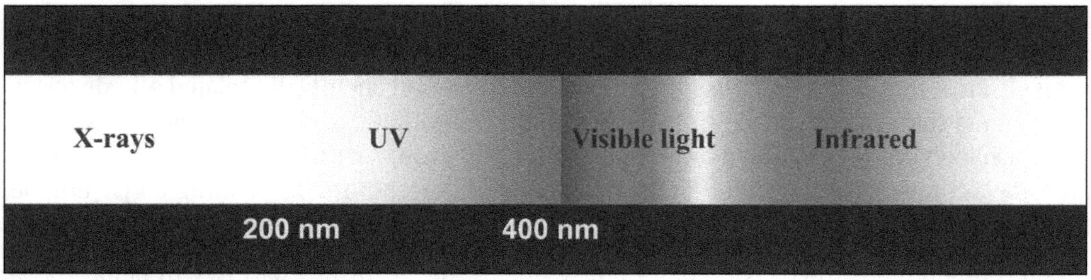

Figure 5.2: High-energy end of the electromagnetic spectrum

short times. The economic and environmental efficiency of the process has been the driver to develop and adapt UV and EB curable formulations for the purpose of coating flat metallic work pieces. The UV technology has been incorporated into the first coil coating lines.

UV is part of the electromagnetic spectrum, more energetic than visible light. Wavelengths of UV radiation range from approximately 200 to 400 nm cf. Figure 5.2. UV triggered polymer formation can occur via radical or cationic mechanisms. The radical mechanism involves generation of a sufficient concentration of radicals to be started, and propagates through attack of these radicals on unsaturated organic compounds with carbon-carbon double bonds (C=C) like acrylic acid and its derivatives. A radical UV photo-polymerisation is described schematically by Equation 5.2.

Equation 5.2

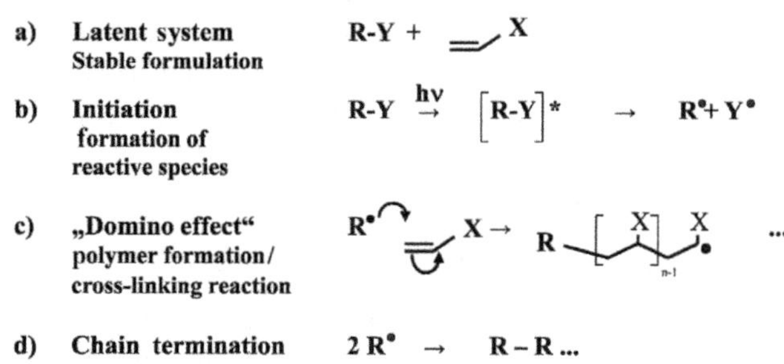

h = Planck's Constant; ν = frequency of electromagnetic radiation

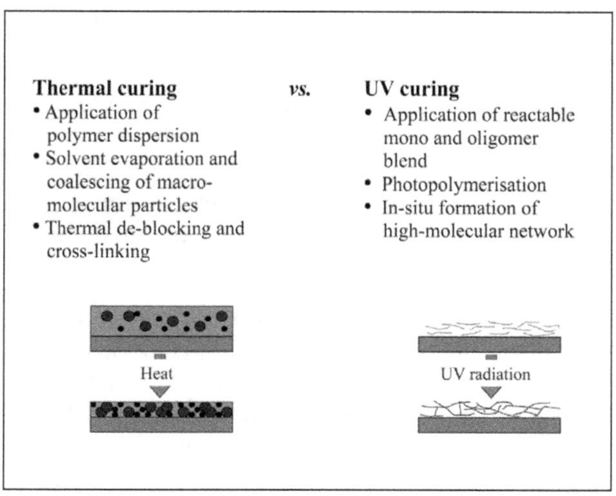

Figure 5.3: Systemic differences between thermal and radiation, e.g. UV curing

UV curable coating chemicals are fundamentally different from conventional, thermally curing systems. Figure 5.3 depicts the systemic differences. Instead of polymers dispersed in a solvent, they consist of a mixture of single organic molecules (monomers or so-called prepolymers) and photo-initiator molecules, R-Y, that are sensitive to UV radiation of suitable energy, E = hν. Being a latent and stable system when kept in the dark, the photoinitiator molecules decompose, via an excited state [R-Y]*, into reactive radicals when exposed to radiation.

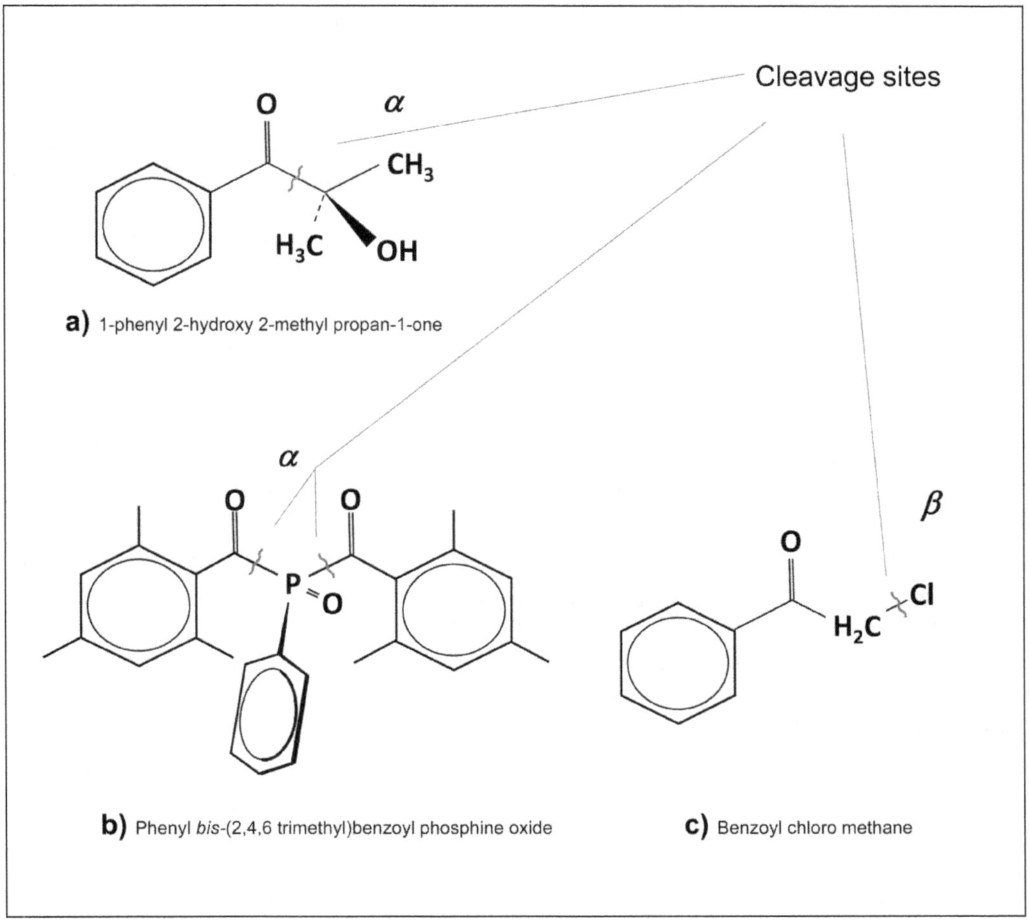

Figure 5.4: Examples of photoinitiators for radical UV photopolymerisation

Common photoinitiators are benzoyl compounds, preferably with hydroxyl groups on the adjacent carbon atom, e.g. benzoyl phenyl hydroxyethane, which will readily decompose under UV irradiation by α-cleavage (Figure 5.4a; Norrish Type I reaction) [14]. Other typical photoinitiators are acyl phosphine oxides (Figure 5.4b) that undergo a cleavage between the carbonyl and the phosphino group under radiation with a wavelength above 370 nm, or benzoyl chloro methane (Figure 5.4c). The radical fragments add to the terminal end of the prepolymer C=C double bond, resulting in an extended radical which in turn reacts with further prepolymer, so that a polymer chain or network develops. The most likely pathway to terminate the polymer growth involves reactive, radical ends of the growing network to combine and form a bond. Other, less desirable, pathways include reactions with other accessible partners. For example, oxygen interferes with reactive centres in a radical reaction. Therefore nitrogen atmosphere is often preferred for UV curing.

The most common prepolymers for radical UV polymerisation belong to the class of acrylic or methacrylic acid esters. The ester can be formed with any hydroxyl function containing compound, cf. Figure 5.5. Useful synthons for ester groups are multi-functional alcohols, amino alcohols and their ethoxylated or propoxylated derivatives, as these allow onward reactions on other terminal functional groups. Polar, and in particular protic, terminal groups enable metal adhesion by virtue of their acidic nature. Examples are acrylated esters of phosphoric and phosphonic acid (Figure 5.6) [15].

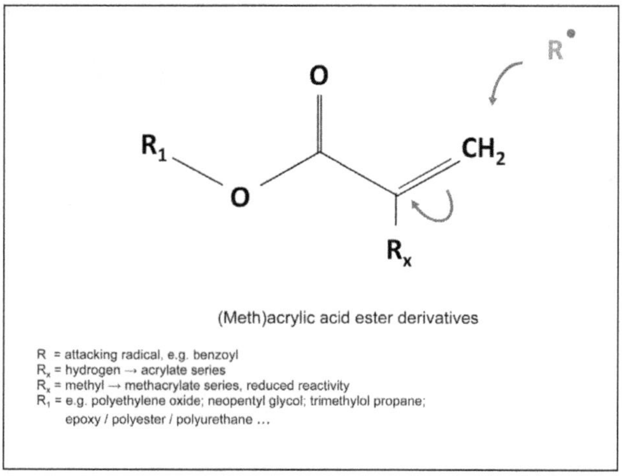

Figure 5.5: Acrylic and methacrylic acid derivatives as versatile precursors for radical UV polymerisation

Multi-functional acrylates allow cross-linking in two ("rope ladder") or three dimensions ("network"), thereby increasing the barrier effect of the film that is formed. Popular synthons are trimethylol propane triacrylate and its ethoxylated derivatives. In principle, acrylate groups can be bonded to any core molecule of suitable chemistry, including polyurethanes, polyesters and epoxies. The chemical features of the resulting polymer correspond to the properties of the prepolymer backbone. Monomeric and polymeric materials with four or even six terminal acrylate groups are commercially available. However with higher cross-link density, the film will also become less flexible. UV cured coatings must be elaborately tailored to balance the chemical and mechanical features of the film according to the specific needs. To this end, the vast variety of acrylate precursors with different backbone chemistry that are commercially available is very useful.

The cationic mechanism proceeds via liberation of protons (H^+) and their attack on electron-rich C=C groups (vinyl, allyl ethers) or cyclic ethers (epoxides, oxetanes). Cationic processes display benefits as well as disadvantages vs. the radical systems. They are usually less prone to shrinkage and brittleness, insensitive to oxygen, and allow "dark" and "shadow cure", as the initiating acid remains to be active after the irradiation has stopped. However, cationic processes are sensitive to moisture, and proceed at lower speed.

Dual cure systems combine both reaction pathways by using precursor materials that carry both reactive groups. Examples for suitable precursor chemistry are oxetane acrylates (cf. Figure 5.7) or glycidyl methacrylate.

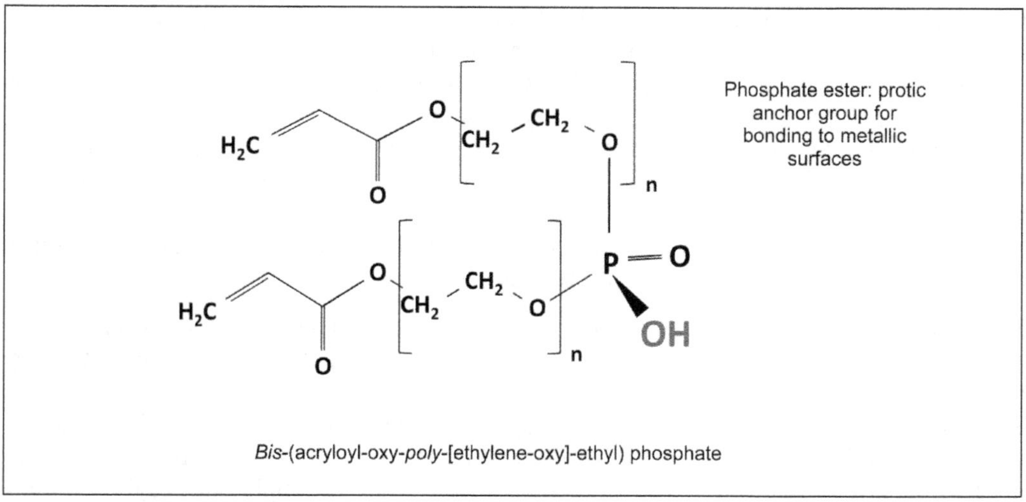

Figure 5.6: Terminal phosphate and phosphonate functionalised acrylates for UV metal coatings

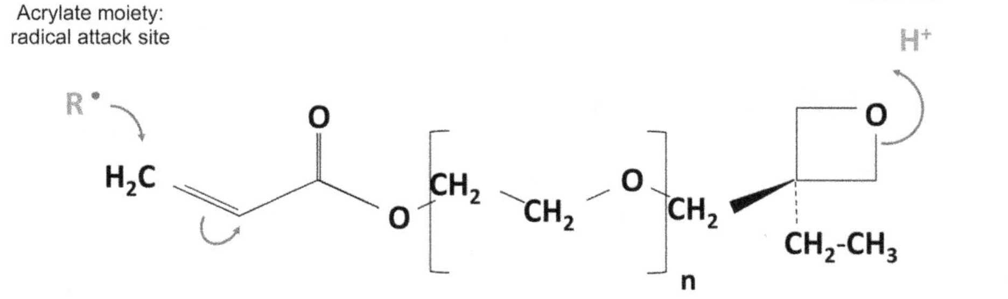

Figure 5.7: Bifunctional monomer for UV dual cure systems

A summary and comparison of the characteristics of the radical and cationic UV coating technology is given in Table 5.1.

EB curing is brought about by accelerated electrons of high energy (90 to 250 keV or higher) that are capable of direct attack on C=C double bonds, triggering radical polymerisation. Therefore, though in principle similarly constituted as the UV curable formulas, EB curable coating chemicals do not contain photoinitiators. A further benefit is the very fast reaction, leaving a very low residual free radical activity, hence no uncontrolled side reactions. Moreover, EB is unobstructed by pigmentation. Currently, EB lines are operated in wood coating (doors, wood fibre board or parquetry) or in printing. The industrial application for metal painting is a focus of development for the near future.

5.2.2.2 Applications

An interesting option for radiation curing emerges from the combination of pretreatment and primer functions in one product. Such a process could be incorporated in a coil coating

Table 5.1: Comparison of UV curing systems

	UV curing, radical	UV curing, cationic
Binder chemistry	prepolymers with polyester, polyurethane, epoxy backbone and acrylate functional groups	vinyl ethers, epoxides, oxetanes
Photoinitiator	alkyl phenones and α-hydroxy phenones; acyl phospine oxides; thiols	aryl iodonium or sulphonium salts
Advantages	fast reaction	reaction completes without irradiation ("dark phase"); formation of hydroxyl groups supports adherence to metals
Limitations	shrinkage, brittleness (cross-linking density)	slow reaction
	sensitive to oxygen	sensitive to moisture
	clearcoats and lightly pigmented lacquers	
	photoinitiator cleavage products are usually not food safe	

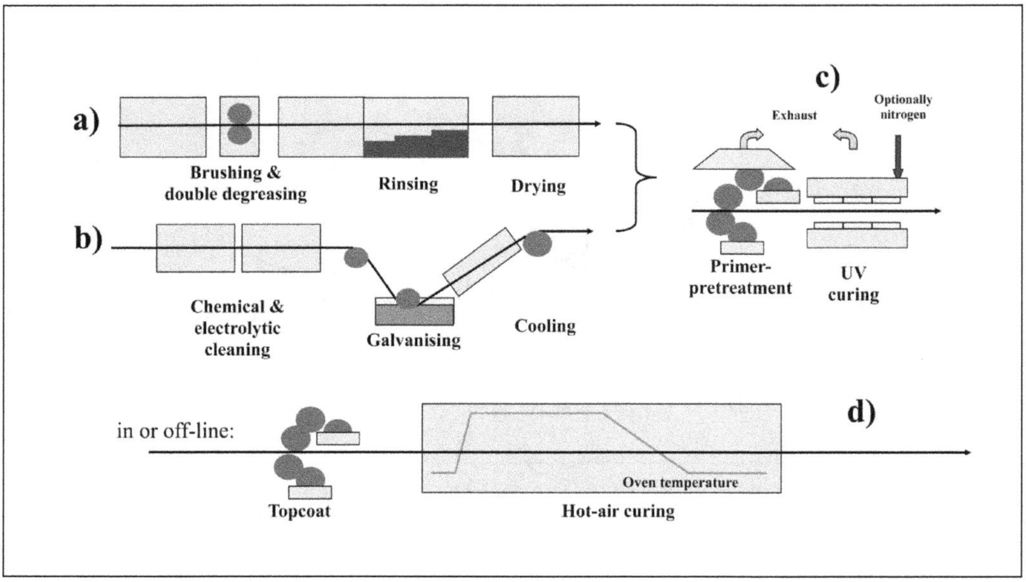

Figure 5.8: Proposed workflow for a UV curable primer pretreatment

where: (a) installation integrated in a coil coating line;
(b) in-line operation at the exit of a hot-dip galvanising line;
(c) adapted roll-coater and UV lamp array; (d) in-line or off-line topcoat operation

line. Equally, equipment for the application and the curing can be installed and operated, in particular, with very low floor space demand and no need for thermal combustion of solvent exhaust, in the exit zone of a coil galvanising line. Figure 5.8 depicts a schematic overview on the workflow options of such an operation [16, 17].

Developmental products for a combined primer pretreatment have been reported elsewhere, yielding cross-linked, non-blocking, adherent films of approximately 2 to 3 μm gauge (specific consumption 3 g/m^2) that provide the required capability for forming, as well as sufficient corrosion resistance in laboratory testing [18].

As another example for the versatility of UV curing, a protective clearcoat for stainless steel coil and sheet (ferritic and austenitic alloys) has gone commercial for domestic appliance uses. The coating is transparent and, being applied in very thin films of less than 2 μm, does not disguise the metallic look. It is claimed to withstand extended exposure to sunlight or heat without discolouration and to prevent staining from fingerprints and contaminations with common domestic cleaners and foodstuffs like mustard, lemon juice etc. Brush and scratch resistance of this coating was also demonstrated by respective tests. Bending and deep-drawing capability are restricted, due to the required mechanical hardness [19, 20].

UV curable coatings are also discussed to be used as sealants for bare, exposed metal areas, like freshly cut edges of sheet. For this purpose, the coatings are applied with thicknesses of 10 to 250 μm [21].

5.2.2.3 Equipment

Common UV lamps are mercury (Hg) vapour lamps that emit a spectrum with a maximum around 365 nm. When UV absorbers are present (e.g. titanium dioxide, TiO_2) in the formulation, the UV lamps may be doted with gallium, indium or iron, to bring about emission spectra with bands above 400 nm wavelength.

High-pressure UV equipment [22, 23] is available with high energy output as is required for the fast and wide-web operation of a coil coating line. Typical lamps have power ratings of 120 to 180 W/cm and offer output densities of 2 W/cm^2, sufficient for high-speed curing up to 100 m/min per lamp for clearcoat systems. For an industrial coil coating line, an array of several lamp rows must be installed that covers the whole width of the strip and provides sufficient radiation energy for the desired line speed. A range of industrial suppliers offer total systems including lamps, power supply and amplifiers, reflectors and shutter systems as well as cooling and air/nitrogen supply. Lamp technology includes electrode plasma discharge lamps and electrode-less, microwave triggered lamps.

Obviously, for industrial line operation, safety precautions regarding radiation protection are a prerequisite. When working without inert atmosphere, the exhaust gas must be delivered outside the plant building to avoid accumulation of ozone in the workplace atmosphere. However, no gas washer is required, as the ozone self-decomposes. Moreover, because there is no solvent involved, the process does not demand an incineration system.

5.3 *Literature*

[1] Goldschmidt, A., Streitberger, H.-J., Basics of Coating Technology, 2nd Edition, BASF Coatings AG, Münster 2007, pp. 327 ff
[2] Manea, M., High Solid Binders, Vincentz, Hannover 2008, p. 106
[3] Müller, B., Rath, W. Formulating Adhesives and Sealants, Vincentz, Hannover 2010, pp. 143 ff
[4] Manea, M., [2] pp. 63 ff, 81 ff, 87 ff
[5] Goldschmidt, A., Streitberger, H.-J., [1] pp. 385 ff
[6] Goldschmidt, A., Streitberger, H.-J., [1] p. 592
[7] Meuthen, B., Jandel, A.-S., Coil Coating, 2nd ed., Vieweg, Wiesbaden 2008, p. 130
[8] Kittel, H., Streitberger, H.-J. (ed.), Lehrbuch der Lacke und Beschichtungen (Coursebook of Paints and Coatings), Vol. 6, 2nd ed., Hirzel, Stuttgart 2008, pp. 286 ff
[9] Manea, M., [2] pp. 50 ff, 125 ff
[10] Kittel, H., Streitberger, H.-J. (ed.), [8] pp. 253 ff, 378 f
[11] Goldschmidt, A., Streitberger, H.-J., [1] pp. 607 ff
[12] Meuthen, B., Jandel, A.-S., [7] pp. 136 f, 300 f
[13] Goldschmidt, A., Streitberger, H.-J., [1] p. 613
[14] Coxon, J. M., Halton, B., Organic Photochemistry, Cambridge Univ. Press, London 1974, pp. 58 ff
[15] Grundmeier, G., Stratmann, M., Adhesion and De-Adhesion Mechanisms at Polymer/Metal Interfaces: Mechanistic Understanding Based on In Situ Studies of Buried Interfaces, Annu. Rev. Mater. Res. 35, 2005, pp. 571 ff, 577
[16] Heylen, M., Franzolin, G., Radiation Curable Coil Coatings: Innovative and Feasible Solutions for the Steel and Aluminium Industry, ECCA Autumn Congress Brussels, Proc., European Coil Coating Association, Brussels 2005
[17] Brown, N., Radiation Curing Systems in Coil Coating, ECCA General Meeting Maastricht, Proc., European Coil Coating Association, Brussels 2006
[18] Sander, J., Koch, M., Picot, P., Réticulation UV - Une nouvelle approche pour le „coil-coating“ (UV Curing - A New Chemical Approach for Coil Coating), Centre d'Actualisation des Connaissances et de l'Etude des Matériaux Industriels (Cacemi), Journée spécialisée, Proc., Paris 2002
[19] anon., Edler Touch für Edelstahl - Neue Oberflächenbehandlungen erhalten die elegante Optik des Edelstahls (Noble Touch for Stainless Steel - New Surface Treatments Preserve the Elegant Aspect...), Stahlmarkt 04, 2007, pp. 50 f; cit. re. Meuthen, Jandel [7] p. 136
[20] Sander, J., Up To Scratch With Novacoat 1000 UV, Steel & Aluminium Newsletter, Henkel KGaA, Düsseldorf 2007
[21] Meuthen, B., Jandel, A.-S.,[7] pp. 227 f
[22] Jung, J., Härtung von Druckfarben und Lacken durch UV-Strahlung (Curing of Printing Inks and Paints by UV Radiation), IST-Metz GmbH, in: Dtsch. Berufsgenossenschaft Energie Textil Elektro Medienerzeugnisse (Institution for Statutory Accident Insurance and Prevention - Energy Textile Electro Media Products, Working Committee for UV printing) BG ETEM (ed.), UV Technology - A Practical Guide for all Printing Processes, BG ETEM, Wiesbaden 2007, pp. 29 ff
[23] Skinner, D., New Development in UV Curing Systems for the Coil Coating Industry, ECCA Autumn Congress Brussels, Proc., European Coil Coating Association, Brussels 2004

6 Mechanism of protection and properties of organic coatings

Jörg Sander

6.1 Measurement of physical properties and influence

Testing of the performance of an organic coating comprises a number of optical, mechanical and corrosion tests. First and foremost, any coating must be properly cured and it must adhere well on its substrate, i.e. the primer must stick to the metal, which is supported by the correct pretreatment, and any subsequent paint layer must adhere on the foregoing coat (intercoat adhesion). While there are specific tests for individual layers, particularly related to their curing and adhesion features, finally the entire finish has to pass laboratory and application tests. Table 6.1 summarises the major tests on the physical features of an organic coating [1].

Table 6.1: Major test methods on the physical features of organic coatings

Test, feature	Norm
Film thickness, gauge	DIN EN 13523-1
Colour	DIN EN 13523-3
Hardness, pencil; scratch resistance Buchholz; resistance to cuts	DIN EN 13523-4 DIN EN ISO 2815
Impact; cracking resistance to fast deformation	DIN EN 13523-5
Bending; coating elasticity, crack resistance adhesive tape; substrate adhesion	DIN EN 13523-7 ditto
MEK Test; curing (cross-linking) and adhesion	DIN EN 13523-11
Heat resistance; mechanical effect of ageing	DIN EN 13523-13
Foil adhesion, protective foils	DIN EN 13523-17

According to Meuthen, Jandel [1]

Apart from cosmetic properties (colour, gloss) and hardness, especially the total thickness of a coating and its adhesive and elastic properties are of interest, as any failure in these areas may result in corrosion.

6.2 Dry film thickness

The thickness (gauge) of an organic coating determines the final aspect, barrier performance, and mechanical properties of the surface finish. High gauges are beneficial to the barrier properties, in particular as an obstacle to the diffusion of water and gas, however usually adversely affect hardness and deformation behaviour. As surface roughness must be taken into account, and any coating will retract from edges due to the action of surface tension, local gauges will inadvertently vary. It is, therefore, important to apply a minimum average coating thickness, in order to ensure sufficient coverage.

Measurement of the dry film thickness [2, 3] is often done by intrusive methods, e.g. by cutting a wedge-shaped scribe or drilling a conical bore into a coating, or by preparing a

Jörg Sander et al.: Anticorrosive Coatings

ISBN 978-3-86630-911-1

cross-section and measuring the thickness directly under a microscope. On the micrometre or even smaller scale, electron microscopy or focused ion beam techniques are used. To cope with the local thickness variation, several single measurements must be performed and averaged to obtain a statistically reliable result. Another method uses differential weighing of coated samples before and after paint stripping. The paint can be removed by wiping with methyl ethyl ketone or immersing in warm dimethyl formamide. All other, non-destructive methods rely on intrusive tests for calibration.

Ferromagnetic substrates like carbon steel can be inspected by electromagnetic measurement. The magnetic flow is influenced by the thickness of the non-magnetic coating, thereby resulting in a decrease of response voltage by induction. Eddy current measurements form the principle of similar techniques for diamagnetic substrates (aluminium, copper, brass, austenitic stainless steel). Both methods are subject of norm pr DIN EN 13523-1.

Continuous measurement of the dry film thickness is possible by back-scattering of β-particles (electrons) released from a radioactive krypton-85 source (^{85}Kr). The emission of secondary electrons is dependent on the thickness (DIN EN ISO 3543). Another method uses the propagation and reflection of heat at interfaces. The heat is induced by carbon dioxide LASER irradiation and monitored by an infrared sensor (DIN EN 15042-2). Due to their extremely fast response, both methods are preferably used in coil coating.

Coating thicknesses may vary between a few micrometres (corrosion protection primers, CPP, or backing coats in coil coating) and several hundred micrometres in severe corrosion protection in water engineering and shipbuilding cf. Chapters 2.1.3 and 6.5.2.

6.3 Adhesiveness

6.3.1 Role of adhesion and factors of influence

Adhesion [4, 5] of an organic coating is its most important property, as any coating that does not adhere will not serve any purpose. Adhesion of an organic coating must ensure that the coating will remain in its place despite any intentional deformation (flanging, profiling, bending), shear or torque force (cutting, drilling, honing, screw fastening) during the manufacturing process of a work piece, dynamic load by stretching, compression and torsion or also accidental deformation (impact, stone chipping) [6] during service life.

Adhesion, however, is also a precondition for a long lasting coated product to serve in a corrosive environment. In particular for coil coating, where deformations of the sheet metal are extreme to obtain the final shaped work piece (e.g. appliance equipment housing, car body parts made from preprimed sheet, or food cans and lids and crown corks [7]), insufficient adhesion and elasticity may cause weak points prone to corrosion by the appearance of cracks, even though no immediate delamination occurs. In addition, mechanical stress can also be induced by volume changes due to hydrothermal stress (cf. Chapter 6.6.1).

Table 6.2: Selected chemical bond strengths

Chemical bond	Binding energy (kJ/mol)
Al – O	512.1
C – O	1076.5
Fe – O	408.8
P – O	596.6
Zn – O	284.1

According to Weast, Astle [10]

The explanation of adhesion is often reduced to describing a mere mechanical interlocking of a coating in the cavities on a rough substrate surface. However, while a certain micro-roughness of the surface is certainly helpful, adhesion is much more the effect of van-der-Waals attraction forces between the substrate and molecules of the coating. As these forces apply in the nanometre range, a precondition is the proper wetting of a surface to allow contact (cf. Chapter 3.1.3). Here again, the importance of proper cleaning and pretreatment of the surface becomes obvious [8]. Adhesive forces include dipole and inductive attraction with a binding energy level of 5 to 25 kJ/mol, and hydrogen bridge bonds that can form between protic molecules (hydroxy, mercapto, amino functions) and metallic/oxidic surfaces, representing the 50 kJ/mol level. Chemical bonds like metal-to-oxygen bonds are much stronger, i.e. several 100 kJ/mol, hence desirable as adhesion promoters. Table 6.2 shows the strengths of a few selected chemical bonds that occur in organic coating of a metal [9, 10].

In terms of the chemical nature of the binder, thermoplastic binders like polyvinyl chloride, PVC, provide a very good adhesion which, together with their chemical inertness, makes them suitable for application in the food and beverage can sector. Also polyesters and polyurethanes and their hybrids are acknowledged to give good metal adhesion and formability. For example, they form the standard resin base for coatings in the manufacture of utility and railway vehicles or aircraft. Epoxies display very good primary adhesion, however restricted formability due to their inherent brittleness [11, 12].

6.3.2 Measurement of adhesion and elasticity

6.3.2.1 Industrial methods

In the industrial practice [13 - 15], adhesion tests comprise direct assessments of the adhesive performance or application tests that simulate practical use. The former class are pull-off or torsion shear tests. These are carried out by gluing a circular or ring punch onto the test surface with an adhesive selected to provide a stronger bond than adhesion and cohesion forces expected for the test specimen, then tearing off the punch from the test panel by perpendicular pulling or by radial twisting. The applied force is measured and the damaged area inspected for the aspect and pattern of the rupture.

For the latter class of tests, the cross-hatch test is most common, where two perpendicular sets of cuts are applied on the coated surface, the cuts being regularly spaced by 1 or 2 mm (coatings below or above 60 µm thickness, respectively), so that a grid of small squares is formed in the centre area. The adhesion is assessed by fixing an adhesive strip on to the test grid, pulling it off and inspecting the grid for the number of removed squares. A code Gt 0 to Gt 5 is defined by norm (DIN EN ISO 2409) for the assessment and classification of the failure. Other tests comprise peel (e.g. ASTM D-1876-08) [16], stone chip and "coin scratch".

Concerning adhesion of a coating under deformation, industrial tests usually involve standardised forming. Common tests comprise bending around cylindrical or conical mandrels or T-bending, i.e. the bending of sheet material back on itself. For the mandrel tests, the minimum numerical radius is measured that can be applied before the coating flakes (adhesion) or shows fissures (elasticity). In the T-bend test, the inner radius is expressed in multiples of the sheet thickness, i.e. a panel bent back on itself has a bending value of 0T, when one sheet of the same thickness is slit in, a value of 0.5T and so forth (DIN EN 13523-7); it must be noted that the corresponding American norm (ASTM D-4145-83) sets a different definition that results in double T values.

The deep drawing capability of coil coated sheet is assessed using the Erichsen dome test (Erichsen cupping). A cup is slowly formed in the test specimen, using a punch with a 20 mm spherical tip, and the deformation denoted (in millimetres dome height) that can be applied before paint delamination occurs on flanks or top of the dome. Often the deformation is combined with a cross-hatch, and the adhesion assessed by taping. Ideally, no delamination occurs until rupture of the metal itself (DIN EN 13523-6). Other deep drawing tests comprise the drawing of cylindrical or quadrangular cups, simulated food or beverage cans etc., and inspection for forming failures.

Shock deformation is assessed by ball impact tests etc., using a weight with a small ball tip that is dropped on the test surface. The test result displays the energy, given as weight by fall height, sometimes also converted in Joules, [J], that can be applied before paint delamination is observed (DIN EN 13523-5; DIN EN ISO 6272).

Propagation and reflection patterns of heat or sound in thermographic or (ultrasonic) echo methods that are used for thickness measurements (cf. Chapter 6.2) can be evaluated to detect adhesion failures. These methods, however, do not give exact numerical values for adhesion strength [17].

6.3.2.2 Laboratory methods

A very versatile scientific method to obtain topographic and surface force data in a microscopic scale is the atomic force spectroscopy, AFM. AFM is based on the principle that a cantilever tip is deflected upon approach towards and removal from a surface, due to van-der-Waals attraction and repulsion forces. By monitoring the cantilever motion, force-distance diagrams are obtained. Using an oscillating cantilever, a true topographic map is derived. The method is applicable in gaseous and liquid environment, thus enabling the study of adhesion phenomena on electrolyte covered metallic or oxidic surfaces. This method is so accurate as to allow studies of the effect of electric field and pH changes in the electrolyte on attraction forces, or single-molecule adhesion investigations [18].

Equation 6.1: $$\sigma = \frac{F}{A} = E\,\varepsilon$$ Hooke´s Law

where σ = tension;
F = force;
A = cross-section;
ε = elongation (as fraction Δl of the initial length l);
E = modulus (material specific)

Elasticity [19] is described by Hooke's Law (Equation 6.1), that is followed by any real material like a polymeric coating only for very small deformations. Above a critical value, it can be considered that the polymers of a resin are disentangled and stretched, but can rearrange and obtain a new conformation upon strain relaxation (visco-elastic flow). Upon overstretching, the material will suffer irreversible conformational changes and finally bond breaking and rupture.

Duroplastic materials have two regimes of elastic deformation, separated by the glass transition temperature T_g, and being characterised by two different moduli. Below T_g, the polymer film is glassy and rigid (high modulus) and displays brittleness. Above T_g, elasticity is much higher (low modulus). While part of the conformational changes that have occurred at low temperature may be reverted when the coating is heated above its T_g (cf. Chapter 6.6.1) [20], fissures and cracks are irreparable once they have occurred. Deformation of a coated work piece therefore is preferably done at temperatures above T_g.

Elasticity can be measured macroscopically on free polymer films by monitoring the elongation or torsion versus force and temperature. To obtain visco-elastic data, dynamic mechanical methods are available. On a microscopic scale, the deflection of a stressed material can be monitored by AFM.

6.4 Permeation through organic coatings

The role of water uptake in weathering and corrosion processes has already been touched in Chapter 2.2.1. Any coating acts as a semi-permeable wall to water vapour and liquid, which will inadvertently penetrate the coating. It will also serve as a carrier for electrolytes and corrosive gases [21]. To assess water uptake, gravimetric, optical (e.g. attenuated total reflection infrared spectroscopy, ATR-IR) and electrochemical, in particular capacitance and electrochemical impedance spectroscopy (EIS, cf. Chapter 7.2.3), methods have been used [22]. As the dielectric constant of water is substantially higher than for most organic matter the water uptake results in an increase of the total capacitance. It can be monitored over time, as well as in its relative extent, and the percentage of water present in the coating can be calculated [23, 24].

Water uptake has substantial effects on the plastification of a polymer film and on adhesion, as shown by a comparison of water uptake and peel test results of an epoxy amine adhesive as an example [25]. ATR-IR inspection confirmed that the water uptake reaches a saturation level that is dependent on the relative humidity (rH). For the epoxy amine, saturation is achieved after 48 h. At high humidity (rH = 97 %), the water uptake is almost double the value observed at low humidity (rH = 82 %). This coincides with a loss of adhesion. Peel results were reduced by a factor of 3 between low and high humidity regime. The water is stored in the bulk coating, which is indicated by the absence of a separated water film at the interface, the saturation level of 2 to 2.5 % water in the coating, and values of the diffusion coefficient in the order of $2 \cdot 10^{-9}$ cm/s, both regardless of whether an intact or a defective coating was investigated [26].

The rate of water uptake quite normally depends on the chemical nature of the polymeric backbone. In a comparison, a polyvinyl chloride paint showed very low, an acrylic coil coating paint a medium, and an automotive acrylic melamine clearcoat system a relatively high uptake [27].

6.5 Corrosion protective performance

6.5.1 Titanium and zirconium fluoro complex based pretreatments

In coil coating, chromium-free pretreatment chemicals based on aqueous solutions of titanium (Ti) and/or zirconium (Zr) fluoro complexes (cf. Chapter 8.2.7.1) have been commercial since more than a decade. They have been reported to provide sufficient corrosion protection and adhesion to subsequent topcoats. Figure 6.1 shows a façade part clad with coil-coated V-groove aluminium panels after 10 years service life. The panels, along with sun blinds, prepared with a chromium-free no-rinse pretreatment and top-coated with high-durable polyester, are mounted on a south-facing wall, covering an area of about 1,200 m². The façade is exposed to sun, wind and rain in an industrial site, and also encountering mechanical damage around loading areas, but does not show any corrosion even at critical points like punch-holes, bolt and screw joints or dripping edges [28].

Strong correlation and comparability between chromium-free and chromate pretreatments in various outdoor and accelerated corrosion tests have been shown for a chromium-free architectural pretreatment for steel, galvanised steel and Galfan substrates. Figure 6.2 displays some laboratory results, showing the maximum creepage values from the scribe mark as were reported under commercial polyester coil coating systems in comparison to standard chromate spray and no-rinse pretreatments. In the same paper, 360 h salt spray results

Figure 6.1: Chromium-free coil coated aluminium V-groove sheet Source: Henkel AG & Co.KGaA [28]

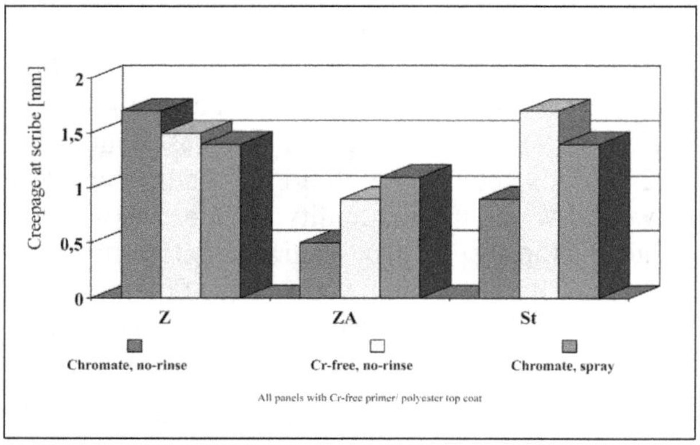

Figure 6.2: Laboratory testing on corrosion resistance
Source: ECCA 1999, [29]

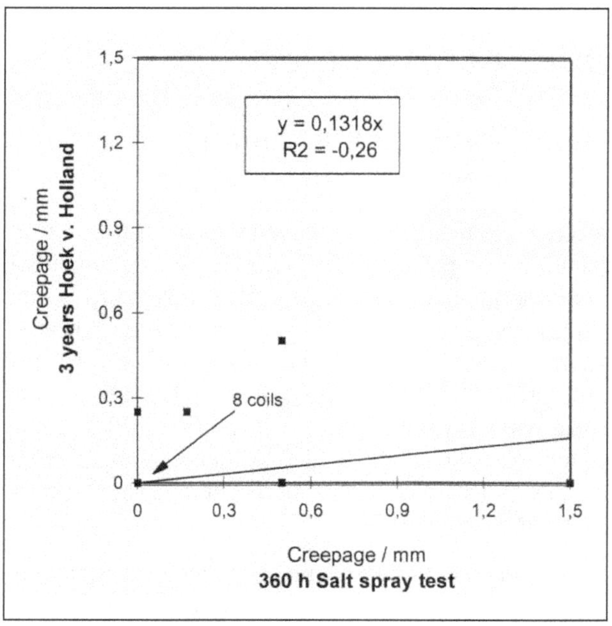

Figure 6.3: Creepage values for chromate and chromium-free pretreatments (coils from industrial production, 360 h salt spray vs. 3 a outdoor exposure, Hook of Holland, scribe to zinc substrate) Source: ECCA 1999, [29]

were shown to correlate well with 3 years outdoor exposure results from Hook of Holland, comparing 13 coil coated samples with chromate and chromium-free pretreatments that gave a maximum creepage of 0.5 mm. Fully equivalent anticorrosive performance between the standard chromate and the chromium-free systems was confirmed in cyclic corrosion testing including VDA and prohesion tests (see Figure 6.3). Moreover, the chromium-free pretreatment appeared to provide good process stability in terms of the distribution of creepage and T-bend adhesion values (see Figure 6.4) [29].

A series of experiments using the height-regulated "scanning Kelvin Probe" technique (HR-SKP, cf. Chapter 7.2.4.2) was carried out to get some insight in the anticorrosive mechanism. The results are shown in Figure 6.5. Electro-galvanised panel samples were cleaned and etched following a standard laboratory process, covered with a protective tape across one end (forming the y-axis), then, perpendicular to it, along the x-axis, pretreated on half of the remaining bare surface. The entire specimen was then coated with a model clearcoat. Upon removal of the protective tape, the specimen displayed a treated and an untreated area adjacent to each other, both coated with the clearcoat and having a sharp border to a bare metal surface.

One part of the bare surface (left area in the schematic of Figure 6.5) was encased with small stripes of resin, to form a small container which was filled with 0.5 M sodium chloride solution.

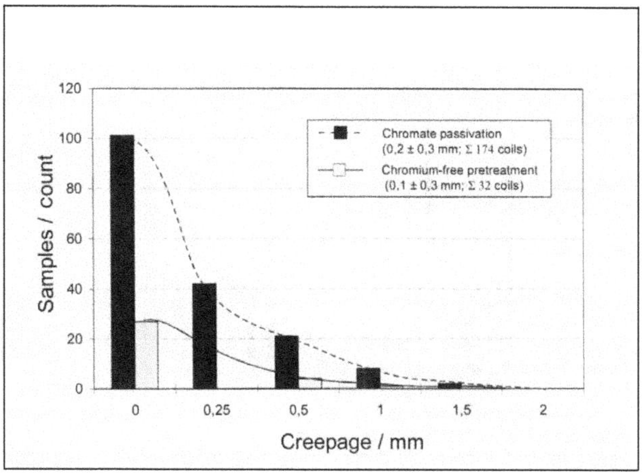

Figure 6.4: Creepage distribution on polyester coated coil with chromate and chromium-free pretreatments (coils from industrial production, 360 h salt spray, scribe to zinc substrate) *Source: ECCA 1999, [29]*

The HR-SKP device was used to scan the area 10 mm along the x-axis into the coating. It monitors the electrical potential underneath the paint film and its propagation from the defect with time, delivering a vivid picture of the advancing corrosion. It was shown that the potential above the untreated zinc surface exhibits the typical Zn/Zn^{2+} standard potential (E_{pot} = -0.76 V) and the difference in potential between the actively corroding defect area and the intact clearcoat/metal interface is around 0.5 V. Contrary to this, the potential of the pretreated area is shifted to more cathodic values. The potential step is therefore only around 0.15 V, making the overall corrosion reaction much less thermodynamically favourable. Consequently, the delamination front remains stable over a long time, hence indicating almost no propagation of the corrosion. Independent results from another source converse with these findings, suggesting the improvement resulting from band gap widening in a Ti/Zr oxide layer that effectively suppresses oxygen reduction [31].

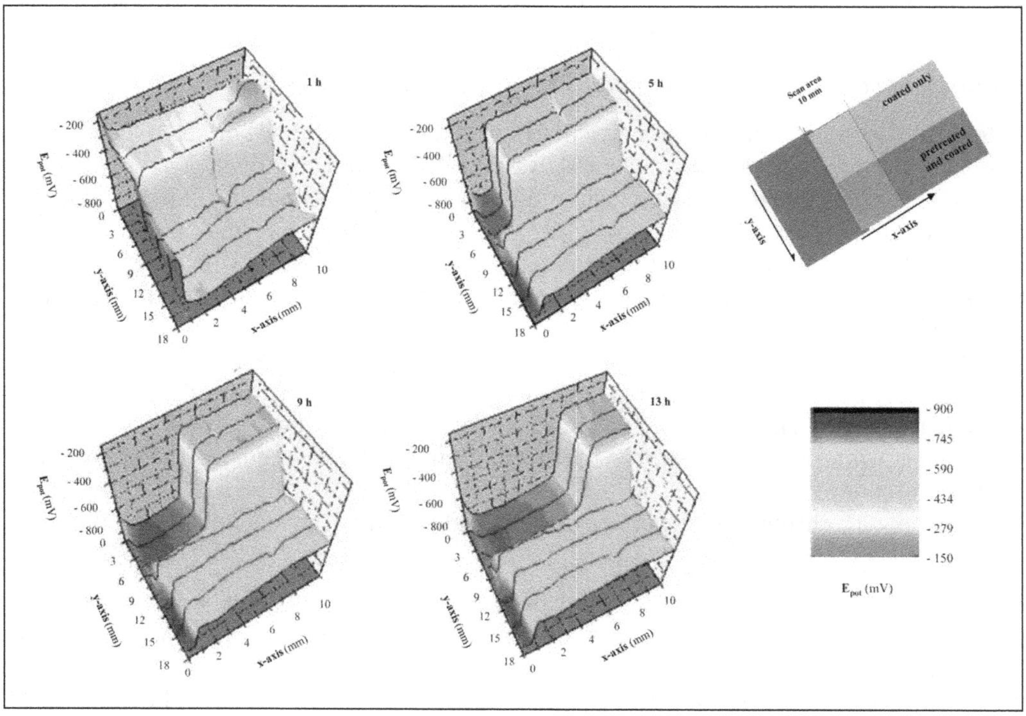

Figure 6.5: Electrochemical delamination: Model experiments studying a chromium-free conversion treatment with the height-regulated "Scanning Kelvin Probe" *Source: Elsevier [30]*

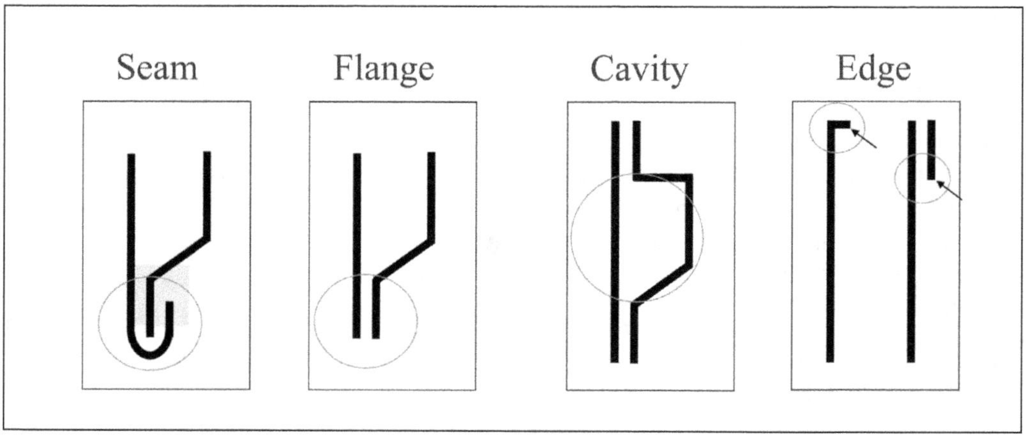

Figure 6.6: Corrosion risk areas in automotive construction *Source: Chemetall GmbH, 2009; [32]*

6.5.2 Weldable corrosion protection primer for automotive sheet

For the manufacture of car bodies, electro-galvanised sheet is one of the substrates that are used for the outer skin and hang-on parts. Particularly in doors and in the suspension frame, flanges and boxes form risk areas that are prone to corrosion, due to the poor accessibility of these areas for the normal pretreatment and painting process (cf. Chapters 2.1.3; 8.2.3.2). Examples are depicted in Figure 6.6. Using a corrosion protection primer (CPP) for these parts, the risk of rusting-through is reduced substantially, thus helping to extend warranty periods. Cost-driving secondary protective measures, like cavity sealing and seam sealing, virtually become redundant. The principle of these CPP coatings is illustrated in Figure 6.7. Table 6.3 summarises the features that characterise CPP for the automotive application.

Features of corrosion protection coatings for automotive sheet

- Improved corrosion protection particularly in box section, flanges or seams, helping to reduce secondary protection measures (wax flooding, seam sealing etc.)
- Suitable for all common welding techniques (spot, stud, projection, LASER, metal-active gas welding)
- Compatible with common cleaning and phosphating processes in car making
- Compatible with cathodic electro-dip paint in terms of conductivity and resistance to high pH at the interface
- Improvements in dynamic crash properties and vehicle safety
- Improved friction and wear properties to reduce lubricant in the press shop
- Compatible with common adhesives and sealants in car making
- Available work range in deep drawing substantially widened, reducing the risk of folds or ruptures
- Combine with chromium-free pretreatment on all galvanised steel and aluminium substrates

Typical CPP are based on polyester, epoxy or polyurethane resin blends. Apart from corrosion protection pigments, they also contain high portions of conductive pigments like zinc dust or graphite. The products of the so-called first generation comprise up to 70 % of Zn dust. They are applied with dry film gauges of 2.5 to 4.5 μm. Commercial CPP are applied on electro or hot-dip galvanised coil that is pretreated with chromium-free, Ti/Zr based processes, as described in the previous chapter [32 - 34].

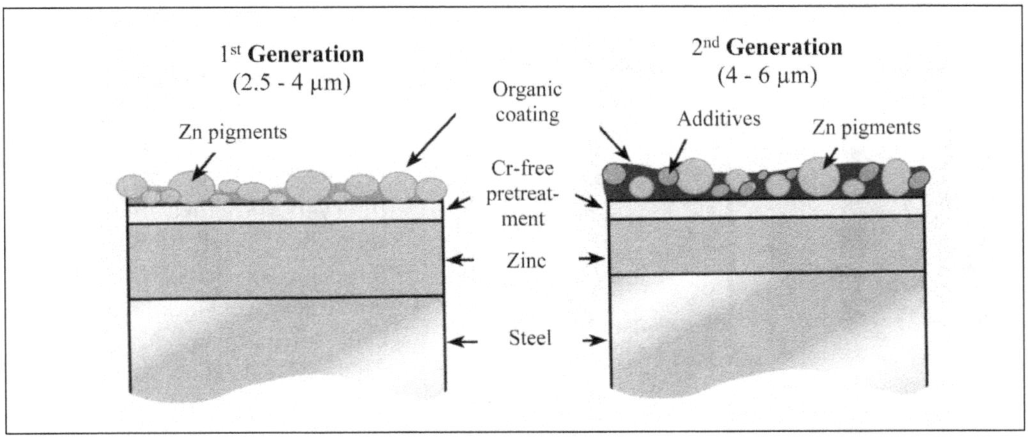

Figure 6.7: Corrosion protection primers for automotive sheet *Source: [33]*

Recently issued products are reported to be suitable for multiple substrates, as have become particularly common in the premium class segments of vehicle manufacturers. Aluminium/steel couples in the automotive mixed-metal construction form neuralgic points, where overlapping metal sheet is tightly joined in flanges by adhesive bonding or welding [35].

Figure 6.8 gives an impression of the performance of a modern CPP for car bodies. Test specimens were prepared to simulate the situation in a car door. As substrates, cold-rolled steel (CRS), 5 µm electro-galvanised (ZE) and ZE with CPP coatings of the 1st and 2nd generations at their respective coating gauges, 4.5 and 3.2 µm, were used. Sheet material was treated according to a standard car paint shop operation, including alkaline cleaning, tri-cation Zn phosphating and cathodic electro-dip coating. The interior and exterior sheets were joined by combined hem flanging and spot welding to form "mini doors", and then subjected to a cyclic corrosion test according to norm 621-415 of the German Automobile Associa-

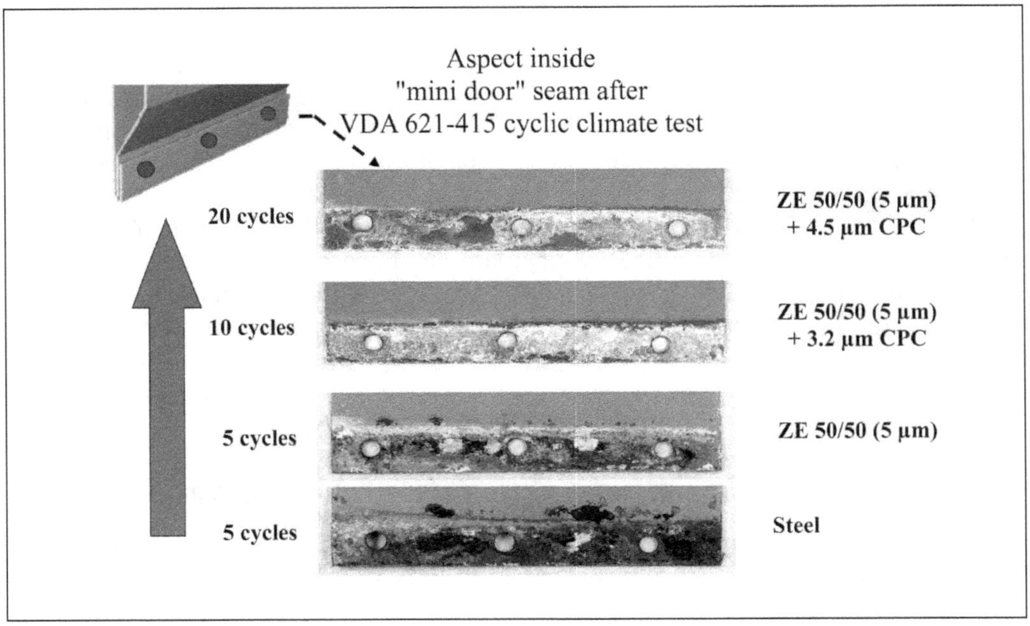

Figure 6.8: Corrosion protection primers: Performance in cyclic climate testing acc. to test norm VDA 621-415 (German Automobile Association) *Source: Chemetall GmbH, [36]*

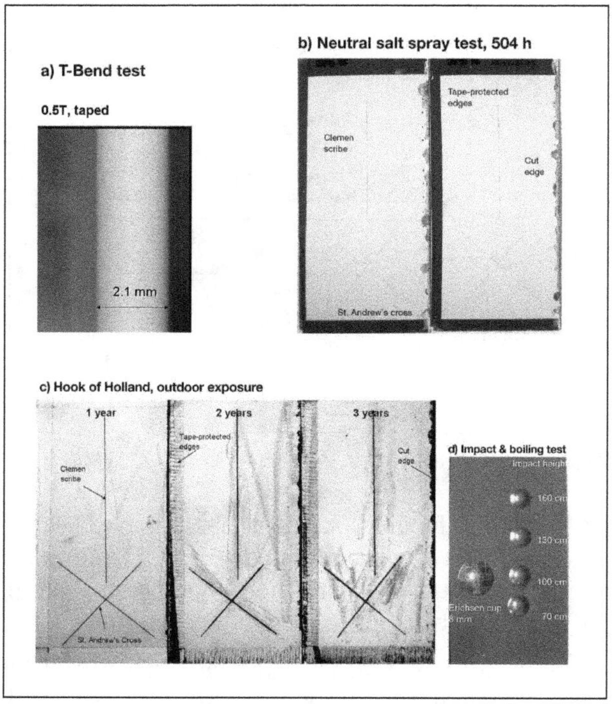

Figure 6.9: Anticorrosive and adhesion performance tests on a thermally curing 2-in-1 primer pretreatment (5 μm) with commercial liquid polyester topcoat (20 μm). (a) T-bend adhesion; (b) neutral salt spray (504 h); (c) outdoor exposure Hook of Holland, 1 - 3a; (d) cross-hatch/Erichsen dome test and Reverse impact *Source: Henkel AG & Co. KGaA [37]*

tion (VDA) for several weeks. This test comprises one-week cycles of consecutive salt spray, humidity (condensation) and drying phases (cf. Chapter 7.2.3.3). Samples were removed from the test in weekly intervals, the joined areas opened, inspected and photographed. Superior corrosion resistance of 20 cycles (weeks) is achieved by CPP in this test. While CRS and ZE substrates show severe corrosion (red rust) after only 5 weeks, the CPP coated samples are only slightly corroded (white rust) when thinly coated (3 μm), or even entirely unaffected with the higher CPP gauge (4.5 μm) [36].

6.5.3 Thermally curing 2-in-1 primer-pretreatment

A waterborne primer-pretreatment was reported elsewhere [37] that combines a conventional chromium-free pretreatment chemistry and a dispersion of a thermally drying acrylic resin in a single-package formulation. In the coil coating process, the product is roll coater applied with a liquid film of around 10 μm thickness, then dried in a hot air furnace or an infrared dryer with a metal temperature of below 100 °C to form a 5 μm primer coating. A line speed of up to 160 m/min is reported from industrial trials. Performance data were provided that match standard industrial requirements for indoors and outdoors architectural uses when combined with commercial polyester topcoats. Adhesion was tested using T-bend, impact and cupping tests. The coating passed 0.5T bending without paint delamination, and little cracking on the bent shoulder. Sufficient ball impact test results were reported, with around 16 J impact energy, as well as Erichsen cupping to 8 mm dome height, combined with a cross-hatch and 15 minutes immersion in boiling deionised water. Both tests were passed without flaking. Corrosion testing included the neutral salt spray test, 360 h QUV-B weathering, and three years outdoor exposure in Hook of Holland (see Figure 6.9).

6.5.4 Chromous based pretreatments – "chromiting"

The overall anticorrosive performance of pretreatments based on trivalent chromium, Cr (III), is claimed to be equivalent to that of chromate standard conversion coatings, however it is very much dependent on the gauge of the chromous oxy-hydroxide layer and its microstructure. EIS studies on galvanised specimens that were obtained after increasing immersion times in the bath solution showed that the impedance increased, having a maximum

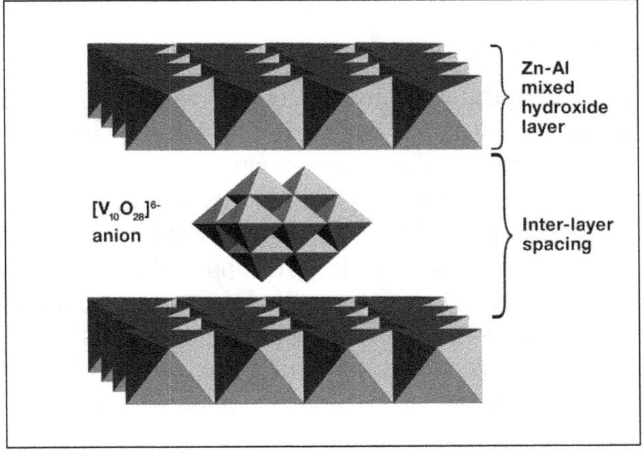

Figure 6.10: Schematic of a hydrotalcite-vanadate structure
Source: NACE International [42]

between 30 and 60 s, of approximately squared level of the untreated zinc specimen at the low-frequency end of spectrum. In this optimum region, the Bode plots exhibit three time constants which are interpreted as the coating forming three intact layers of different composition that serve as effective oxygen barriers [38]. Upon further extension of the immersion time, cracks appear in the coating. Neutral salt spray resistance shows almost no creepage on aluminium and hot-dip galvanised steel and only little creepage from the scribe (< 1 mm) on cold-rolled steel after 600 h. As reported, also the acetic acid salt spray on aluminium shows only little creepage. In constant climate and boiling tests or in "acid rain" cyclic humidity testing, no blisters nor delamination are observed. However, no duration is disclosed for these tests. Ball impact, taping, Erichsen cupping and thorn bending also show good adhesion and flexibility [39, 40].

6.5.5 Active pigments, ion exchangers and scavengers

The influence of non-chromate active pigments in a model epoxy coating was investigated by potential measurements and EIS after salt spray and immersion in sodium chloride solution. Though zinc chromate standard pigmentation showed the best initial data in electrochemical evaluation, the differences diminished over exposure time, due to the solubility and leaching of the chromate. Zinc phosphate, and in particular zinc phospho-molybdate and zinc-calcium phospho-molybdate proved to be viable non-toxic alternatives. Water permeability of free films did not reveal any differences between chromate and its replacement pigments. Thermogravimetric analysis (TGA) showed the beneficial effect of all pigments as compared with a pigment-free model varnish, thus indicating some stabilising interaction [41].

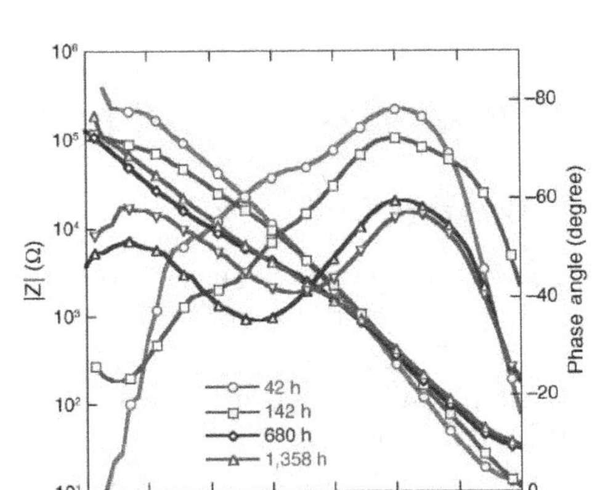

Figure 6.11: EIS Bode plots of hydrotalcite-vanadate pigmented epoxy coatings subject to salt spray exposure; near-linear curves represent the resistance/frequency plot (left ordinate), wave-shaped curves correspond to the phase angle/ frequency plots (right ordinate). *Source: NACE International [43]*

A coprecipitate of zinc and aluminium hydroxide and metavanadate VO_3^- can be synthesised from alkaline solution. As verified by X-ray diffraction, the reaction results in a hydrotalcite structure that consists of layers of mixed Zn-Al hydroxide species, whose inter-layer spaces are populated by decavanadate anions $[V_{10}O_{28}]^{6-}$ as illustrated in Figure 6.10 [42].

From there the decavanadate can be released slowly by way of ion exchange, resulting in equilibrium or steady-state concentrations of vanadate much below its solubility. The steady state is achieved after approximately 900 h of contact with a chloride solution. Simultaneously, Zn^{2+} ions are released. Both ionic species were shown to reduce the current density on AA 2024 alloy ($AlCu^4Mg^1$) in polarisation experiments, i.e. suppress the corrosion. When admixed to an epoxy coating, and applied on AA 2024, the coating showed a protective behaviour. Figure 6.11 represents the resistance (Bode) and phase angle plots vs. the voltage frequency, as taken in repetitive EIS measurements of specimens that were subject to salt spray. The bell-shaped phase angle curve shows a maximum at high frequencies, which however is reduced with time, while a second maximum develops at the low-frequency end.

This resembles an additional capacitance that arises from direct electrolyte/metal contact and the respective double-layer that is formed at the exposed interface. On the other hand, the Bode plot does not change much with time. This means that despite the access of electrolyte to the metal surface, no effective dissolution takes place, i.e. active inhibition is displayed. However, the performance is less pronounced than with strontium chromate [43].

6.5.6 UV-curable primer-pretreatment

UV technology has been adapted to develop a combined 2-in-1 primer-pretreatment as mentioned in Chapter 5.2.2.2. The liquid, flowable single-package formulation was reported to be applicable by roll-coater, at coil speeds of 100 m/min. The developmental formulation was tested in pilot line trials on galvanised steel, Galfan and Galvalume. Based on titanium-organic compounds with acrylic acid functional groups as corrosion inhibition system, it renders corrosion protection under commercial polyester topcoats with creepage levels of below 2 mm from the scribe after 504 h salt spray test. Sufficient adhesion was demonstrated with T-bend (0.5T without flaking), reverse impact and Erichsen cupping tests. For adaptation to the industrial use, further improvements are being sought including higher application speed, enhanced corrosion resistance for harsh industrial/marine environments and versatility under a variety of liquid or powder topcoats [44, 45].

6.6 Degradation and ageing

6.6.1 Weathering

In service life, a coated work piece is weathered, i.e. it is subject to manifold random ageing influences: The UV spectral part of sunlight, combined with oxygen, brings on photochemical reactions. Temperature changes and all forms of precipitation result in mechanical stress by swelling and shrinking or even hydrolytic or thermal decomposition. Chemicals, like deicing salts, "acid rain" (sulphur dioxide, SO_2), volatiles, solvents or fuel, or natural substances, like tree gum or mould all have their detrimental effect. All these climatic factors combine with mechanical loads like static or dynamic deformations, thus leading to a multiple burden.

Degradation by UV radiation, hydrolysis and thermal destruction, in the presence of oxygen, follows different pathways, but basically, all these reactions lead to a reduction in molecular weight, thus impairing the mechanical and protective properties of the coating. This becomes apparent first by gloss reduction, fissures and chalking, later by corrosion and eventually spalling of the entire coating.

Generally, oxidation under UV radiation (photo-oxidation) involves the formation of hydroperoxides that further decompose to yield shorter carbonyl compounds and other by-products under chain cleavage. In this way, ethers and esters undergo the formation of hydroperoxides on the carbon atom that bears the C–O bond. Further decomposition forms the lower carbonyl compound and, by further oxidation, eventually liberates carbon dioxide, CO_2. Urea resins are oxidised in a similar way to form carbamides and, finally, free urea compounds, CO_2 and water. Radiation cleavage of acrylates may occur in the main chain or in side branches of the polymer. In polyvinyl chloride, PVC, the C–Cl bonds are homolytically cleaved by the radiation, finally resulting in chain length reduction and the generation of CO_2 and hydrogen chloride. Chain shortening and, on the other hand, after-curing induced by UV radiation simultaneously reduce elasticity and cohesion [46].

In a comparative study, different automotive clearcoat systems with variations in their binder (polyacrylate or polyester polyol), hardener, i.e. hexamethylene or isophorone diisocyanates (HDI or IPDI) and stabiliser (with or without HALS additions, cf. Chapter 4.5) were investigated regarding their weathering resistance. Apparently, the polyacrylate polyol is much more prone to photo-oxidative degeneration than the polyester system. HDI on its own appeared to be the more effective cross-linker for the polyester; for the acrylic system, a blend of HDI with IPDI was preferable. HALS showed a beneficial effect, decelerating the degradation, however less pronounced for the acrylic urethane [47, 48].

The interdependence between degradation by weathering and susceptibility to corrosion was shown in a combined study of the EIS data and performance in salt spray tests of various coatings on galvanised steel after 2,000 h exposure to QUV-A (cf. Chapter 7.2.3.4). The organic coatings consisted of polyester or polyurethane primer (5 to 15 μm) and different polyester-melamine topcoats (20 μm). In the majority of cases, blistering occurred on the unscribed surface that was unrevealed without the UV conditioning. The salt spray resistance correlated with the impedance values measured in the low-frequency range, indicating the resistivity at the metal/polymer interface. Good performers with regard to blistering also had a particularly high insulating protection. The performance observed in the combined exposure roughly coincides with gloss retention results, but does not necessarily reflect the durability classification of the paint systems as obtained only from UV resistance measurements [49].

Also hydrothermal stress can result in a mechanical damage. For instance, with polyester-urethane and epoxy-polyamine systems, a reduction in cross-link density of more than 50 % was observed, though the polymer strands themselves did not seem to break down, as the average molecular weight between cross-links grew by about the same factor. The degradation of the epoxy system occurred at about the double rate as with the polyester-polyurethane, and it also suffered a substantial reduction of the glass transition temperature (T_g). Moreover, when combined with humidity, the degradation of epoxy was enhanced [50]. Furthermore, esters can be degraded by saponification. Thermal hydrolysis of urethanes leads to the formation of alcohols and N-substituted carbamic acids, which further degrade to the lower amine while liberating CO_2 [51].

The susceptibility to blistering of coil applied organic coatings on galvanised steel was also demonstrated by EIS studies after cyclic humidity testing comprising wet and dry cycles at 25 °C/50 % rH and 50 °C/95 % rH over four weeks. Critical polyurethane and durable polyester primer were investigated. The polyurethane system showed significant roughing and a locally dropping potential (about 100 to 120 mV) in the SKP mapping after exposure to humidity, in accordance with the blisters that were observed visually. EIS revealed the resistivity in the low frequency end to be reduced by a factor of >10. There was, however, no indication of corrosive activity going on, so that the resistivity decrease is interpreted

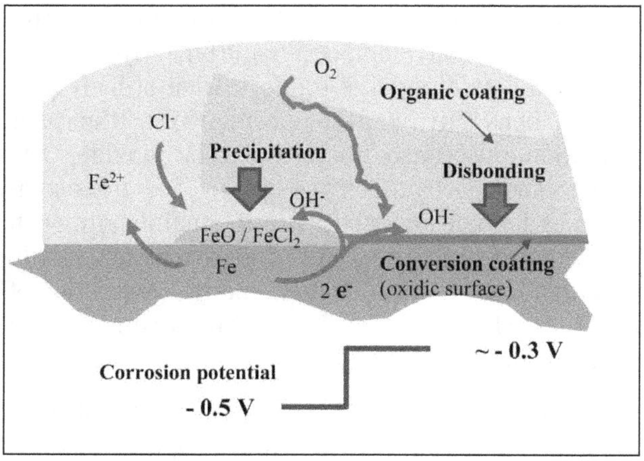

Figure 6.12: Reactions at an iron/polymer interface that result in delamination and creepage

in terms of displacement of the polymer from the metal surface by water molecules. The polyester system was much less critical [52]. Similar results are also reported for cathodic electro-dip primers that were subjected to hydrothermal ageing in cyclic tests.

EIS after hydrothermal ageing in cycles between 55 °C/84 % rH and 23 °C/40 % rH again showed the decrease of the pore resistance with ongoing exposure. This electrical weakening correlates with higher residual stress in the coating. The stress increase was by the factor 2 for two different coatings. Hence, ageing resulted in an irreversible swelling, thereby generating more interstices and rendering the coating more prone to water uptake upon immersion [53].

6.6.2 Electrochemical degradation

6.6.2.1 Cathodic disbonding: oxygen reduction

Apart from mechanical forces, the delamination of organic films from a metallic substrate occurs via two major different pathways, i.e.

- water uptake, swelling and displacement of adsorptively bonded (hydrogen bridge bond) polymer moieties
- dissolution of metal/polymer valence bonds by chemical (hydrolysis, saponification) or electrochemical reactions (corrosion)

The latter interaction will lead to a separation of electrical charges at a metal/polymer interface, either by hydrolysis (e.g. dissolution of metal soaps) or by the corrosion reactions. Combined HR-SKP (cf. Chapter 7.2.4.2.1) and atomic force microscopy (AFM) confirmed that the electrochemical delamination process often takes place before the polymer-to-metal bonds are weakened by swelling and hydrolysis through the ingressing electrolyte.

The ease of electron transfer depends on the conductive properties of the metal surface which is covered by native oxide (cf. Chapters 2.1.1, 2.1.2). If this oxide is an insulator (e.g. SiO_2, Al_2O_3, MgO) the transfer will be decelerated. Oxygen reduction, on the other hand, leads to the generation of anions that may attack both the native oxide (e.g. dissolution of zinc or aluminium oxides by formation of zincate or aluminate), and saponifiable bonds between the metal and the polymer.

Cathodic disbonding prevails on conductive surfaces, like that of steel as shown in Figure 6.12. At the delamination front, the potential, as can be measured by SKP, jumps up from a low, cathodic level, typical for a bare steel surface that is immersed in an electrolyte, to the anodic value in the region of the intact coating. Here, the reaction is controlled by the conductivity of the iron oxide layer that diminishes with the conversion of Fe from the divalent to the trivalent state. Ions that ingress from the defect site into the oxide layer reconstitute the conductivity and accelerate the corrosion again. In effect, the delamination front propagates with the diffusion of ions along the interface. Other investigations indicate that the potential front propagates in correlation with a pH gradient that represents the active oxygen reduction, while the eventual delamination may be kinetically retarded when bond-breaking reactions are slow [54].

Simultaneously, oxygen, during its reduction, may induce the degradation of the polymer film itself as depicted in Equations 6.2 a to d. These describe a set of reactions that involve radical intermediate species like oxygen radical anions that have been confirmed to occur under these conditions. They may abstract hydrogen from C–H bonds, generating free organic radicals that undergo onward reaction with molecular oxygen, O_2. The formation of organic peroxides **1** that further decompose to form carbonyl compounds **2** (aldehydes and ketones) and water is therefore a likely reaction path [55].

Equation 6.2

$$R^1R^2CH{-}H + {}^{\bullet}O{-}O^- \rightarrow R^1R^2\dot{C}H + H{-}O{-}O^-$$

$$R^1R^2\dot{C}H + O{-}O \rightarrow R^1R^2CH{-}O{-}O^{\bullet}$$

$$R{-}H + {}^{\bullet}O{-}O{-}CHR^1R^2 \rightarrow R^{\bullet} + \underset{\underline{1}}{H{-}O{-}O{-}CHR^1R^2}$$

$$\swarrow$$

$$H_2O \quad + \quad R^1{-}C(R^2){=}O \quad \underline{2}$$

6.6.2.2 *Anodic disbonding: filiform corrosion*

Underneath paint films that display a high barrier effect towards oxygen and moisture penetration, a corrosion process is observed that advances in threadlike filaments (filiform corrosion, FFC). FFC has been an issue in the European architecture since the early 1980s [56]. In this kind of corrosion reaction, de-adhesion is anodic-driven. While it preferentially takes place on aluminium or magnesium surfaces that form insulating oxides, it has also been reported for galvanised steel. FFC does not appear in a condensation climate, i.e. at dew point. Its maximum occurrence and speed are observed at relative humidity around rH = 80 %. The mechanism is explained via an aeration gradient that results from retarded oxygen supply. That means that the oxygen reduction takes place in the tail section of the corrosion filament, while its front is characterised by active anodic metal dissolution [57, 58].

Several resin systems were investigated in terms of their resistance against FFC. 40 µm topcoats were applied on AlMg1 architectural aluminium alloy, on top of 1 µm layers of adhesion promoting organic primers. When using anodic films as pretreatment, the amount of FFC was found to decrease exponentially with the thickness of anodic films. Hydrothermal sealing was found to further improve the performance. This is attributed to the roughness of the surface that is increased by this treatment, thereby improving the mechanical interlinking of resin and anodic film. As a general trend, duroplastic primers, e.g. based on hard polyester or cured phenolic resins, appeared to provide a good result in combination with a topcoat based on aromatic polyester polyol and aliphatic diisocyanate hardener. However, the authors make a point in that the performance of coatings may be impaired by cleavage reactions that are caused by the action of hydrochloric acid on ester and urethane bonds, and therefore results of the accelerated filiform test, when incubated with hydrogen chloride vapour, may diverge from practice [59]. The same workgroup, however, explicitly focusing on the comparability of accelerated test and outdoor exposure results in Hook of Holland (cf. Chapters 7.2.2.3; 7.4) in a later investigation, observed similar rankings for various pretreatment/paint combinations in both tests. No clear preference could be given for particular pretreatments or paint systems. Good results, i.e. a filiform corroded area of less than 0.5 mm^2/cm in front of a scribe mark, were likewise obtained with chromate or chromium phosphate pretreatment, with 5 µm DC anodic films and a chromium-free, hexafluoro zirconate based pretreatment. Paint systems comprised chromate containing polyvinylidene difluoride (PVDF), polyester powder and liquid paints, polyester primer/ polyurethane topcoat and epoxy primer/PVDF powder combinations [60].

For rolled aluminium, literature states a particularly critical susceptibility for FFC. Investigations of this behaviour studied the influence of the rolling and subsequent annealing processes. It is believed that the particular electrochemical activity in the magnesium (Mg) containing AA 3005 and 8006 alloys (AlMg0.5Mn; AlFe1.5Mn0.5) is caused by the formation of a reactive layer just underneath the superficial 1 µm layer of grain-refined structures that is effected by the rolling itself. The presence of enrichments of the alloying elements in the surface layer may cause a negative shift of the electrochemical potential.

Lower alloyed aluminium like 1000 series alloys (i.e. <100 ppm Mg) shows a less critical behaviour. As a solution to prevent FFC on these substrates, the authors recommend the complete removal of the critical surface layer by caustic pickling and strong alkaline cleaning, respectively, or by DC anodic treatment [61 - 63]. Also from other sources, reported laboratory and long-term exposure data indicate that DC anodising with film gauges of 6 µm appears to be the most robust pretreatment [64, 65].

Chromating and standard chromium-free pretreatments give a good performance against FFC, with the exception of high-alloyed copper and zinc containing alloys [66]. Driven by the high operational cost of anodising, the toxicology of chromate and the missing absolute performance safety of standard chromium-free pretreatments, alternatives are still being sought, of which the phospho-molybdate technology seems to have some potential [67].

6.7 Literature

[1] Meuthen, B., Jandel, A.-S., Coil Coating, 2nd ed., Vieweg, Wiesbaden 2008, p. 175

[2] Goldschmidt, A., Streitberger, H.-J., Basics of Coating Technology, 2nd edition, BASF Coatings AG, Münster 2007, pp. 342 ff

[3] Meuthen, B., Jandel, A.-S., [1] pp. 180 ff

[4] Goldschmidt, A., Streitberger, H.-J., [2] pp. 372 ff

[5] Meuthen, B., Jandel, A.-S., [1] pp. 51 ff, 74, 94 f

[6] Goldschmidt, A., Streitberger, H.-J., [2] pp. 715 f

[7] Kittel, H., Streitberger, H.-J. (ed.), Lehrbuch der Lacke und Beschichtungen (Coursebook of Paints and Coatings), Vol. 6, 2nd ed., Hirzel, Stuttgart 2008, pp. 101 f, 384, 495

[8] Goldschmidt, A., Streitberger, H.-J., [2]pp. 452, 451 f, 663, 688 f

[9] Barrow, G. M., Physikalische Chemie (Physical Chemistry), vol. 3, 3rd ed., Bohmann-Vieweg, Vienna 1977, pp. 335 f

[10] Weast, R. C., Astle, M. J. (ed.), Handbook of Chemistry and Physics, 60th ed., CRC Press, Boca Raton 1981, pp. F-220 ff

[11] Kittel, H., Streitberger, H.-J. (ed.), [7] p. 495

[12] Meuthen, B., Jandel, A.-S., [1] p. 95

[13] Goldschmidt, A., Streitberger, H.-J., [2] pp. 373 ff, 391 ff

[14] Meuthen, B., Jandel, A.-S., [1] pp. 189 ff

[15] anon., Internationale Qualitätsrichtlinien für die Beschichtung von Bauteilen aus Aluminium (International Quality Standards for the Coating of Work Pieces), GSB AL 631, ed. 07/2009, Gütegemeinschaft für die Beschichtung von Bauteilen (GSB International), Schwäb. Gmünd 2009, pp. 32, 35 ff

[16] anon., ASTM D1876 - 08 Standard Test Method for Peel Resistance of Adhesives (T-Peel Test), Am. Soc. f. Testing and Materials (ASTM) Intern., W. Conshohocken PA, USA, 2010; www.astm.org/Standards/D1876.htm; (28-03-2010; 11:00 h)
[17] Goldschmidt, A., Streitberger, H.-J., Handbuch Lackiertechnik (Handbook Coating Technology), BASF Coatings AG, Münster 2002, pp. 396 ff
[18] Grundmeier, G., Stratmann, M., Adhesion and De-Adhesion Mechanisms at Polymer/Metal Interfaces: Mechanistic Understanding Based on In Situ Studies of Buried Interfaces, Annu. Rev. Mater. Res. 35, 2005, pp. 584 ff
[19] Goldschmidt, A., Streitberger, H.-J., [2] pp. 385 ff
[20] Shi, X., Fernando, B. M. D., Croll, S. G., Concurrent physical aging and degradation of crosslinked coating systems in accelerated weathering, J. Coat. Technol. Res. 5, 2008, pp. 299 ff
[21] Kittel, H., Streitberger, H.-J. (ed.), [7] p. 390
[22] Lewis, O. D., A Study of the Influence of Nanofiller Additives on the Performance of Waterborne Primer Coatings, PhD Thesis, Loughborough Univ., Loughborough 2008,, pp. 77 ff
[23] Vlasak, R., Klüppel, I., Grundmeier, G., Combined EIS and FTIR-ATR study of water uptake and diffusion in polymer films on semiconducting electrodes, Electrochim. Acta 52, 2007, pp. 8075 ff
[24] Lewis, O. D., [22] pp. 88 ff
[25] Posner, R., Giza, G., Vlasak, R., Grundmeier, G., In-situ electrochemical Scanning Kelvin Probe Blister Test studies of the de-adhesion kinetics at polymer/zinc oxide/zinc-interfaces, Electrochim. Acta 54, 2009, pp. 4837 ff
[26] Vlasak, R., Klüppel, I., Grundmeier, G., [23]
[27] Hardcastle, H. K., Meeks, W. L., Considerations for characterizing moisture effects in coatings weathering studies, J. Coat. Technol. Res. 5, 2008, pp. 181 ff
[28] Lienkamp, P. (ed.), Sander, J., Körner, T., Kleinschmidt, T., More than 10 years of experience with chrome-free pre-treatment of aluminium – Perfect protection against corrosion, Press release, Henkel AG & Co. KGaA, Düsseldorf 2008, cit. Int. Aluminium. J. 84 (11), 2008, p. 36 ff
[29] Androsch, F. M., Stellnberger, K.-H., Wolpers, M., Jandel, L., Drescher, D., Sander, J., Seidel, R., Chromate-free Coil Coating and One Year of Production Experience, ECCA General Meeting Monte Carlo, Proc., European Coil Coating Association, Brussels 1999, Chapter 3
[30] Klimow, G., Fink, N., Grundmeier, G., Electrochemical Studies of the Inhibition of Cathodic Delamination of Organically Coated Galvanised Steel by Thin Conversion Films, Electrochim. Acta 53, 2007, pp. 1290 ff, 1291
[31] Stoulil, J., Thierry, D., Structure and corrosion properties of Ti-Zr based conversion coatings applied on hot-dip galvanised steel, NACE Corrosion Congress 2009, Atlanta GA, USA, Proc., NACE International, Houston TX, USA 2009
[32] Schinzel, M., Advanced Corrosion Protection of Automotive Body Sheet - A Challenge for Coil Coating, ECCA Autumn Congress Brussels, Proc., European Coil Coating Association, Brussels 2009
[33] Meuthen, B., Jandel, A.-S., [1] p. 76
[34] Sander, J., Novel Chromium-free Surface Treatments for Coil, ECCA General Meeting Budapest, Proc., European Coil Coating Association, Brussels 2001
[35] anon., Corrosion Protection in Light-Weight Construction, APT Aluminium News 5, 2009, p. 4
[36] Schinzel, M., [32]
[37] Sander, J., Novel Surface Treatment of Metal Strip in the Coil Coating Process, Millennium Steel 2009, pp. 146 ff, 148 ff
[38] Chang, Y.-T., Wen, N.-T., Chen, W.-K., Ger, M.-D., Pan, G.-T., Yang, T.-C.-K., The effects of immersion time on morphology and electrochemical properties of the Cr(III)-based conversion coatings on zinc coated steel surface, Corr. Sci. 50, 2008, pp. 3494 ff
[39] Pohl, J., Möller, R., New Cr (III)-based pretreatment – A simple, all-purpose, environmentally friendly process, JOT International Surf. Technol. (1) 2009, pp. 2 ff
[40] Volk, P., Echte Alternative – Chrom(III)-haltige Vorbehandlung für Aluminium (True Alternative – Cr[III] Containing Pretreatment ...), Metalloberfläche 60, 2006, pp. 24 ff
[41] de Lima-Neto, P., de Araújo, A. P., Araújo, W. S., Correia, A. N., Study of the anticorrosive behaviour of epoxy binders containing non-toxic inorganic corrosion inhibitor pigments, Prog. Org. Coatings 62, 2008, pp. 344 ff
[42] Mahajanam, S. V. P., Buchheit, R. G., Characterization of Inhibitor Release from Zn-Al-$[V_{10}O_{28}]_6$-Hydrotalcite Pigments and Corrosion Protection from Hydrotalcite-Pigmented Epoxy Coatings, Corrosion 64, 2008, p. 231
[43] Mahajanam, S. V. P., Buchheit, R. G., [42] p. 238
[44] Sander, J., Koch, M., Picot, P., Réticulation UV - Une nouvelle approche pour le „coil-coating“ (UV Curing - A New Chemical Approach for Coil Coating), Centre d'Actualisation des Connaissances et de l'Etude des Matériaux Industriels (Cacemi), Journée spécialisée, Proc., Paris 2002
[45] Sander, J., UV Curable Coil Coating, Symposium Innovative Surface Treatments for Steel and Aluminium Coil Coating, Neuss, Proc. Henkel AG & Co.KGaA, Düsseldorf 2004
[46] Goldschmidt, A., Streitberger, H.-J., [2] pp. 406 ff

[47] Krüger, S., Wachtendorf, V., Rauth, W., Klimmasch, T., Krüger, P., Screening the weathering stability of automotive coatings by chemiluminescence, J. Coat. Technol. Res., 5, 2008, pp. 17 ff
[48] Krüger, S., Brademann-Jock, K., Rauth, W., Klimmasch, T., Krüger, P., Analysis of weathering effects in automotive coatings using thermoluminescence, J. Coat. Technol. Res., 5, 2008, pp. 11 ff
[49] Mallégol, J., Poelman, M., Olivier. M.-G., Influence of UV weathering on corrosion resistance of prepainted steel, Prog. Org. Coatings 61, 2008, pp. 126 ff
[50] Croll, S. G., Shi, X., Fernando, B. M. D., The interplay of physical aging and degradation during weathering for two crosslinked coatings, Prog. Org. Coatings 61, 2008, pp. 136 ff
[51] Goldschmidt, A., Streitberger, H.-J., [2] p. 410
[52] Nazarov, A. Prosek, T., Thierry, D., Application of EIS and SKP methods for the study of the zinc/polymer interface, Electrochim. Acta 53, 2008, pp. 7531 ff
[53] Olivier, M.-G., Romano, A.-P., Vandermiers, C., Mathieu, X., Poelman, M., Influence of the stress generated during an ageing cycle on the barrier properties of cataphoretic coatings, Prog. Org. Coatings 63, 2008, pp. 323 ff
[54] Huang, M.-W., Allely, C., Ogle, K., Orazem, M. E., A Mathematical Model for Cathodic Delamination of Coated Metal Including a Kinetic pH-Porosity Relationship, J. Electrochem. Soc. 155, 2008, pp. C 279 ff
[55] Grundmeier, G., Stratmann, M., [18] pp. 597 ff
[56] Haagen, H., Rihm, K.-H., Filiformkorrosion auf organisch beschichtetem Aluminium (Filiform Corrosion on Organically Coated Aluminium), FARBE UND LACK 96, 1990, pp. 509 ff
[57] Grundmeier, G., Stratmann, M., [18] pp. 604 ff
[58] Heinrich, M., Gaszner, K., Schuler, T., Korrosion von beschichtetem Aluminium (Corrosion of Coated Aluminium), Metalloberfläche 49, 1995, pp. 802 ff
[59] Haagen, H., Gaszner, K., Scheck, K., Vorbehandlungsschichten und Filiformkorrosion bei organisch beschichtetem Aluminium (Pretreatment Layers and Filiform Corrosion…), FARBE UND LACK 97, 1991, pp. 306 ff
[60] Heinrich, M., Gaszner, K., Schuler, T., [58] pp. 802 ff , pp. 804 f
[61] Leth-Olsen, H., Nisancioglu, K., Filiform Corrosion of Aluminium Sheet. I. Corrosion Behaviour of Painted Material, Corr. Sci. 40, 1998, pp. 1179 ff, 1183
[62] Leth-Olsen, H., Afseth, A., Nisancioglu, K., Filiform Corrosion of Aluminium Sheet. II. Electrochemical and Corrosion Behaviour of Bare Substrates, Corr. Sci. 40, 1998, pp. 1195 ff
[63] Leth-Olsen, H., Nordlien, J. H., Nisancioglu, K., Filiform Corrosion of Aluminium Sheet. III. Microstructure of Reactive Surfaces, Corr. Sci. 40, 1998, pp. 2051 ff
[64] Pietschmann, J., Gardein, R., Filiform Corrosion and Results of 10 Years Natural Weathering, Galvanotechnik 106, 2008, pp. 1764 ff
[65] Pietschmann, J. E., Pfeifer, H., Filiformkorrosion auf organisch beschichtetem Aluminium, Teil III (Filiform Corrosion on Organically Coated Aluminium, Pt. III), Aluminium 70, 1994, pp. 82 ff
[66] Rudolf, A., Kaiser, W.-D., Filiformkorrosion - Einfluss der Legierung und der Vorbehandlung (Filiform Corrosion - Influence of Alloy and Pretreatment), JOT 42, 2002, pp. 98 ff
[67] Rauscher, G., Neue chromfreie Vorbehandungsverfahren für Aluminium (New Chromium-free Pretreatment Processes), JOT, 2005, pp. 54 ff

7 Testing of organic coatings

Jörg Sander

7.1 Performance testing

In addition to mechanical testing, the assessment of endurance of a coated metallic work piece requires certain tests that provide information on the anticorrosive performance of the coating. This information must be available before such a work piece enters the market place. Therefore, to enable performance forecasts, a number of accelerated corrosion tests have been designed.

7.2 Accelerated corrosion tests

7.2.1 Overview

A long-term service life is simulated by a number of accelerated corrosion tests [1 - 5], in order to assess the effect of corrosive stress in a time lapse. Typical corrosion tests comprise salt spray, boiling and condensation tests, constant and cyclic climate tests. Weathering includes combined humidity and irradiation tests (visible light and ultraviolet radiation, UV), that help in assessing resistance of the coating against photochemical degradation (chalking, discolouration, gloss reduction). Coil and sheet material that is used for appliance manufacture is tested on its resistance to water soak and detergents, next to tests regarding the resistance to e.g. domestic cleaners, foodstuffs, dyes and colourants. Coatings for the canning industry, e.g. for beverage cans and food containers, are often subject to additional tests regarding their porosity (e.g. copper sedimentation after immersion in acidulated copper salt solution) and their sterilisability or retortability in pressurised cooker tests. Table 7.1 gives an overview on accelerated tests related to corrosion.

7.2.2 Constant climate tests

7.2.2.1 Salt spray

The salt spray test provides data on the corrosion resistance of coated work pieces in general. The coating is scratched with a defined tool (scribe mark), and subjected to salt fog from a solution that contains 50 g/l sodium chloride (NaCl). Steel substrates are usually tested in a neutral salt fog, while aluminium (Al) is subjected to an atmosphere that is additionally acidulated with acetic acid (pH ≈ 3.2, AASS test). In some cases, e.g. automotive applications like wheels, this is even enhanced by adding copper sulphate (CASS test). The usual test temperature is 35 °C. Test specimens are removed from the test chamber after the agreed test time and inspected for blistering and for adhesion loss (creepage) progressing from the scribe and, when applicable, from cut edges. Typically, 336 to 1,008 h (2 to 6 weeks) resistance is required for steel substrates, including galvanised steel, with

Jörg Sander et al.: Anticorrosive Coatings

ISBN 978-3-86630-911-1

only small defects being accepted. The most stringent requirements are made for Al sheet for aircraft, i.e. a maximum creepage of 1.25 mm (0.05 in) from the scribe is accepted after 3,000 h AASS.

7.2.2.2 Constant climate, humidity

For constant climate testing, coated specimens are scribe-marked and may be bent or deformed in another way. After storage in a test compartment under constant humidity and temperature, they are inspected regarding blistering and other apparent visual changes.

7.2.2.3 Filiform corrosion

The filiform corrosion test is mainly carried out on Al work pieces. It focuses on the threadlike corrosion pattern that may occur in chloride containing, reduced-humidity corrosive environments. Scribe marked work pieces are stored over fuming hydrochloric acid (HCl, 36 %) for an incubation period, then kept under constant atmosphere with controlled humidity and temperature (80 % relative humidity [rH], 40 °C). The specimens are inspected for the number and length of threads that have developed. Again the aircraft industry claims the highest performance, i.e. 1,000 h resistance with max. 2 mm thread length. Steel based materials are tested after incubation in NaCl solution (EN ISO 4623).

Table 7.1: Major test methods on the corrosion performance of organic coatings

Test, feature	Norm
Constant climate tests	
Salt spray – neutral	EN ISO 9227 NSS (DIN 50021), ASTM B117, EN 13523-8
Salt spray – acetic acid	EN ISO 9227 AASS
Salt spray – copper acetate	EN ISO 9227 CASS
Humidity	DIN 50017 KK, pr EN 13523-25
Condensation	EN ISO 6270-1 and 2, pr EN 13523-26
Filiform	ISO 4623-2, EN ISO 4628-10, EN 3665
Boiling, water soak	EN 13523-9
Cyclic climate tests	
Humidity/Temperature cycle tests	
- Humidity	DIN 50017 KFW, pr EN 13523-25
- Tropical test	analogous, ambient to 40 °C
- Humidity-freeze test	analogous, -5 °C to ambient
Humidity in SO_2 atmosphere	EN 13523-13 (DIN 50018)
Cyclic salt spray/humidity tests	
- Automotive industry	VDA 621-415, Renault ECC1 D172028, VW PV1210, Volvo VCS1027,149, other OEM specifications
- Prohesion test	ASTM G85,A5
Cyclic immersion test	ISO 15710, EN 3212
Weathering tests	Xenon test, VDA 612-430 etc.
Outdoor exposure	EN 13523-19

According to Meuthen, Jandel [1]; Kittel [2]; Goldschmidt, Streitberger [4]; LeBozec et. al. [5]

7.2.2.4 Condensation

The condensation test is focused on the resistance of coatings under elevated temperature and permanent humidity. Specimens are stored at 100 % rH and 40 °C, and then checked for blisters and visible deterioration.

7.2.2.5 Boiling, water soak

Water soak tests usually comprise immersion in deionised water at 40 °C, however may be extended to boiling. In case of sheet material, it is usual to apply a deformation by bending or cupping. Test specimens are checked for blisters and other visual changes, as well as for adhesion (by taping), if applicable, in deformed areas. Test times range from 15 min to several hours and are subject to agreement.

7.2.3 Cyclic climate tests

7.2.3.1 Cyclic humidity

All tests described above for constant climate conditions can be combined with repetitive cyclic variations in temperature, humidity, additional chemical or radiation load, etc.

Cyclic humidity and condensation testing may involve temperature cycles between ambient and 40 °C (tropical test), or freezing and ambient (humidity-freeze test). Humidity tests are also combined with cyclic exposure to sulphur dioxide containing atmosphere (SO_2, DIN EN 13523-13), to simulate "acid rain" industrial environments.

7.2.3.2 Prohesion

The so-called prohesion test was invented to cope with the issue that results obtained in the harsh neutral salt spray on galvanised steel do not correspond very well with reality. Samples are subjected to a cyclic test comprising salt spray with a diluted, slightly acidulated NaCl solution, with intermediate intervals of drying. Though it was first intended for use on steel substrates, the test is also used for aluminium. Test times are up to 2,016 h (12 weeks) on Al materials for the aircraft industry [6].

7.2.3.3 VDA test

In particular, the automotive industry has standardised cyclic tests that involve salt spray, humidity and open-air storage, to obtain a more realistic representation of outdoor results. A full cycle usually lasts a week, however the order and duration of the individual load and relief periods vary with the specific standard. Standard 621-415 of the German Automobile Association (VDA) is generally accepted as a guideline, though there are also internal norms by individual manufacturers like Daimler, Ford, General Motors, Renault, Volkswagen, or Volvo. A minimum of 10 cycles (10 weeks) is usually specified as test duration for cathodic electro-dip primer coats, whereas the full finish must withstand longer periods.

7.2.3.4 UV test, weathering

Irradiation with UV is primarily used to assess the photochemical resistance of the binder system, however may also give a good account on the overall resistance of the entire coating against natural weathering, when combined with humidity (VDA 621-430). Xenon lamps or special UV lamps are common. The latter provide a smaller spectral segment, either UV-A (spectral maximum at 340 nm, simulating the short wave solar radiation) or UV-B (313 nm, more energetic, preferred to simulate extremely solarised climate). Performance requirements range up to 2,000 h or more.

7.3 Electrochemical testing

7.3.1. General remarks

Apart from phenomenological assessment of the corrosion of coated metals, as is obtainable by accelerated and outdoor exposure testing, modern corrosion science has developed a lot of designs for an experimental access to this problem. Though being still far from providing a true prediction, modern tools are powerful aids in understanding of corrosion and the delamination mechanisms it induces. As corrosion is predominantly an electrochemical phenomenon, electrochemical testing plays an important role in these investigations. However recently, scientists have focused on combining analytical methods, that comprise e.g. optical (infrared and Raman, in particular surface-enhanced and total-reflection methods), electron-optical and atomic force methods, to complement the electrochemical tests. Increasingly often, experimental setups are created to enable real-time, in-situ investigations [7]. Selected electrochemical methods will be described hereunder.

7.3.2. Electrochemical potential

7.3.2.1 Standard potential

Electrochemical potentials apply for any kind of chemical reaction involving free flow of electrons, and therefore being linked with an electric current. Such reactions are commonly referred to as red-ox reactions. The voltage measured in a bridge circuit when no electric current is observed (galvanostatic method) is defined as the electrochemical potential. By convention, it is given versus a standardised counter electrode, most commonly the standard hydrogen electrode, whose potential is set to be 0 V. Electrochemical potentials expressed in this way are called standard potentials. The more oxidisable the metal, the lower (more negative) is its electrochemical potential. On the other hand, a high

Table 7.2: Standard electrochemical potentials of selected elements and compounds
Oxidation occurs more readily with lower potential

Description	Electrode reaction in aqueous solution	Standard potential E_0 at 25 °C (V)
Pt | F_2 | F^-	F_2 (*g*) + 2 e^- → 2 F^-	2.87
Au | $AuCl_4^-$, Cl^-	$AuCl_4^-$ + 3 e^- → Au + 4 Cl^-	1.00
Ag | Ag^+	Ag^+ + e^- → Ag	0.7991
Pt | Fe^{2+}, Fe^{3+}	Fe^{3+} + e^- → Fe^{2+}	0.771
Pt | H_2 | H^+	2 H^+ + 2 e^- → H_2	0.0000
Pb | Pb^{2+}	Pb^{2+} + 2 e^- → Pb	-0.126
Fe | Fe^{2+}	Fe^{2+} + 2 e^- → Fe	-0.4402
Zn | Zn^{2+}	Zn^{2+} + 2 e^- → Zn	-0.7628
Al | Al^{3+}	Al^{3+} + 3 e^- → Al	-1.662
Mg | Mg^{2+}	Mg^{2+} + 2 e^- → Mg	-2.363
Pt | MnO_2 | MnO_4^-	MnO_4^- + 2 H_2O + 3 e^- → MnO_2 + 4 OH^-	0.588
Pt | O_2 | OH^-	O_2 + 2 H_2O + 4 e^- → 4 OH^-	0.401
Pt | H_2 | OH^-	2 H_2O + 2 e^- → H_2 + 2 OH^-	-0.8280

According to Barrow [8]

(positive) electrochemical potential of a metal indicates what is called a noble metal which is resistant to (atmospheric) corrosion. A selection of standard potentials is given in Table 7.2 [8].

7.3.2.2 *Cyclo-voltammetry*

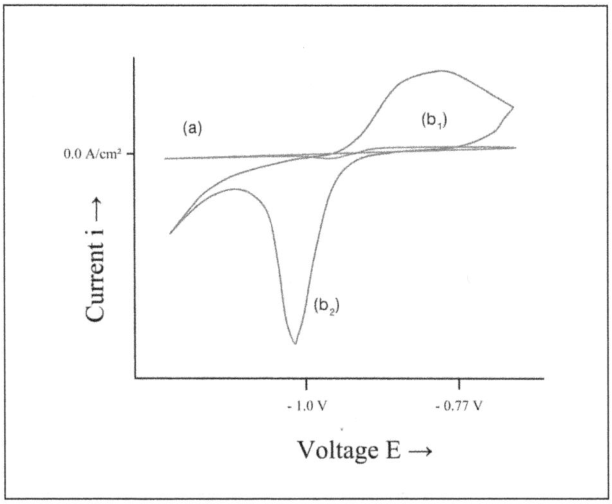

Figure 7.1: Cyclovoltammograms on galvanised surfaces; (a) native oxide; (b$_1$) freshly cleaned Zn surface - oxidative branch, (b$_2$) reductive branch *Source: Fink et al. [10]*

Red-ox systems can be investigated by a dynamic method that is carried out by scanning the voltage (E) up and down in the expected range of the electrochemical reaction to occur, and monitoring the current (i) that is measured in response. The method is commonly referred to as cyclo-voltammetry (CV). Negative currents indicate a reduction process, while positive currents represent an oxidation. The area (E · i) enveloped by a curve over its base line is an indication of the reaction turnover. Repetition of the voltage sweep results in a set of curves that form a loop. In case of ideally reversible reactions, the loops are closed and do not alter their shape. In real systems, reactions are more likely to be irreversible, i.e. products are removed from the equilibrium, for example by onward reactions, and in consequence, the curves will be transitory. Figure 7.1 shows a typical set of voltammograms as can be obtained on a technical zinc (Zn) surface. Curve (a) represents the situation on Zn covered with native oxide which inhibits the red-ox reactions (cf. also Chapter 7.3.3) hence almost no current is measured over the voltage sweep. Curve (b) shows the typical loop of metallic, freshly etched Zn that enables the oxidation (b$_1$) of the metal and the reduction (b$_2$) of its ions to occur [9, 10].

7.3.3 *Electrochemical impedance spectroscopy*

Electrochemical impedance spectroscopy (EIS) can be used to monitor any electrochemical activity that occurs at the paint/substrate interface upon corrosion and degradation of metal-to-paint bonds. In terms of its electrical behaviour, a coating serves as a barrier that displays both capacitance and resistivity features which can be simulated by an "equivalent circuit". In a simplistic model, such a circuit contains a resistor (considering that the ionic transport through the electrolyte is slower than the free flow of electrons, hence forms a resistance to an electric current), followed by a loop with a second resistor and a capacitor in parallel. This latter resistance is attributed to the retardation of any molecule or ion, when it passes through the pores and interstices of the coating (pore resistance), while the capacitance is interpreted in terms of the dielectric behaviour of the coating that separates metal and surrounding electrolyte. As soon as electrochemical activity starts at the metal/coating interface, the situation is getting more complex. For simulation, additional components must be incorporated into the equivalent circuit, to account for charge transfer, double layer and similar phenomena, as illustrated for an example in Figure 7.2 [11].

EIS measurements are carried out by applying AC voltage to a coated test specimen, which, by nature of the mixed resistor/capacitor behaviour of a coating, results in a responding electric current that alternates with a phase difference. The phase angle usually shows

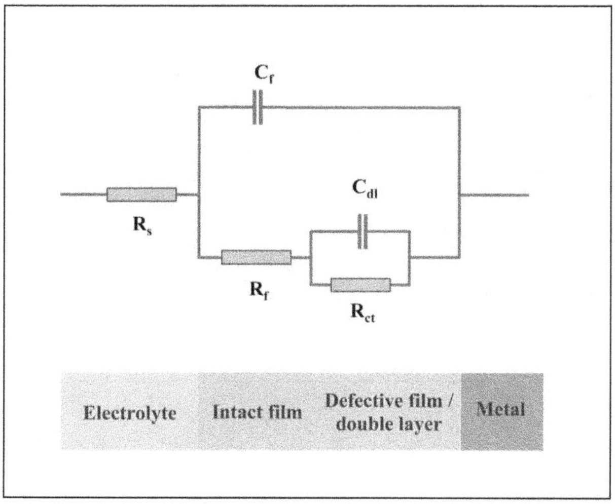

Figure 7.2: Schematic: EIS equivalent circuit; R_s = resistance of solution, R_f = film (pore) resistance, R_{ct} = charge-transfer resistance (electron transport through interface); C_f = film capacitance, C_{dl} = double-layer capacitance; according to Lewis [11]

a maximum coinciding with the inflection point of a drop in the impedance level (Bode plots). By Fourier transformation, the impedance can be separated into real and imaginary components which are plotted against each other in a diagram (Nyquist plot).

A simple, intact coating usually shows a semi-circle in this plot, its radius being an indication of the overall protective power. For example, a technical galvanised steel surface displays a large semi-circle (i.e. higher protection) when the native oxide layer, that mainly consists of mixed oxy-hydroxide of the alloying element aluminium, is left intact (solvent cleaning). An alkali-cleaned surface exposes more oxy-hydroxide of the base metal, zinc, which is more reactive, hence resulting in a smaller semi-circle in the Nyquist plot [12]. Film degradation that is induced by electrochemical activity is indicated by shifts and deformations in the plots of repetitive EIS cycles. In particular, in the Nyquist plot, the appearance of film defects results in the introduction of a second semi-circle (i.e. a second time constant representing the additional resistor/capacitance circuit loop), often camouflaged in real spectra by complex enveloping curves.

7.3.4. Electrochemical techniques with high spatial resolution

7.3.4.1 Scanning vibrating electrode

Local corrosion currents at a metal/electrolyte boundary can be detected and mapped with a high spatial resolution, i.e. on a micrometre scale, by the scanning vibrating electrode technique (SVET). This technique is capable of making in-situ studies without intrusive action. Its principle is the measurement of currents induced by ionic species that migrate in the course of local corrosive reactions. The locations of anodic and cathodic events can be clearly separated by the method. The lateral resolution is several 10 µm, and current densities of less than 5 $\mu A/cm^2$ can be detected.

For example, the method was used to study the protection mechanism on freshly cut exposed edges of hot-dip galvanised steel [13]. In another application of the method, the effect of the composition of Al alloyed galvanised (Galfan, ZA) layers on the corrosion performance was assessed [14].

7.3.4.2 Height-regulated Scanning Kelvin Probe

7.3.4.2.1 General technique

The inhibiting effect of conversion pretreatments can be studied by high-resolution height-regulated "Scanning Kelvin Probe" (HR-SKP) experiments [15 - 18]. The HR-SKP allows the precise measurement of the local distribution of electrical potentials on a metal surface underneath an organic coating. Thereby, changes in the oxide structure and the ingression of ions become detectable. The local resolution in the x- and y-direction is approximately 15 µm. With repetitive measurements, it is thus possible to obtain a view of how corrosion and delamination develop and move forward over time. The height regulation (z-axis) has a precision of below 200 nm. With the HR-SKP, it is possible to distinguish between the weakening of the structures by moisture (wet delamination), and red-ox (corrosive) reactions. Figure 7.3 shows a schematic of the experimental array of the HR-SKP.

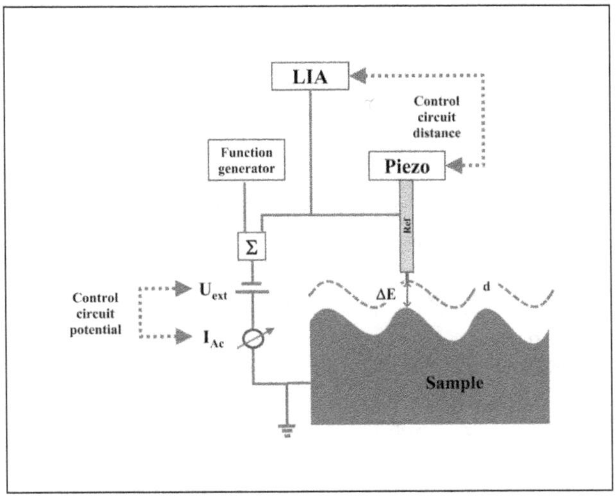

Figure 7.3: Schematic view of the height-regulated Scanning Kelvin Probe (HR-SKP)
LIA = Lock-in Amplifier; Ref = Reference Electrode with capillary tip; d = tip distance over sample
ΔE = potential difference, Source: Grundmeier et al. [17]

7.3.4.2.2 *Blister test*

Another test for the investigation of corrosive delamination mechanisms combines the HR-SKP with a setup that allows putting an organic film under electrochemical and mechanical strain at once. The coated test specimen is prepared by drilling a small hole through the substrate from the underside. When an electrolyte is pumped in, delamination will occur, as soon as the hydraulic force exerted on the film by electrolyte exceeds adhesion. Consequently, a blister will form which is monitored with the HR-SKP, simultaneously giving electrochemical and topographic information on the blister growth. The test is referred to as the Scanning Kelvin Probe blister test (HR-SKP-BT). Figure 7.4 shows a schematic.

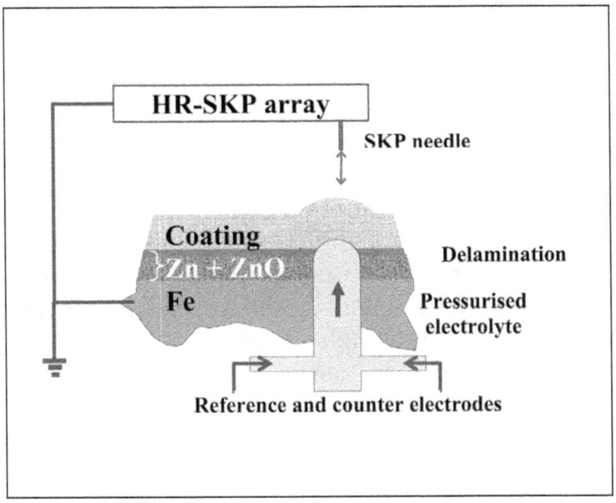

Figure 7.4: Schematic view of the blister test (HR-SKP-BT)
Source: Posner et al.[19]

With this setup, cathodic delamination processes can be studied under variation of atmospheric humidity, electrolyte pressure and polarisation conditions. Combination with data from peel force and attenuated-total-reflection infrared (ATR-IR) spectroscopic inspection of water uptake renders additional useful information about the incorporation of water by the bulk coating and along the interface. Thus it is possible to differentiate between red-ox driven and chemical delamination that is effected by swelling (plastification of the coating, cf. Chapter 2.2.1), hydrolysis and displacement of adsorbed organic binders by the water molecules (cf. Chapter 6.6.1).

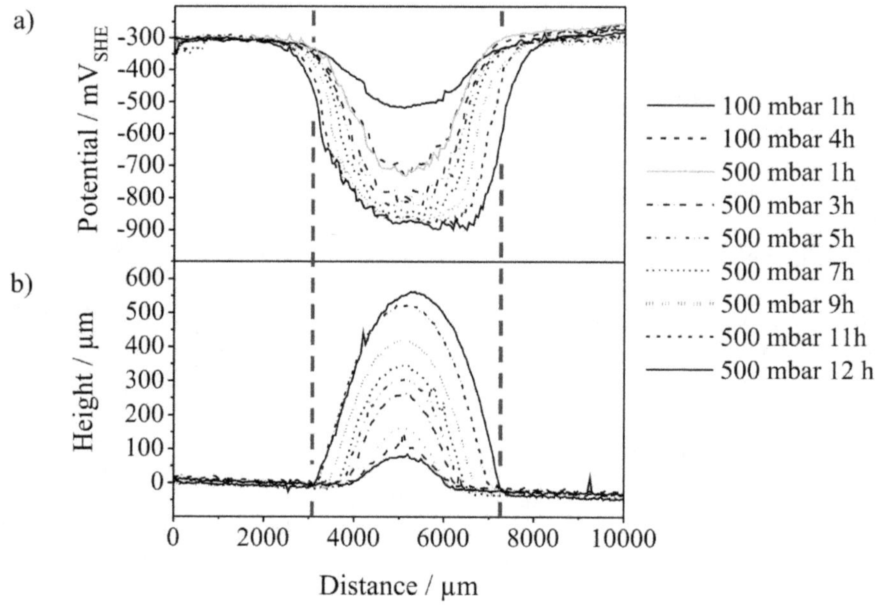

Figure 7.5: Blister test (HR-SKP-BT) on a Zn / ZnO / epoxy amine interface at 500 mbar electrolyte pressure and 97 % rel. humidity; (a) SKP potential profiles; (b) film topography *Source: Posner et al. [19]*

A basic study with electro-galvanised substrate and an epoxy amine adhesive as model coating showed that the potential inside a blister dropped to the zinc standard potential within a day, after which point a steady, linear growth of the corroded area was observed. The corrosion progress was in the order of several 10 µm/h. The delamination rate in high relative humidity (rH = 97 %) was measured two to three times higher than in less humid conditions (rH = 82 %). Topographically, the corroded area spread faster than the blister, though this was apparent only at higher hydraulic strain. Therefore obviously, electrochemical breakdown by oxygen reduction occurs before the onset of mechanical delamination. However it is largely decelerated in low humidity, which is interpreted in terms of the coating not being sufficiently softened and a fluid water film not being present at the interface, thereby suppressing ion transport and slowing down oxygen diffusion to the delamination front [19]. Figure 7.5 shows an exemplary image of potential and topographic profiles of an HR-SKP-BT experiment.

7.4 Outdoor exposure tests

Despite the necessity of simulation by accelerated corrosion tests, any of them can only give a clipped and limited description of the complex reality. Therefore, painted objects are tested in outdoor exposure [20, 21], covering various climates with regard to solarisation, temperature and humidity. This includes extremes, for example Florida, with its set of heat, humidity, high salt load, and high UV irradiation, or highly populated and industrialised areas on the North Sea or North Atlantic coasts, where high salt concentration and atmospheric contamination, e.g. by sulphur dioxide (SO_2), prevail.

Several test sites have been installed that are categorised according to their climate and corrosivity classification according to ISO 12944-2 (e.g. C2 = low; C3 = moderate;

C5 = very strong). The European Coil Coating Association has set up its own test programme EURODES [22], with certified test sites in

- Geleen (Netherlands); continental/industrial; corrosivity class C2 (steel)/C2 (zinc)
- Brest (France); aggressive marine; (C5/C3)
- Hook of Holland (Netherlands); marine/industrial; (C3/C2)
- Lisbon (Portugal); high UV exposure; (C2/C2)

Test panels are placed at defined declinations according to the type of application and its typical load that shall be investigated, i.e. for the European sites

- 5° South (low pitched roofing)
- 90° North (cladding)
- 45° South (sunshine)

Other European exposure sites include regional facilities, e.g. in Sweden (Bohus Malmön), Southern France (Hendaye), Great Britain (Sandwich, Kent) or Italy (Fusina, Venice) [23]. Minimum accepted exposure periods in European exposure sites are two to three years.

Due to its very high solarisation and subtropical climate, the Florida test facility is a preferred test site also for European manufacturers and users, to assess the durability of coatings. Exposure in this site, located near Daytona Beach at 29° northern latitude, is done in 5° and 45° South declination, with durations of up to five years [24, 25].

Outdoor exposure might be combined with salt contamination that is applied by weekly spraying of NaCl solution onto the test specimens. Corrosion progress is usually accelerated by a factor of three to four, as compared to natural exposure, i.e. results may be available within six months rather than years. The automotive industry often uses combined test programmes that include humidity and salt spray testing, and exposure in the driving range [26].

7.5 Literature

[1] Meuthen, B., Jandel, A.-S., Coil Coating, 2nd ed., Vieweg, Wiesbaden 2008, p. 193

[2] Kittel, H., Streitberger, H.-J. (ed.), Lehrbuch der Lacke und Beschichtungen (Coursebook of Paints and Coatings), Vol. 6, 2nd ed., Hirzel, Stuttgart 2008, pp. 321 ff

[3] anon., Internationale Qualitätsrichtlinien für die Beschichtung von Bauteilen aus Aluminium (International Quality Standards for the Coating of Work Pieces...), GSB AL 631, ed. 07/2009, Gütegemeinschaft für die Beschichtung von Bauteilen (GSB International), Schwäb. Gmünd 2009, pp. 35 ff

[4] Goldschmidt, A., Streitberger, H.-J., Basic of Coating Technology, 2nd edition, BASF Coatings AG, Münster 2007, pp. 414 ff

[5] LeBozec, N., Blandin, N., Thierry, D., Accelerated corrosion tests in the automotive industry: A comparison of the performance towards cosmetic corrosion, Materials and Corrosion 59, 2008, pp. 889 ff

[6] Yasuda, H. K., Reddy, C. M., Yu, Q. S., Deffeyes, J. E., Bierwagen, G. P., He, L., Effect of Scribing Modes on Corrosion Test Results, Corrosion 57, 2001, p. 30

[7] Grundmeier, G., Stratmann, M., Adhesion and De-Adhesion Mechanisms at Polymer/Metal Interfaces: Mechanistic Understanding Based on In Situ Studies of Buried Interfaces, Annu. Rev. Mater. Res. 35, 2005, pp. 571 ff, 577

[8] Barrow, G. M., Physikalische Chemie (Physical Chemistry), vol. 3, 3rd ed., Bohmann-Vieweg, Vienna 1977, p. 213

[9] Klimow, G., Fink, N., Grundmeier, G., Electrochemical Studies of the Inhibition of Cathodic Delamination of Organically Coated Galvanised Steel by Thin Conversion Films, Electrochim. Acta 53, 2007, pp. 1290 ff, 1291

[10] Fink, N., Wilson, B., Grundmeier, G., Formation of Ultra-Thin Amorphous Conversion Films on Zinc Alloy Coatings, Part 1: Composition and Reactivity of Native Oxides on ZnAl (0.05 %)-Coatings, Electrochim. Acta 51, 2006, pp. 2956 ff

[11] Lewis, O. D., A Study of the Influence of Nanofiller Additives on the Performance of Waterborne Primer Coatings, PhD Thesis, Loughborough Univ., Loughborough 2008, p. 81

[12] Fink, N., Wilson, B., Grundmeier, G., [10] p. 2969
[13] Thébault, F., Vuillemin, B., Oltra, R., Ogle, K., Allely, C., Investigation of self-healing mechanism on galvanized steels cut edges by coupling SVET and numerical modeling, Electrochim. Acta 53, 2008, pp. 5226 ff
[14] Taylor, C. J., Elvins, J., Sullivan, J. H., Worsley, D. A., Corrosion Performance Evaluation of Zn/Al Galvanized Steels Using the Scanning Vibrating Electrode Technique (SVET), Electrochem. Soc. Transactions 13, 2008, pp. 95 ff
[15] Grundmeier, G., Stratmann, M., [7] pp. 590 ff
[16] Grundmeier, G., Schmidt, W., Stratmann, M., Corrosion Protection by Organic Coatings: Electrochemical Mechanism and Novel Methods of Investigation, Electrochim. Acta 45, 2000, pp. 2515 ff
[17] Grundmeier, G., Wapner, C., Stratmann, M., Applications of a New Height Regulated Scanning Kelvin Probe for the Study of Polymer/Metal Interfaces in Corrosive Environments, ICEPAM Conf. Oslo, SINTEF, Trondheim 2004, www.sintef.no/static/mt/norlight/ICEPAM/09-Grundmeier_Max-Planck.pdf; (15-03-2010; 19:12 h)
[18] Klimow, G., Fink, N., Grundmeier, G., [9] pp. 1290 ff, 1291
[19] Posner, R., Giza, G., Vlasak, R., Grundmeier, G., In-situ electrochemical Scanning Kelvin Probe Blister Test studies of the de-adhesion kinetics at polymer/zinc oxide/zinc-interfaces, Electrochim. Acta 54, 2009, pp. 4837 ff
[20] Goldschmidt, A., Streitberger, H.-J., [4] pp. 416 ff
[21] Meuthen, B., Jandel, A.-S., [1] p. 194
[22] anon., Eurodes Programme, Eurodes: Outdoor Exposure, European Coil Coating Association (ECCA), Brussels 2009; www.prepaintedmetal.eu/prg/selfware.pl?id_sitemap=190&language=EN; (15-03-2010; 19:14 h)
[23] Pietschmann, J., Gardein, R., Filiform Corrosion and Results of 10 Years Natural Weathering, Galvanotechnik 106, 2008, pp. 1764 ff
[24] anon., [3] p. 44
[25] anon., Florida Subtropical Exposure Facility, Brochure: Battelle Florida Materials Research Facility, Battelle Memorial Institute, Columbus OH, USA 2010; www.battelle.org/daytona.aspx; (16-03-2010; 16:31 h)
[26] Goldschmidt, A., Streitberger, H.-J., [4] pp. 429 f

8 Chemical conversion treatment

8.1 Substrates

Various metallic materials serve as substrates for industrial painting, however, the most common metal used in construction is steel [1]. Due to its wide distribution on earth and its useful features, in particular strength and ductility, iron (Fe) has dominated human culture and economy for three millennia, making this epoch of human history the "iron age". As homage to the metal, the Atomium, architectural highlight of the 1958 World Exposition in Brussels, Belgium, represents a model of an iron crystal, magnified by 165 billion. The building, originally clad with aluminium sheet, was refurbished in 2006, its outer skin being made now from stainless steel.

The earth's lithosphere contains between 4.5 to 5 % iron, i.e. an existence of two quintillion (10^{18}) tons, mainly as silicates, oxidic and sulfidic ore [2]. Iron is produced in blast furnaces, by smelting the ore with coke and lime under an oxygen-depleted atmosphere, and separating the molten metal from the silicate/carbonate slag. The crude pig iron is then converted to steel by controlled oxidation. Further alloying and annealing processes are used to produce a wide range of steels of different quality [3]. Highly alloyed steels display particular qualities, like stainless steel (alloying elements: chromium, manganese, nickel, molybdenum) or electro-sheet with high magnetic susceptibility (silicon). However, only minor amounts of these materials are subject to painting operations. Low-alloyed, carbon steel, when exposed to the atmosphere, quickly develops a layer of oxide scale, composed of FeO (wuestite), mixed oxide Fe_3O_4 (magnetite) and Fe_2O_3 (haematite), on its surface.

Massive steel parts are manufactured by casting, hot forging, extrusion or hot rolling. Work pieces made via these hot production cycles often develop a thick scale that requires mechanical cleaning and pickling prior to any coating process (cf. Chapter 3.3).

The choice of steel is made with respect to the intended use. Particular qualities and strength properties are required for roll profiling, bending, deep-drawing and other metal working operations. The mechanical and technological specifications of both the base steel and any metallic coatings are subject to norms [4]. Sheet steel of high surface quality is produced by cold rolling, further reducing the gauge of hot-rolled feedstock. Cold-rolled steel sheet is used for indoor applications like furniture, some appliances or lighting. However, when intended for outdoor use, e.g. car manufacture, façade cladding or roofing, most of the material is covered with a zinc (Zn) layer by hot-dip or electro-galvanising, in order to prevent the reddish-brown iron oxide formation (rust).

Zinc coatings serve as a first protective barrier against corrosion. Due to its lower electrochemical potential zinc is less noble than iron [5] (cf. also Chapter 2.1), and therefore corrodes preferentially. However, this corrosion reaction is kinetically retarded by formation of a dense surface layer of mixed zinc oxide, hydroxide and carbonate (cf. Chapter 3.3.3.1). Moreover, these corrosion products are colourless (white rust). The service life of a zinc coating is determined by its thickness (or gauge). Under normal weathering conditions, the gauge loss of the coating has been reported in the order of 0.4 µm/a (2.5 to 3.5 g/[m²a]) [6].

Jörg Sander et al.: Anticorrosive Coatings

ISBN 978-3-86630-911-1

Table 8.1: Substrates

Ferrous	Non-ferrous
Steel, stainless steel	Wrought aluminium (alloys)
Metallic-coated steel	Cast aluminium (Si, Cu)
- Galvanised, alloy-galvanised zinc	Magnesium
- Electro-galvanised, alloy-galvanised zinc	Cast zinc, zinc sheet
- Aluminised	Others (copper, brass etc.)
- Tinplate	

On hot-dip galvanised sheet metal, a zinc coating is produced in fast, continuous coil lines, immersing cleaned, annealed steel in a zinc melt. The zinc melt may contain aluminium (Al), as well as small amounts of other alloying elements, e.g. silicon (Si), lead or antimony. Specified alloy-galvanised steel sheet is known as "Galfan", ZA (5 % Al), or "Galvalume", AZ (55 % Al, 1.6 % Si). After the zinc melt, the strip is cooled down by pressurised air (air knife) in order to control the Zn thickness. Nowadays, a spangle-free, finely crystalline surface is state of the art. More recently, ternary systems with magnesium and aluminium, that provide a corrosion resistance superior to ZA coatings [7], have entered the marketplace.

Electro-galvanised sheet is produced in a continuous coil operation by electro-deposition of Zn or Zn alloys from an electrolyte solution onto the clean, annealed and temper-rolled steel strip. The Zn layer gauge can be adjusted by choice of the parameters of the electrolysis. Tables 8.1 and 8.2 summarise important substrates for industrial painting.

Aluminium [8 - 10] (Al) is the most common metal in the outermost crust of the earth, making up for 8 % of this so-called "Si-Al" layer. Even less noble than zinc, Al is found almost exclusively in its oxidised state in silicate and carbonate ore, i.e. clay and bauxite minerals. Bauxite contains approx. 50 % Al oxides. Being discovered in the early 19th century, elementary Al took five decades before being produced on an industrial scale. Today's improved process involves hot extraction of Al oxide from the mineral with sodium hydroxide solution, from which Al hydroxide, gibbsite, $Al(OH)_3$, is precipitated upon cooling, and filtered off. Dehydration at 1.100 °C in fluidised bed furnaces renders pure Al oxide, Al_2O_3. Elemental Al is derived from the oxide by electrolysis in molten cryolite, $Na_3(AlF_6)$. The process requires a high amount of energy, i.e. up to 16 kWh per kilogramme of Al. The recycling of Al scrap takes much less energy, making the material that is recovered (secondary Al) an important

Table 8.2: Most common substrates for coil coating

Substrates				Uses		
Base metal	**Coating**	**Gauge [µm]**	**Symbol***	**Architectural**	**Appliance**	**Transport**
Cold-rolled steel (CRS)	none			indoor		
	electro galv. (EG)	3 to 7.5	ZE			x
	hot-dip (HDG)	7 to 30	Z	indoor, outdoor	x	x
	Galfan	7 to 30	ZA	indoor, outdoor	x	x
	Galvalume	15 to 30	AZ	outdoor		
Aluminium				indoor, outdoor	x	x

** System of symbols cf. Meuthen, Jandel [4]*

source for the aluminium industry. In 2007, the world production was 38 Mt of primary and 9.6 Mt of secondary Al.

Despite its low standard potential, technical aluminium is resistant to corrosion, since it forms dense surface oxide layers under normal atmospheric conditions. These layers inhibit ongoing structural corrosion, as long as the metal surface is neutral and no contact corrosion with more noble metals is possible. Hence, corrosion phenomena on Al surfaces are prominent with alloys that contain copper or when acidulated conditions apply, i.e. with the so-called filiform corrosion (cf. Chapter 6.6.2.2). A multiplicity of Al alloys is used for different technical purposes. Table 8.3 shows typical alloys and their applications [11, 12].

Al alloys for casting [13, 14] often contain high percentages of silicon (Si), copper (Cu) and magnesium (Mg), plus smaller amounts of additional elements, in order to obtain materials of particular workability and strength. Typical alloys for dye-casting contain 8 to 12 % Si, up to 3 % Cu and/or up to 10 % Mg. Upon cleaning and pickling with strong alkalis, these alloying elements cause dark deposits on the work piece surface, e.g. enrichments of elementary Si and metasilicate, Na_2SiO_3, that require hydrofluoric acid (HF) pickling for their removal. Alternative measures include the depletion of surfaces of cast work pieces in Si, in order to avoid the highly toxic and hazardous HF.

8.2 Chemicals for pretreatment

8.2.1 General remarks

Adhesion of a coating is highly dependent on the type of substrate and its proper cleaning and conversion treatment (pretreatment) [15 - 18] by an adjusted process, that also provides an improved corrosion resistance and durability. The painting operation, therefore, includes preceding process steps for cleaning and pretreatment. In immediate succession paint is applied. Especially when a flexible substrate is coated, or the painted goods are subject to dynamic stress during their service life, or when the entire paint coat is thin (e.g. <35 μm), the paint coat often comprises a separate primer and a topcoat, sometimes even additional layers to mediate stress distribution or to add protective features to the coated surface (scratch resistance, anti-scaling, anti-fouling properties etc.). The primer is applied to

Table 8.3: Aluminium alloys and typical uses

Alloy	AA[a] Key No.	Features	Uses
Al 99 % min	1xxx		Foil, sheet, litho sheet
Cu	2xxx	+ mechanical – corrosion protection	Aircraft industry
Mn	3xxx	+ mechanical ~ deep-drawing	Food packaging, architecture, radiators
Si	4xxx	+ casting	Engineering, heat exchangers
Mg	5xxx	+ strength after cold forming	Food packaging, automotive, transport, architecture
Mg, Si	6xxx	+ strength	Automotive, architecture, transport
Zn	7xxx	– corrosion protection	Aerospace, radiators
Others	8xxx		Foil (Fe), aircraft industry (Li)

[a] *The Aluminium Association Inc.*

enhance corrosion resistance and to render improved adhesion, while the topcoat provides the required colour, gloss and texture.

The prominent function of pretreatment and primer is to provide corrosion resistance, if one is to generate a long-lasting product with an aesthetic appeal and high durability in all environmental conditions. Pretreatment and primer preclude or delay creepage of organic coating once it is damaged or punctured. After thorough cleaning (cf. Chapter 3), the chemically clean, wettable surface is extremely reactive and therefore needs to be rendered inert immediately, by coating with a thin, non-metallic layer that simultaneously serves as passivation and adhesion promoter for the subsequent paint coat. The typical specific weight of these coatings ranges from 50 to 500 mg/m^2, gauges of 20 to 100 nm are common.

Most conversion coatings are applied in spray or immersion processes. Alkaline and chromate passivation or phosphate processes have been widely used. These processes require a fresh water rinse to remove excess chemicals and by-products. Often, post-rinses are used to enhance the corrosion resistance. Chromate or chromium-free post-rinses are available.

As an example, Table 8.4 summarises prior-art pretreatment technology as it has been used in the coil coating industry. In particular, hot-dip galvanised (Z) strip is commercially pretreated either with alkaline passivation or acidic, chromate passivation processes, while cold-rolled steel (St) is usually treated with iron phosphating solutions; all of these pretreatments, however, commonly are completed by chromate post-rinses.

After any pretreatment, the work piece surface must be completely dry before being processed to the paint section. Indirectly fired hot-air, infrared or induction dryers are state of the art.

8.2.2 Alkaline passivation

Alkaline passivation [19, 20] has been the predominant pretreatment process for hot-dip galvanised coil. The treatment solution comprises caustic soda, iron (Fe) and cobalt (Co) compounds and solubilising complexants. Optionally nickel ions (Ni) could be present. In the course of the reaction, a zinc oxide coating is formed that incorporates the Fe and Co (or Ni) ions in a spinel-type lattice. The alkaline passivation exhibits outstanding adhesion, thus allowing extreme deformation, however requires a post-rinse to obtain sufficient corrosion protection. Typical specific coating weights range between 0.2 and 0.4 g/m^2. Key reactions of the film formation are illustrated by Equations 8.1 to 8.3.

Equation 8.1 $2\,Co(OH)_2 + 0.5\,O_2 \rightarrow CoO(OH) + H_2O$ Co oxidation

Equation 8.2 $2\,CoO(OH) + Zn(OH)_2 \rightarrow Zn(Co_2O_4)\downarrow + 2\,H_2O$ Zn,Co spinel

Equation 8.3 $2\,Fe^{3+} + Co^{2+} + 8\,OH^- \rightarrow Co(Fe_2O_4)\downarrow + 4\,H_2O$ Co,Fe spinel

Table 8.4: Conventional coil coating pretreatments for architecture, white goods etc.

Process	Substrates[a]	Remarks
Alkali passivation	Z, ZA, AZ, ZE	Chromate post-rinse
Yellow chromate	Z, ZA, AZ, ZE, Al	Chromate post-rinse[b]
Green chromate	Al	Foodgrade
No-rinse process	Z, ZA, AZ, ZE, Al, St	Chromate containing
Iron phosphate	St	Chromate post-rinse

[a] *Abbreviations: Z: Hot-dip galvanised, ZA: Galfan, ZE: Electro-galvanised, AZ: Galvalume (AluZinc), Al: Aluminium, St: Cold-rolled steel*

[b] *Except aluminium*

The conversion layer is completed by a post-rinse that contains chromium (III) and chromate ions. Alkaline passivation is suitable for all galvanised substrates, including the aluminium alloyed, Galfan and Galvalume, despite earlier publications [21]. More recently, the use of heavy metals like Cr, Ni, that form toxic compounds, and in their wake, Co, has become a matter of concern. As a result, chromium-free post-rinses have been introduced. In Europe, some development efforts on the alkaline passivation process have been directed towards the replacement of the heavy metals by magnesium (Mg) [22], however these developments have not taken ground. Instead, the single-stage, no-rinse chromium-free pretreatments (cf. Chapter 8.2.7) have proliferated.

8.2.3 Phosphating

8.2.3.1 Iron phosphating

The term "phosphating [23]" is used for two different classes of pretreatment, i.e. iron phosphating, and zinc phosphating. Iron phosphating [24, 25] comprises the treatment of steel surfaces with acidic sodium phosphate solution (dihydrogen phosphate). Iron is dissolved by the attack of the acid, and a thin, amorphous iron phosphate layer is formed. The thickness of the conversion coating is increased by the use of oxidants like molybdate, chlorate or organic nitro compounds that inhibit the formation of hydrogen, thereby accelerating the net reaction.

Equation 8.4 $6\ Fe + 3\ NaClO_3 + 18\ H^+ \rightarrow 6\ Fe^{3+} + 3\ NaCl + 9\ H_2O$

Equation 8.5 $Fe^{3+} + H_2PO_4^- \rightarrow FePO_4 \downarrow + 2\ H^+$

Commercial iron phosphating processes usually contain surfactants, which makes them suitable for the use as cleaners. The process is often carried out in two stages, with intermediate and final rinsing, where the first active bath is adjusted for the cleaning step, e.g. by addition of surfactants, and the second active bath is operated at a lower pH value by addition of phosphoric acid. The process can be operated in immersion or spray lines. The conversion coating ranges at specific weights between 0.2 and 1 g/m^2. Iron phosphating renders a basic level of corrosion protection. To enhance the performance, the final rinse may be carried out with a sealant, including chromate chemicals.

8.2.3.2 Zinc phosphating

Zinc phosphating [26 - 28] is state of the art in the pretreatment of car bodies or appliance housing. It is particularly well suited for the common process of car manufacture. Car bodies are constructed of a variety of substrates, mostly hot-dip galvanised or electro-galvanised steel, but also including cold-rolled steel, and aluminium. After cleaning and phosphating, the entire car body is coated with a cathodic electro-dip primer. Zinc phosphate pretreatments yield crystalline coatings of 0.5 to 4 g/m^2 that mainly consist of hopeite, $Zn_3(PO_4)_2 \cdot 4\ H_2O$, but also contain manganese (Mn) and nickel (Ni) doted phosphates, in order to improve the stability vs. high pH that occurs during the electro-dip painting (tri-cation phosphate technology). More recent developments use magnesium (Mg) or calcium (Ca) phosphates. Zinc phosphating baths further contain e.g. nitrite or hydroxyl amine as accelerators. Common Mn and Ni contents in the coating are 4 and 0.5 %, respectively. On ferrous surfaces, iron containing phosphate crystals, phosphophyllite, $Zn_2Fe(PO_4)_2 \cdot 4\ H_2O$, are deposited. Quality zinc phosphate coatings require a controlled crystal size that can be effected by application of e.g. titanium phosphate slurries that provide seeding nuclei for the crystal growth [29]. Usual crystal sizes range from 1 to 3 µm. Smaller crystals enable thinner, yet denser phosphate coatings for better formability.

Equation 8.6 $Zn + 2\ H^+ \rightarrow Zn^{2+} + H_2 \uparrow$

Equation 8.7 $3\ Zn^{2+} + 6\ H_2PO_4^- \rightarrow Zn_3(PO_4)_2 \downarrow + 4\ H_3PO_4$

The application of Al has proliferated due to the low density of this material and the weight and resulting fuel savings implied by its use. It is often found as construction material for hang-on parts like hoods, trunk lids and doors, but also in the suspension and axle assembly. As free Al^{3+} ions would inhibit the precipitation of Zn phosphate, they must be masked by appropriate measures. Therefore, fluoride containing anions are present in multi-metal phosphating compositions. While complex fluorides are preferred, e.g. hexafluoro silicic or hexafluoro titanic acid, some free fluoride is necessary which is preferably supplied by ammonium hydrogen fluoride [30]. The free fluoride concentration can be measured by a fluoride ion sensitive electrode.

8.2.4 Chromating

Historically, the unique properties of chromium have been the key factor of both pretreatments and paints. Chromium, in its oxidation states +VI and +III, serves as an electrochemical couple that can inhibit most corrosive reactions on a metal surface [31]. Therefore, most pretreatments have comprised at least one processing step with chromate containing chemicals to obtain the necessary corrosion resistance of the final product. Typical reactions for the formation of a chromate conversion layer on a galvanised steel surface are described as depicted in Equations 8.8 to 8.10. The surface reaction is characterised by acidic attack of zinc and subsequent film formation by partial reduction of chromic acid. Owing to these reactions, the acid is consumed and zinc dissolves. By using an immersion or spray technology and removing solution from the bath by drag-out and overflow, the composition of the bath would slowly change with operating time. In practice, it is therefore necessary to correct the pH and the ingredients of the bath by an appropriately formulated replenisher which is fed continuously or in short intervals, to maintain the correct composition.

Equation 8.8 $Zn + 2\ H^+ \rightarrow Zn^{2+} + 2\ [H]$

Equation 8.9 $2\ H^+ + 6\ [H] + 2\ CrO_4^{2-} \rightarrow Cr_2O_4^{2-} + 4\ H_2O$

Equation 8.10 $Zn^{2+} + Cr_2O_4^{2-} \rightarrow Zn(Cr_2O_4) \downarrow$

Similar considerations apply for cold-rolled steel or aluminium substrates. Pretreatments for aluminium have been extensively reviewed, in order to understand the principles that would help to identify a suitable chemistry for chromate replacement [32].

Equation 8.11 $Al + 2\ H_2O \rightarrow AlO(OH) \downarrow + 3\ [H]$

Equation 8.12 $CrO_4^{2-} + 3\ [H] + PO_4^{3-} + 5\ H^+ \rightarrow Cr(PO_4) \downarrow + 4\ H_2O$

Equations 8.11 and 8.12 depict the situation for a so-called green chromate coating, i.e. the chromous phosphate layer that precipitates from a phosphoric acid containing chromate pretreatment solution and is virtually free of chromium (VI) (chromate). Since chromous (chromium [III] containing) compounds are almost insoluble and non-toxic, such conversion coatings have been used for decades on Al surfaces, as a replacement for "yellow" chromium chromate treatments, and in particular as pretreatments in the food-related sector (canning, i.e. food containers, beverage cans and lids).

8.2.5 Anodising of aluminium

While chromate containing chemicals have been in use for decades, the discussion on environmental and health risks and according legislative action have driven the coating industry's search for „greener", more environmentally favourable solutions. It is expected that the implementation of the new chemical policy REACH will further boost the replacement of the chromate processes that are technically proven, but hazardous to human health and environment (cf. Chapter 8.2.10) [33]. Additional economic factors are also driving the need for further coating developments.

Anodising offers an approach to chromium-free treatment of aluminium surfaces. It is often carried out to provide a stand-alone, dense oxide coating that is left uncoloured or is coloured with inorganic or organic dyes that settle in the pores of the anodic film before these pores are closed by hydrothermal sealing. However, thin-layer anodising processes have been adapted to serve as pretreatments prior to organic coating.

The anodising process is carried out by immersing the work pieces in an acidic bath, and subjecting them to an electrolytic treatment with asymmetric alternating or pulsed direct current. The aluminium surface is converted to aluminium oxide, Al_2O_3, at a typical film building rate of 3 μm/min. The oxide forms a honeycomb structure, with pores in the centre of the cells. For the normal anodising process, the most commonly used electrolyte is sulphuric acid, and anodising is extended to allow the formation of a 20 μm layer, which is subsequently subjected to a hydrothermal sealing treatment in order to close the pores. Contrary to this, when used as pretreatment prior to painting, anodic films are only thin (3 to 8 μm), and they are left unsealed. While historically, phosphoric acid was often used as the electrolyte, to obtain more flexible, less brittle anodic films, today the DC/sulphuric acid process is preferred, the operational parameters and obtainable quality properties being subject to industry specification [34].

Particularly for aircraft and military equipment, however, chromate still is the specified standard in terms of high durability under severe conditions. The required performance is prescribed in a series of military specifications (Mil Spec). For these uses, anodising is carried out in chromic acid, hence resulting in a mixed Al, Cr oxide coating [35, 36].

8.2.6 "Chromiting"

The term "chromiting" refers to a class of passivates and pretreatments that have been developed on the basis of chromous (Cr [III]) chemicals. Typical, highly acidic pretreatment baths contain chromous nitrate or sulphate, optionally up to 2.5 g/l cobalt (Co), plus certain complexants [37, 38]. The process is reported to generate dense, nano-scale oxidic coatings on various substrates like cold-rolled steel, galvanised steel, and aluminium, providing good corrosion resistance in salt spray testing. The process is aimed at replacing iron and tri-cation zinc phosphating in multi-metal lines, substantially reducing the sludge level. It is claimed to be particularly suited as pretreatment prior to coating with polyester and epoxy/polyester powder paints [39].

8.2.7 Chromium-free pretreatment

8.2.7.1 Titanium and zirconium fluoro-complex technology

In anticipation of a more stringent environmental protection policy, chromium-free pretreatment processes for architectural applications have been developed and introduced to the market since the early 1990s. This applies to the coating of single work pieces (batch painting) e.g. window frame profiles and façade parts, as well as to the industrial painting of metal coil (coil coating) that is used for large wall claddings or roofing, but also for garage doors or sun blinds and roller shutters [40].

Having been introduced in the batch painting process of aluminium extrusion in 1996, chromium-free technology nowadays is standard in this market in Europe. It has become the favourite choice whenever a new painting line is designed, or major renovations imple-

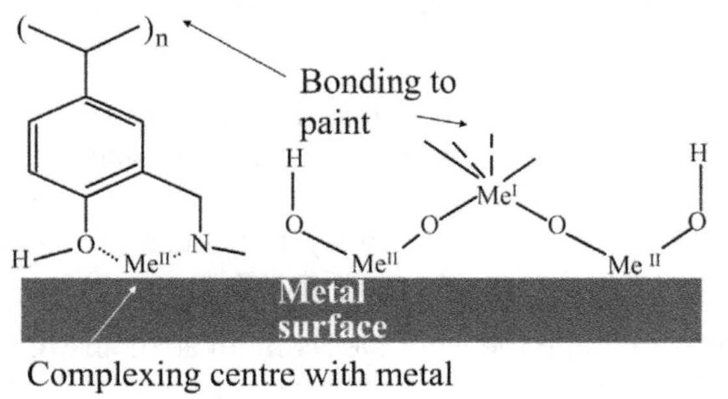

Figure 8.1: Potential bonding structures of chromium-free pretreatments

mented in existing lines, and has been awarded approvals from several European wide operating quality associations.

For coil coating of aluminium, another chromium-free process was introduced in the same period, comprising the so-called no-rinse application. Chromium-free treated aluminium coil stock has been in use for over 10 years, for wall cladding and sun blinds. The coil material was top-coated with a single-layer, high-durable polyester paint. For steel and galvanised steel surfaces, chromium-free products have been in industrial use since 1998.

Table 8.5 represents the alternative, chromium-free pretreatment processes that have become available in coil coating. Here, the chromium-free no-rinse pretreatment has become particularly important, as it is a multi-substrate process, and prevents high amounts of waste water. Strong correlation and comparability between chromium-free and chromate pretreatments in various outdoor and accelerated corrosion tests have been reported [41].

Commonly, the chromium-free pretreatments are based on aqueous solutions of non-toxic transition metal complexes and compounds and special water-soluble polymers. The mechanisms of bonding and corrosion resistance are attributed to a metal oxide layer that is formed in an initial reaction on top of the treated metal surface and then covered by the organic matrix. Figure 8.1 shows a corresponding schematic. Thermal and calorimetric experiments, micrographic methods and spectroscopy of the layers are consistent with this picture [42].

Chromium-free conversion coatings have gauges in the order of a few 10 nm, corresponding to Ti or Zr build-up values between 1 and 30 mg/m^2. The key reactions during the formation of such chromium-free conversion coatings can be sketched as in Equations 8.13 and 8.14:

Equation 8.13 $Zn + 2\ H^+ \rightarrow Zn^{2+} + 2\ [H]$

Equation 8.14 $(Ti, Zr)\ F_6^{2-} + H_3PO_4 + Zn^{2+} + 3\ H_2O$
$\rightarrow Zn\ (OH)_2 \cdot (Ti, Zr)\ (OH)\ (PO_4) \downarrow + 6\ HF$

A distinctive layer is reported that was analytically characterised using Auger and photo electron spectroscopy combined with argon sputtering technique. Figure 8.2 shows the

Table 8.5: Chromium-free coil coating pretreatments for architecture, white goods etc.

Process	Substrates*	Remarks
Alkali passivation	Z, ZA, AZ, ZE	Chromium-free post-rinse
Chromium-free pretreatment	Z, ZA, Al	Spray and dip, optionally: chromium-free post-rinse
No-rinse process	Z, ZA, AZ, ZE, Al, St	Chromium-free
Iron phosphate	St	Chromium-free post-rinse

** Abbreviations: Z: Hot-dip galvanised, ZA: Galfan, ZE: Electro-galvanised, AZ: Galvalume (AluZinc), Al: Aluminium, St: Cold-rolled Steel*

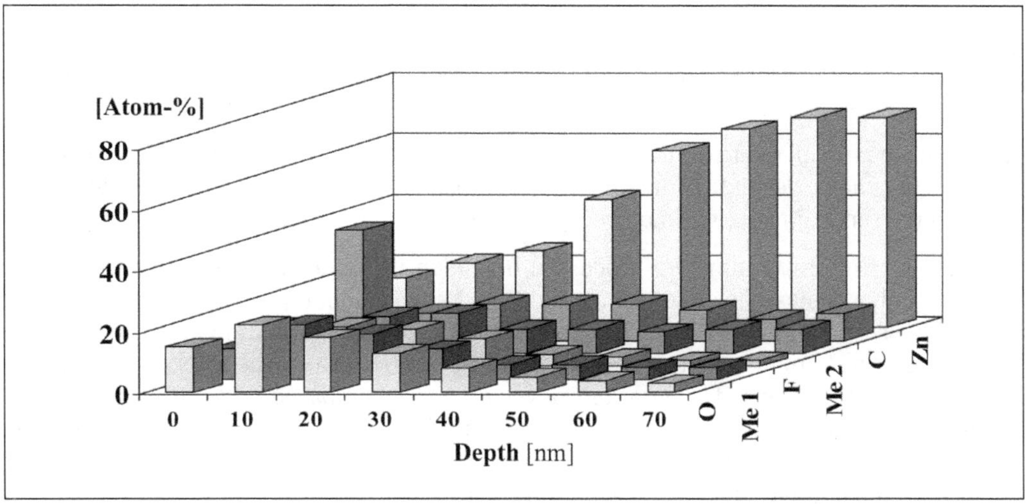

Figure 8.2: Depth profile of a chromium-free pretreatment on galvanised steel

Source: European Coil Coating Association 1999; Androsch et al. [41]

depth profile obtained by this method. Zinc, originating from the substrate, is found in all parts of the layer, albeit in higher concentration close to the interface between metal and pretreatment. This also applies to alloying elements of the galvanising layer, like aluminium. The transition metals (Me 1 and 2), and fluorine (F) that are ingredients of the pretreatment, are generally distributed throughout the layer. The concentration of carbon (C), indicating the presence of the organic polymer, increases sharply close to the outer surface.

Chromium-free pretreatments of the kind described above are used for a variety of substrates, such as aluminium, zinc, zinc alloys and cold-rolled steel. In particular for coil coating application, the chemical solution is adjusted to a no-rinse process, i.e. applied at an appropriate dilution onto the clean, dry strip surface where it immediately reacts off to form transparent, uniform layers upon drying at temperatures above 50 °C. Neither chemicals in excess nor products of side reactions are left, hence no rinsing is required after pretreatment.

Currently, pretreatment of Zn-Al-Mg substrate is carried out with unaltered pretreatment processes that have been in use for state-of-the-art galvanised substrates. Adaptations are

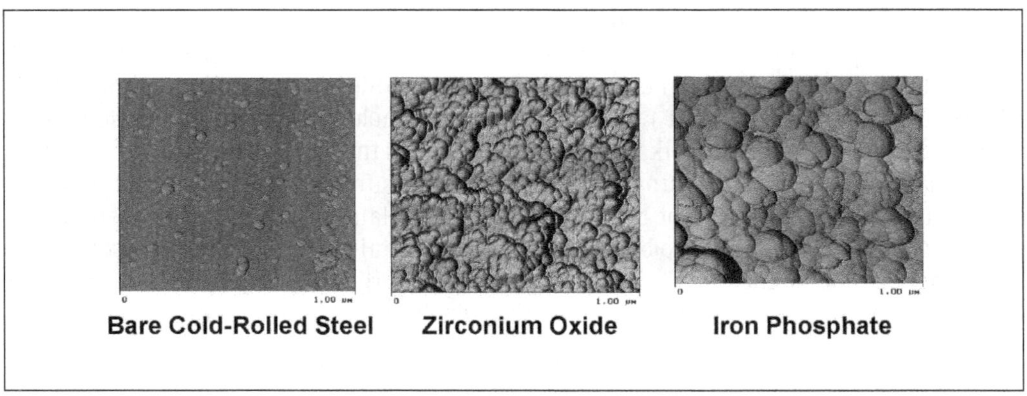

Figure 8.3: Atomic force microscopy (AFM) images of bare, zirconium oxide coated and alkali phosphated cold-rolled steel (CRS)

Source: Rogers Publishing Ltd.; cf. Fristad [43]

under way to improve on certain features that are impacted by the physics and surface chemistry of the new substrate that tends to be less flexible and more sensitive to moisture than ordinary galvanised material. For the substrate composition and properties itself, there is yet no standard set.

In the European coil coating market, chromium-free pretreatments have only recently gained some share, despite the technical acceptance they have achieved. Since 1998 they have grown from 0.3 to approximately 5 % of the pretreatment use.

Another class of chromium-free processes has been introduced recently for which the absence of phosphates is claimed as an additional environmental benefit [43, 44]. These are generally described to form thin (20 to 100 nm) amorphous zirconium oxide (ZrO_2) films. "First generation" products of this class are reported to perform at the level of iron (alkali) phosphate, while a "second generation" development is compared with zinc phosphates, in particular under cathodic electro-dip primers. The target use for this latter range of products is the replacement of tri-cation zinc phosphate pretreatments in the automotive original equipment manufacturers' (OEM) market.

8.2.7.2 Other chromium-free pretreatments

Apart from the transition metal complex technology, other approaches for chromium-free pretreatment have been investigated, including cerate, phospho-molybdate [45, 46] and cobalt amine [47, 48] processes, as well as processes based on so-called self-assembling molecules (SAMs) [49]. The latter are linear hydrocarbon molecules with two terminal functional groups, preferably phosphonate groups that have a bonding capability to both the metal and the subsequent paint coat. Owing to their molecular structure, SAMs assemble to form a monolayer that has a thickness in order of the hydrocarbon chain length, i.e. 1.5 to 2 nm. Due to their complexity, these processes currently have found only limited applications, e.g. in the automotive wheel manufacture and in the aircraft industry [50]. Another approach to Cr-free pretreatments is opened up by the so-called hybrid coatings.

8.2.8 Hybrid pretreatment coatings

As new developments, aqueous pretreatment processes have been introduced that combine organic and inorganic chemistry, and are therefore referred to as hybrid coatings. Hybrid coatings are described in greater detail in Chapter 10. However two examples shall also be outlined in this current context, as they incorporate typical pretreatment features.

8.2.8.1 Silane/siloxane coatings

One class of these coatings employs the formation of silane/polysiloxane layers [51, 52] from a sol-gel process [53]. Typical representatives of this class of compounds are terminal aminopropyl triethoxy silane (γ-APS), $H_2N\text{-}(CH_2)_3\text{-}Si(OC_2H_5)_3$ or *bis*-(triethoxysilyl) ethane (BTSE), $(C_2H_5O)_3Si\text{-}(CH_2)_2\text{-}Si(OC_2H_5)_3$. The technology requires a hydrated metal oxide surface as obtained by alkaline cleaning. The further process involves controlled hydrolysation of the $Si\text{-}OC_2H_5$ functions, bonding to the surface via metal-O-Si links, and polymerisation/cross-linking of the affixed hydroxy-silane moieties to form a polysiloxane. Siloxane coatings can also be coupled with the metal surface by plasma or chemical vapour deposition technology [54, 55]. The functionalised organic groups protrude from the surface, becoming available for bonding with the subsequent paint coat.

8.2.8.2 Combined thermal processes for primer-pretreatment

The combination of inorganic and organic, polymeric materials can also render coatings with additional features apart from adherence and passivation, for instance, fingerprint resistance, lubrication, or priming capability. Such processes allow add-on quality and changes in the manufacturing chain, like omission of extra lubrication in the press shop or in-line application of extra coats and varnishes.

In particular for precoated coil, the idea of combining pretreatment and primer functions in one single stage in the coil coating process has been around for a number of years. By employing this principle, workflow changes are rendered possible, like doing away with one of the currently separate treatment steps, or making the respective treatment station available for other purposes, i.e. multi-layer coatings.

The advantageous impact of a combined primer-pretreatment on the operating economy of compact coating lines for coils and sheet has been reported very recently [56]. As described in the literature [57], a primer-pretreatment may incorporate up to five different technology steps with their related economic and environmental benefits at once. These include low toxicity by the omission of chromium (particularly Cr VI) compounds, environmental benefits by the avoidance of organic solvents and by low curing temperatures, a maximised yield by a high solids/reactables content, the obsoletion of a separate conventional priming step, as well as the possibility of in-line priming in galvanising installations. With an aqueous primer-pretreatment and a powder topcoat an entirely solvent-free coating process becomes feasible.

Embracing both the pretreatment and priming steps, such a combined process is described that can be integrated at exit of a coil galvanising line. Even cleaning that is normally the mandatory first stage in a conventional coil coating line can be omitted. For the first time in the coil coating history, the new primer-pretreatment enables process speeds in galvanising and coating to be matched. The process is expected to go fully commercial soon.

8.2.9 Surface preparation of other substrates – copper alloys, white metal, magnesium, stainless steel

For copper and brass, and similar non-ferrous metals, a conversion treatment is usually omitted. Special care must be taken to use mild cleaning conditions in order to avoid uncontrolled oxide formation and discolouration. Alkalinity, temperature and also duration of cleaning must be carefully adjusted. Aged surfaces that have already developed a brownish colouration must be pickled prior to painting. Environmentally friendly solutions comprise sulphuric acid and stabilised hydrogen peroxide, instead of nitric acid and/or fluorides. For white metal (tin and its alloys) mild cleaning forms a sufficient preparation of the surface.

Historically, pretreatment of magnesium and its alloys comprised chromating or anodising procedures that were adapted from aluminium processing. More recent processes also use chromium-free chemistry that has been derived from pretreatments for aluminium. Free fluoride usually must be present in order to achieve sufficient etching of the metal surface.

Stainless steel surfaces develop a chromous oxide (Cr_2O_3) layer that has a very low wettability. Work pieces made of stainless steel therefore require a strong alkaline cleaning and subsequent treatment with a nitrate/fluoride etch or a chromium-free, hexafluoro titanate based pretreatment immediately prior to the painting operation, in order to achieve proper adhesion.

8.2.10 Environmental considerations

As mentioned in Chapter 8.2.5, the European Union (EU) has issued strict legislation in order to control and restrict environmental and health hazards related to the use of chemicals. Some of the related laws and directives have their worldwide counterparts, like the End of Life Vehicles Directive (ELV), the Waste Electrical and Electronic Equipment Directive (WEEE) and the Regulation on Hazardous Substances (RoHS), which have resulted in an effective restriction of the use of chromates and other toxic substances in the respective target markets. On top, the EU has now brought into force a new codex whose acronym, REACH, stands for registration, evaluation, authorisation and restriction of chemicals [58]. This policy rules the use of all chemical substances that are manufactured in or imported into the European Union and is governed by the EC Regulation 1907/2006. Under the provisions of this Regulation, all chemicals that are commercially used above certain threshold volumes require official authorisation and therefore may be subject to assessments of their exposure risks to health and environment. Chemicals of very high concern, e.g. CMR-labelled (carcinogenic, mutagenic and toxic to reproduction), and highly toxic substances like chromium (VI) (chromate) compounds are likely to be phased out wherever possible or effectively banned.

Waste water that is generated in a chromating operation (treatment baths and rinses) must be processed via a detoxification of chromate ions, e.g. by reduction with sodium bisulphite in acidic solution. After neutralisation with lime and precipitation, the concentration of heavy metals that are present, like chromium (III), but also zinc, aluminium and others from the substrate, as well as fluoride can be reduced below the required legal limits.

Table 8.6: Safety data regarding labelling (examples)

Chromium-free pretreatment	**Chromate containing pretreatment**
Hazard Symbols: C Corrosive	Hazard Symbols: T Toxic C Corrosive N Environmentally harmful
Hazardous substance: Hexafluoro zirconic acid	Hazardous substance: Cr(III) chromate Cr(VI) oxide
R clauses: R 34: Causes burns	R clauses: R 45: May cause cancer R 22: Hazardous when swallowed R 35: Causes severe burns R 43: Sensibilisation possible upon skin contact R 50/53: Very toxic for aquatic organisms, may have long-term harmful effects on waters
S clauses: S 26: Upon eye contact, immediately flush thoroughly with water. Get medical attention S 36/37/39: During work, wear suitable protective clothing, gloves, eye/face protection S 45: In case of an accident or when feeling sick, immediately see a doctor (if possible, show this label)	S clauses: S 53: Avoid exposition – get special instructions before use S 26: Upon eye contact, immediately flush thoroughly with water. Get medical attention S 36/37/39: During work, wear suitable protective clothing, gloves, eye/face protection S 45: In case of accident or when feeling unwell, immediately see a doctor (if possible, show this label)
National regulations (Germany): WGK = 2, water polluting product. Classification according to Mixing Rule VwVwS May 17th, 1999	National regulations (Germany): WGK = 3, highly water polluting product. Classification according to Mixing Rule VwVwS May 17th, 1999

The detoxification process, however, also builds up a considerable sulphate level. If limits for direct effluent discarding apply, a further reduction of the sulphate content is necessary, e.g. by precipitation with calcium aluminate.

Contrasting to this, waste water treatment for chromium-free processes is simplified to a treatment with lime, to ensure precipitation of the commonly used titanium or zirconium compounds, along with metal ions that originate from the substrate. In both cases the sludge that is obtained must be dehydrated and deposited in a landfill site. The cost for treatment and disposal of a typical chromating process is five to eight times as much as with a chromium-free process. Also labelling instructions are advantageous for a chromium-free process, owing to its much lower hazard potential. Table 8.6 gives examples of the respective safety data of two typical processes.

Solvent-free coating products as described in Chapter 8.2.6.2, moreover, do away with expensive installations for combustion and thermal destruction of exhausts from curing ovens, and substantially reduce fuel consumption and VOC and CO_2 emissions.

8.3 Application of pretreatments

8.3.1 Immersion and spray treatment

For the application of pretreatments in single work piece operations, immersion or spray techniques are used as have been described in Chapter 3 for cleaning. A further technique that has been developed for single work piece operation is "floating" which involves the use of a low-pressure spray curtain.

Small parts are often processed in rotating drums that are dipped entirely into the active and subsequent rinse baths. Larger parts are usually processed in baskets or fastened on a rack and dipped into the treating baths by a crane. Particularly sophisticated and complex parts like entire car bodies are transported through a series of treatment baths by a conveyor system. These conveyors can be combined also with pitch or even rotating installations, to make sure that all parts of the metal surface, including bowl-shaped and box sections, are properly contacted by the treating fluid. For example, patent DE 19950892 describes such a conveyor system for car bodies [59].

Work pieces that have a comparably simple geometry, like flat parts, sheets, rods or profiles, can be treated in spray application, providing that all parts of the surface are properly covered. Such work pieces are fixed to a rack or conveyor chain and transported through a series of spray booths. Most conveniently, the several compartments form a tunnel through which the work pieces are moved with a uniform speed. By the similar principles that have been explained for the cleaning operation, spray pretreatment can be used with lower concentrations of the active bath, as compared to the immersion technique. However, spray pressures are usually lower than in the cleaning operation, as no turbulent action is required to remove excess solution and by-products of the pretreatment reactions. In some cases, e.g. when treating car bodies, immersion and spray techniques are used in combination.

Considerable care must be taken to avoid excess drag-out, as has already been described in Chapter 3 for the cleaner case. Therefore, the operation in immersion lines or in installations with separate spray booths must observe suitable drip-off times, optionally supported by a brief spray with fresh or deionised water. In tunnel systems, the construction of the spray tunnel must be adapted to avoid transfer or mixing of fluids between adjacent active or rinse tanks. Concerning the overflow control and the economy of rinses, the same principles apply as for the cleaner operation.

8.3.2 Pretreatment application for coil: spray/squeeze, spray-cell, roll-coating

Immersion and spray techniques are used also for the pretreatment of coil. To immerse the coil into the treatment bath, dip rolls are used. A number of dip rolls can be mounted in a way that the strip forms several loops in the bath, in order to extend the available treatment time. The speed of the travelling strip is sufficient to bring on the required exchange of reactands from the surface. Immersion lines have been common for the so-called green chromating (chromous phosphate) pretreatment of aluminium coil.

For certain treatments for steel and galvanised steel coil, the so-called spray-cell technique has been in use. Such equipment is constructed as a virtually enclosed compartment that allows a high excess of a concentrated pretreatment solution to be sprayed onto the coil at the cell entrance, and staying there for a reaction time of several seconds, before the excess and reaction by-products are removed from the coil surface by a squeeze roller. Large amounts of by-products generated by this technique, however, result in a fast contamination of the active bath, and consequently in a poor economy, frequent bath dumps and downtime for remaking, as well as high pollution.

Ordinary spray technique, applying lower concentrations of the chemicals over the whole distance of the spray compartment has been more convenient and eco-efficient, also saving operation time by bath life control measures.

The drag-out of chemicals is restricted by squeeze rollers at the exit of every active or rinse stage. When properly maintained, the wet film can be minimised to a few millilitres per square metre with these rollers. To avoid damage of their surface, rollers have to be lifted off the strip automatically when the welded or stitched joint between two coils passes, and return to their squeeze position immediately after.

Over the last twenty-five years, coil coating pretreatments have been adapted to the preferred application method for no-rinse technology. As mentioned before, this involves the application of a precisely controlled amount of pretreatment chemicals to the aluminium, steel or zinc-coated steel strip surface. This can be brought about by a roll-coater. The acidic pretreatment chemicals react with the metallic surface, and the following film formation takes place by drying the strip immediately (dry-in-place principle). No reaction side-products can therefore change the process chemistry and a following water rinse is not necessary. Consequently, the main advantages are found in the efficiency of chemical usage and in substantial waste water savings.

Roll-coaters have been constructed in different ways. When adapted to application of waterborne pretreatment chemicals, they are often also referred to as chemcoaters. In principle, the liquid is taken from a tray by a first roller and transferred via another roller (or a set of rollers) onto the surface of the moving coil. While transferring the liquid, the wet film thickness is controlled by means of the nip distance and pressure, and by the relative tangential speeds of rollers. Two or three-roller coaters are common, where the additional roller (metering roll), which is often mounted between the first roller (pick-up roll) and the final roller (applicator roll), helps in maintaining a uniform wet film across the coater. Roller surfaces usually are made from high-grade or hard chrome-plated stainless steel. For aqueous pretreatment solutions, that display the low viscosity of water, the pick-up roll may be ground to a certain roughness or even engraved, in order to facilitate a sufficient amount of chemical to be taken from the reservoir tray. The applicator roll is clad with a synthetic rubber blanket that has to be acid-proof (i.e. made of neoprene etc.), due to the chemistry of the pretreatment solution.

Apart from the roll-coater technique, also simpler, less expensive equipment has been used, involving the controlled spreading of a small amount of the pretreatment liquid that is applied onto the strip a short distance before squeeze rollers, by flooding, low-pressure spraying or nip-feeding. Ideally, these rollers are motor-driven, in order to avoid aquaplaning effects, and better control the resulting wet film.

8.4 Literature

[1] anon., Eisenkristall - Atomium in Brüssel wieder eröffnet (Iron Crystal - Atomium Re-opened in Brussels), BauNetz Media, Berlin 2006; www.baunetz.de/meldungen/Meldungen_Atomium_in_Bruessel_wieder_eroeffnet_22792.html; (15-03-2010; 21:31 h)

[2] Sander, J., Über die Reaktivität von Lithium-trimethylferrat (II) gegenüber organischen Substraten (On the Reactivity of ...vs. Organic Substrates), Diss., Münster University, Münster 1986, p. 5

[3] Lehmann, V., Der moderne Hochofenprozess (Modern Blast-Furnace Process), Naturwissenschaften im Unterricht Chemie, vol. 4, Friedrich-Verlag, Seelze 1993, p. 4

[4] Meuthen, B., Jandel, A.-S., Coil Coating, 2nd ed., Vieweg, Wiesbaden 2008, p. 47 ff

[5] Barrow, G. M., Physikalische Chemie (Physical Chemistry), vol. 3, 3rd ed., Bohmann-Vieweg, Vienna 1977, p. 213

[6] Townsend, H. E. (ed.), Outdoor Atmospheric Corrosion, Am. Soc. f. Testing and Materials (ASTM) Intern., W. Conshohocken PA, USA, 2002, p. 219

[7] Prosek, T., Larché, N., Vlot, M., Goodwin, F., Thierry, D., Corrosion Performance of Zn-Al-Mg Coatings in Open and Confined Zones in Conditions Simulating Automotive Applications, Materials and Corrosion 60, 2009, publ. pend.

[8] Schäfer, J. H., Stoffstrom Aluminium - vom Bauxitabbau bis zum Recycling (Material Flow Aluminium - From Bauxite Mining to Recycling), Gesamtverband der Aluminiumindustrie GDA, Düsseldorf 2008; www.aluminium-ressourceneffizienz.de/media/Stoffstrom_Aluminium.pdf; (15-03-2010; 21:35 h)

[9] anon., Aluminium, GDA, Düsseldorf 2009, www.aluinfo.de/index-php/alu-lexikon.html; (15-03-2010; 21:38)

[10] anon., About Aluminium - Production Process, Europ. Aluminium Assoc., Brussels 2009; www.eaa.net/en/about-alumium/production-process/; (15-03-2010; 21:44 h)

[11] anon., About Aluminium - Alloys, Novelis Painted Products, Novelis Deutschland, Göttingen 2010; www.novelis-painted.com/about_alumium/alloys/index_eng.html; (15-03-2010; 21:40 h)

[12] Cornell, R., Bhadeshia, H. K. D. H., Aluminium-Silicon Casting Alloys, Univ. of Cambridge Materials Sc. & Metallurgy, Cambridge, UK 1999; www.msm.cam.ac.uk/phasetrans/abstracts/M7-8.html; (15-03-2010; 21:45 h)

[13] anon., ISO 3522:2007, Aluminium and aluminium alloys - Castings - Chemical composition and mechanical properties, International Organization for Standardization (ISO), Geneva 2007

[14] Blecher, A., Grasme, D., Hinüber, H., Husmeier, F., Meyer, B. D., Röhricht, G., Sander, J., Beschichten von Aluminiumguß - Lackieren (Coating of Cast Aluminium - Painting), Worksheet, Dtsch. Forschungsges. f. Oberflächenbehandlung DFO, Düsseldorf 1992, p. 5

[15] Androsch, F. M., Stellnberger, K.-H., Wolpers, M., Jandel, L., Drescher, D., Sander, J., Seidel, R., Chromate-free Coil Coating and One Year of Production Experience, ECCA General Meeting Monte Carlo, Proc., European Coil Coating Association, Brussels 1999

[16] Schinzel, M., Chrome-free Systems for Coil Coating, ECCA Autumn Congress Brussels, Proc., European Coil Coating Association, Brussels 2006

[17] Meuthen, B., Jandel, A.-S., [4] pp. 50, 53 ff

[18] Sander, J., Chromatfreie Systeme für Coil Coating - Stand der Erfahrungen (Chromate-Free Systems for Coil Coating - Status...), Eur. Coil Coating Assoc. - German Group, Workgroup Proc., Solingen 2002, cit. re. Schinzel [16]

[19] Meuthen, B., Jandel, A.-S., [4] p. 53 f

[20] Sander, J., Mady, R., Korrosionsschutz durch Feuerverzinken - neue Entwicklungen beim Herstellen, Passivieren und Lackvorbehandeln von schmelztauchverzinktem Bandstahl (Corrosion Protection by Hot-Dip Galvanising...), Metalloberfläche 43, 1989, p. 329 ff

[21] Ries, C., Metalloberfläche 41, 1987, p. 81, cit. re. Sander, Mady [20]

[22] Riesop, J., Krautbauer, H. (inv.), Alkaline Strip Passivation, Patent Nos. DE 19733972, US 6346295, Henkel KGaA, 1997 and 2002

[23] Meuthen, B., Jandel, A.-S., [4] p. 55

[24] Brands, K. D., Krause, M., Mayer, B., Molz, T., Willer, A. (inv.), Iron Phosphatisation Using Substituted Monocarboxylic Acids, Patent No. EP0760870, Henkel KGaA 1995

[25] Fristad, W. E., Saad, K. I. (inv.), Phosphate conversion coating concentrate, Patent No. US6706123, Henkel KGaA 2003

[26] Narayanan, T. S. N. S., Surface Pretreatment by Phosphate Conversion Coating - A Review, Rev. Adv. Mater. Sci. 9, 2005, pp. 130 ff

[27] Satoh, N., Effects of heavy metal additions and crystal modification on the zinc phosphating of electrogalvanized steel sheet, Surf. Coatings Techn. 30, 1987, pp. 171 ff
[28] Ogle, K., Tomandl, A., Meddahi, N., Wolpers, M., The alkaline stability of phosphate coatings I: ICP atomic emission spectroelectrochemistry, Corrosion Sc. 46, 2004, pp. 979 ff
[29] Tegehall, P.-E., Colloidal titanium phosphate, the chemical activator in surface conditioning before zinc phosphating , Colloids and Surfaces 42, 1989, pp. 155 ff
[30] Sienkowski, M. L. (inv.), Zinc Phosphating Process And Composition With Reduced Pollution Potential, Patent Publ. No. WO/2001/032953, Henkel Corp. 2001
[31] Androsch, F. M., et al., [15] Chapter 3
[32] Kendig, M. W., Buchheit, R. G., Corrosion Inhibition of Aluminum and Aluminum Alloys by Soluble Chromates, Chromate Coatings, and Chromate-Free Coatings, Corrosion 59, 2003, p. 379ff
[33] Sander, J., Novel Surface Treatment of Metal Strip in the Coil Coating Process, Millennium Steel 2009, pp. 146 ff, 148 ff
[34] anon., Internationale Qualitätsrichtlinien für die Beschichtung von Bauteilen aus Aluminium (International Quality Standards for the Coating of Work Pieces...), GSB AL 631, ed. 07/2009, Gütegemeinschaft für die Beschichtung von Bauteilen (GSB International), Schwäb. Gmünd 2009, pp. 32-35
[35] anon., Anodic Coatings for Aluminum and Aluminum Alloys, Military Spec., Mil-A-8625F(1), Dept. of Defense, Defense Logistics Agency, Document Automation & Production Service, Philadelphia PA, USA 2003; www.dodssp.daps.dla.mil/...
[36] anon., Inspection Of Aluminum Alloy Parts, Anodizing Process For, Military Spec. Mil-I-8474C Not 2, Dept. of Defense, Defense Logistics Agency, Document Automation & Production Service, Philadelphia PA, USA 1998; www.dodssp.daps.dla.mil/...
[37] anon., Product Informations, Surtec GmbH, Zwingenberg 2009; www.chromiting.surtec.com; (15-03-2010; 21:48 h)
[38] Volk, P., Echte Alternative - Chrom(III)-haltige Vorbehandlung für Aluminium (True Alternative - Cr[III] Containing Pretreatment...), Metalloberfläche 60, 2006, pp. 24 ff
[39] Pohl, J., Möller, R., New Cr (III)-based pretreatment - A simple, all-purpose, environmentally friendly process, JOT International Surf. Technol. (1) 2009, pp. 2 ff
[40] Lienkamp, P. (ed.), Sander, J., Körner, T., Kleinschmidt, T., More than 10 years of experience with chrome-free pre-treatment of aluminium - Perfect protection against corrosion, Press release, Henkel AG & Co. KGaA, Düsseldorf 2008, cit. Int. Aluminium. J. 84 (11), 2008, p. 36 ff
[41] Androsch, F. M., et al., [15] Figures 18 - 19, discussion
[42] Sander, J., Novel Surface Treatment of Metal Strip in the Coil Coating Process, Millennium Steel 2009, pp. 146 ff, 148 ff
[43] Fristad, W. E., Phosphate-free Pretreatments for all Substrates, Coatings Magazine (11) 2008, pp. 14 ff
[44] Saji, V. S., Thomas, J., Nanomaterials for corrosion control, Current Science 92, 2007, pp. 51 ff
[45] Rauscher, G., Neue chromfreie Vorbehandungsverfahren für Aluminium (New Chromium-free Pretreatment Processes...), JOT, 2005, pp. 54 ff
[46] Jatsch, A., Rauscher, G., Verbesserung des Korrosionsschutzes von Aluminium durch neuartige chromfreie Vorbehandlungsmethoden vor der Lackbeschichtung, (Improvement of the Corrosion Protection of Aluminium...), Galvanotechnik 101, 2003, pp. 674 f
[47] Schriever, M. P. (inv.), Non-chromated oxide coating for aluminum substrates, Pat. US 5411606, The Boeing Comp. 1995
[48] Schriever, M. P. (inv.), Non-chromated oxide coating for aluminum substrates, Pat. US 6432225, The Boeing Comp. 2002
[49] Pfaff, D., SAM bewährt sich in der Praxis (SAM Proves Itself in Practice), JOT 2003, pp. 22 f
[50] Pietschmann, J., Alternativen zur Chromatierung - Individual- statt Universallösung, (Alternatives to Chromating...), JOT 2007, pp. 42 ff
[51] Van Ooij, W. J., Replacement of Chromates by Silanes - An Overview, ICEPAM Conf. Oslo, SINTEF, Trondheim 2004, www.sintef.no/static/mt/norlight/ICEPAM/01-van-Ooij_Cincinnati.pdf; (15-03-2101; 21:50 h)
[52] Franquet, A., van Schaftinghen, T., de Graeve, I., Laha, P., Le Pen, C., Terryn, H., Vereecken, J., Silane Coating of Metal Substrates, ICEPAM Conf. Oslo, SINTEF, Trondheim 2004, www.sintef.no/static/mt/norlight/ICEPAM/03-IDeGraeve_Brussels.pdf; (15-03-2010; 21;52 h)
[53] Meuthen, B., Jandel, A.-S., [4] p. 57
[54] Lebelt, P., Gehrke, J., Hallmeier, K. H., Reuther, H., Dörfel, I., Qualitative und quantitative Analysen dünner Funktionsschichten für das Kleben durch XPS-, AES-, REM- und TEM-Untersuchungen (Qualitative and Quantitative Analyses of Thin Functional Coatings...), Galvanotechnik 107, 2009, p. 930 ff
[55] Kaiser, W.-D., Reinhard, G., Tiller, H.-J., Göbel, R. (inv.), Verfahren zur Oberflächenvorbehandlung von Metallen für die katodische Elektrotauchlackierung (Process for the Surface Pretreatment of Metals for the Cathodic Electro-Deposition Coating), Patent DE 4012086A1, Inst. Mat. Sc. Appl. Technique GmbH, 1991

[56] Bielefeld, F.W., de Zeeuw, A., How Can the Steel Industry Generate Add-on Business in the Post-finishing Market?, ECCA Autumn Congress Brussels, Proc., European Coil Coating Association, Brussels 2009
[57] Sander, J., [42] pp. 148 f
[58] anon., Regulation (EC) No 1907/2006 of the European Parliament and of the Council of 18 December 2006 concerning the Registration, Evaluation, Authorisation and Restriction of Chemicals (REACH), establishing a European Chemicals Agency, amending Directive 1999/45/EC and repealing Council Regulation (EEC) No 793/93 and Commission Regulation (EC) No 1488/94 as well as Council Directive 76/769/EEC and Commission Directives 91/155/EEC, 93/67/EEC, 93/105/EC and 2000/21/EC, Official Journal of the European Union 49, 2006, L 396; eur-lex.europa.eu/…
[59] Krizek, J. (inv.), Vorrichtung zur Behandlung von Gegenständen (Device for the Treatment of Objects), Patent DE 19950892, Eisenmann GmbH, 1999

9 Organic coatings for maintenance

Jörg Sander

9.1 Surface tolerant coatings

9.1.1 General considerations

The term "surface tolerant coating" [1] (STC) is often used when referring to maintenance and repair coatings in shipbuilding, pipeline construction, offshore construction and other water engineering. In these fields of application, damaged areas usually have to be repair-coated on the spot, i.e. they are treated manually rather than in shop processes. STCs hence require tolerance in various aspects. For example, they must be applicable in environments where residual moisture by condensation or spray cannot be avoided or where residual old coatings or rust cannot be entirely stripped. In short, STCs must adhere, cure and perform under less than ideal conditions. STCs have been applied for decades [1, 2], but the achievable results are of variable quality and highly dependent on the preparation of the surfaces and the way of application [3]. The development of standards for testing the surface tolerance of coatings is a subject of recent publications, with the aim of identifying critical parameters that can be used to evaluate the coating performance [4].

As a convenient method for surface preparation in an open construction site, ultra-high pressure water jet cleaning is recommended (cf. Chapter 3.2). This, however, will leave residues of the foregoing, aged coatings, and rust on the surface, as well as it causes new, flash rust on the exposed metal, and obviously the water is unlikely to dry off completely. Regardless of the apparently good adhesion (that must be proven by inspection tests), rust and old paint may be loosened by the action of overcoating, causing the whole repair attempt to fail. The compatibility of old and repair paint systems must be assured, in terms of binder chemistry, curing type, wettability, tension levels and degree of deterioration. Trialling of the repair coat on a small test area is therefore recommended, following Standard ASTM D 5064-95.

9.1.2 Surface tolerant coating materials [5, 6]

To cope with residual rust, imperfectly cleaned surfaces, and old, weathered, micro-cracked coatings, an STC should provide free, low-molecular binder. The binder furthermore must provide sufficient flexibility to avoid mechanical stress between old and new paint coatings. The choice and amount of solvent must be suitable for the purpose, i.e. old paints must not be weakened, and adhering rust must not be dissolved while the STC is applied. Finally, STCs must be applicable at low temperatures and high humidity. Historically, this scope has been matched by alkyd paints with red lead ($Pb_2[PbO_4]$) pigmentation, although limitations were found with these systems in terms of water sensitivity and chemical resistance of the long oil alkyds. Moreover, the lead compound is being phased out due to its toxicity.

Jörg Sander et al.: Anticorrosive Coatings

ISBN 978-3-86630-911-1

Nowadays, the most common coatings in heavy corrosion protection are high-solids, two-pack epoxies. STCs of this class can be applied in thick layers in single operation (100 to 200 µm dry film thickness), and provide the required initial adherence and resistance in dew point conditions. Curing is mostly achieved by diisocyanate agents, rendering a high degree of cross-linking, hence forming high barriers against humidity and oxygen. The choice of binders allows UV reflection and dissipation of thermal and mechanical strain. The formulations might involve active (zinc phosphate etc.) and lamellar pigments (e.g. aluminium flakes and micaceous iron oxide), and polysiloxanes or fluorinated polymers to add barrier properties. STCs may also contain zinc particles for cathodic protection.

Another option is the use of moisture-curing polyurethanes as binder base which can be applied at up to 98 % relative humidity and temperatures even below the freezing point. As an advantage in handling, moisture-curing polyurethanes are single-pack products.

9.2 Organic coatings on residual rust and old coatings

While STCs are often applied on surfaces that are contaminated by residual rust, it is obvious that this is only acceptable as enforced by the conditions of on-site painting. In proper shop painting processes, corrosion products like rust from larger damage areas must be removed by mechanical cleaning, e.g. grit or dry ice blasting, and manual sanding (cf. Chapter 3.2), if possible, supported by pickling.

Residual rust and flash rust are often treated with so-called rust converters that are based on formulations containing phosphoric acid and/or tannic acid derivatives to remove iron oxide by acid pickle and form stable precipitates of iron phosphate and iron tannate. Tannin based industrial rust converters have been in use for the preservation of iron and steel for half a century [7, 8]. However the usefulness of this method as a preparation for subsequent coating is doubtful. A number of papers show the controversial results and opinions in the scientific community [9 - 15]. As a conclusion, the active ingredients of rust converters are suitable to stabilise residual rust particles and flash rust on steel surfaces, however they do not release the operator from careful abrasive removal of rust, and they by no means provide reliable performance as conversion coatings.

In artisan coating operations, corrosion protection is often provided by coating of the damaged, thoroughly cleaned and derusted areas with a so-called wash primer (cf. Chapter 4.5). Wash primers contain phosphoric acid that enables chemical attack and bonding to the exposed metal surface. The coatings are only thin and do usually not exceed 8 µm, in order to secure good bonding. State-of-the-art products still contain zinc chromate as active corrosion preventive pigment. Accordingly, their use is subject to safety precautions. Admixtures for spray application are only allowed with very low chromate contents below 0.1 % [16]. Some wash primers offer a higher filler pigmentation, which makes them suitable for their use as undercoats with a gauge around 20 µm. So-called reactive undercoats are based on epoxy or polyurethane chemistry. Zinc filled corrosion protection primers (CPP) are often used for larger areas (cf. Chapters 2.1.3; 6.5.2) [17].

Repair coats on automobiles are also manually applied and must not exceed 80 °C for curing, to avoid damage to other car components. Physically drying or reactive paints that cure under catalytic action have been replaced by 2-component systems with polyacrylates or polyesters as base component and isocyanates as cross-linkers. Repair paints are available as solvent-based and waterborne high-solid systems. They cure at room temperature, but curing can be shortened by local application of elevated temperatures, e.g. by infrared lamps. Old paint may be stripped from the damaged area, using paint strippers based on e.g. benzyl alcohol or formic acid [18]. The damages are ground down to the metal, derusted, solvent-cleaned and coated with a corrosion protection primer (CPP) [19, 20]. Cleaning can be done with solvent wiping or by vapour or high pressure spraying with aqueous, mostly neutral or mild alkaline cleaners. It must be performed very carefully and its efficiency controlled by wetting tests (cf. Chapters 3.1.3; 3.3.3).

9.3 Literature

[1] Mühlberg, K., Surface-Tolerant Coatings - Some Experiences, J. Prot. Coat. Lin. Prot. Coat. Europe Technology Publ. 2001, pp. 13 ff, 18

[2] Flick, E. W., Prepaint Specialties and Surface Tolerant Coatings, Noyes Publ., Park Ridge NJ, USA 1991

[3] Boxall, J., Advances in surface-tolerant coatings, Pigment & Resin Technology 19, 1990, pp. 4 ff

[4] Wilds, N., Surface Tolerant Coatings for Offshore Maintenance, NACE Corrosion Congress 2004, New Orleans LA, USA, Proc., NACE International, Houston TX, USA 2004

[5] Gui, F., Hill, D., Brossia, C. S., A Method for the Evaluation of Wet Surface Tolerant Coatings, NACE Corrosion Congress 2008, New Orleans LA, USA, Proc., NACE International, Houston TX, USA 2008

[6] Mühlberg, K., [1] pp. 13 ff, 18

[7] Wilds, N., [4] p. 2

[8] Groysman, A., Corrosion for Everybody, Springer, Dordrecht 2010

[9] Rogers, B. A., The Archeologist's Manual for Conservation, Kluwer Academic, New York 2004, pp. 69 f

[10] Collazo, A., Nóvoa, X. R., Pérez, C., Puga, B., EIS study of the rust converter effectiveness under different conditions, Electrochimica Acta 53, 2008, pp. 7565 ff

[11] Singh, D. D. N., Yadav, S., Role of tannic acid based rust converter on formation of passive film on zinc rich coating exposed in simulated concrete pore solution , Surface and Coatings Technology 202, 2008, pp. 1526 ff

[12] Ocampo, L. M., Margarit, I. C. P., Mattos, O. R., Córdoba-de-Torresi, S. I., Fragata, F. L., Performance of rust converter based in phosphoric and tannic acids, Corr. Sci. 46, 2004, pp. 1515 ff

[13] Barrero, C. A., Ocampo, L. M., Arroyave, C. E., Possible improvements in the action of some rust converters, Corr. Sci. 43, 2001, pp. 1003 ff

[14] Matamala, G., Smeltzer, W., Droguett, G., Comparison of steel anticorrosive protection formulated with natural tannins extracted from acacia and from pine bark, Corr. Sci. 42, 2000, pp. 1351 ff

[15] Favre, M., Landolt, D., Hoffman, K., Stratmann, M., Influence of gallic acid on the phase transformation in iron oxide layers below organic coatings studied with Moessbauer spectroscopy, Corr. Sci. 40, 1998, pp. 793 ff

[16] Kittel, H., Streitberger, H.-J. (ed.), Lehrbuch der Lacke und Beschichtungen (Coursebook of Paints and Coatings), Vol. 6, 2nd ed., Hirzel, Stuttgart 2008, pp. 459 f

[17] Goldschmidt, A., Streitberger, H.-J., Basic of Coating Technology, 2nd edition, BASF Coatings AG, Münster 2007, pp. 739 f

[18] Kittel, H., Streitberger, H.-J. (ed.), Lehrbuch der Lacke und Beschichtungen (Coursebook of Paints and Coatings), Vol. 6, 2nd ed., Hirzel, Stuttgart 2008, p. 318

[19] Goldschmidt, A., Streitberger, H.-J., Basic of Coating Technology, 2nd edition, BASF Coatings AG, Münster 2007, pp. 710 f

[20] Kittel, H., Streitberger, H.-J. (ed.), [18] pp. 452 ff

10 New corrosion protection concepts

Dmitry Shchukin, Ekaterina Skorb

High-value materials in automotive and aerospace industries, for example, require increasingly sophisticated eco-friendly coatings for improved performance, self-healing, and durability. In this respect, recent developments in nanotechnology are most promising [1]. New types of functional nanomaterials could provide the coating not only with enhanced characteristics like anticorrosion properties or external performance but also with additional functionalities which can respond on demand to external factors, thus making the coating "active" on macro and micro-levels. These modifications are important for protective coating systems, especially in consideration of the banning of carcinogenic Cr(VI), Cd, and some organic solvents [2].

Standard anticorrosion coatings developed so far passively prevent the interaction of corrosive species with the metal. This approach requires thicker coatings and does not solve the "cut-edge" problem. The next generation of protective coatings should be much thinner (thus reducing CO_2 emission) and the quantity of raw materials per 1 m^2 of the substrate) and possess self-healing or self-curing effects in scratched (damaged) areas. This will require to develop components of the coatings that react to the impact of external or internal conditions (pH, humidity changes, or distortion of the coating integrity, etc.), and, ideally, combine the self-healing property with other functionalities (e.g. detection, controlled reflection, self-cleaning). In the following chapter active coating systems based on self-responding thin films or nanocontainer-based films are described which have received highest attention in the scientific community in the scientific community in recent years.

10.1 Thin films

Currently, various types of thin films are employed for corrosion protection. There is a number of new progressive concepts:

Self-assembled monolayers which started to be widely developed in the 1990s of the last century (see Chapter 10.1.1);

Conducting polymers (see Chapter 10.1.2) on which corrosion specialists have focused their interest during the last 10 to 15 years,

Biopolymers (see Chapter 10.1.3) which are interesting either as a potential conductive matrix or as passive protective coatings with simultaneous biocompatibility.

Individual use of these techniques for corrosion protection and combination of thin films with nanomaterials enabling the formation of **nanocomposites** is also an area of growth in today's corrosion science. Another type of well-known and widely used coatings are **galvanic coatings**; however, these are most conservative and only a few reports dedicated to their modification can be found in literature [3], mostly related to the incorporation of nanoparticles and polymer microcontainers into the coating (see Chapter 10.2.1). Due to the metallic structure of the coating matrix, it is difficult to ensure the mechanical integrity

Jörg Sander et al.: Anticorrosive Coatings

ISBN 978-3-86630-911-1

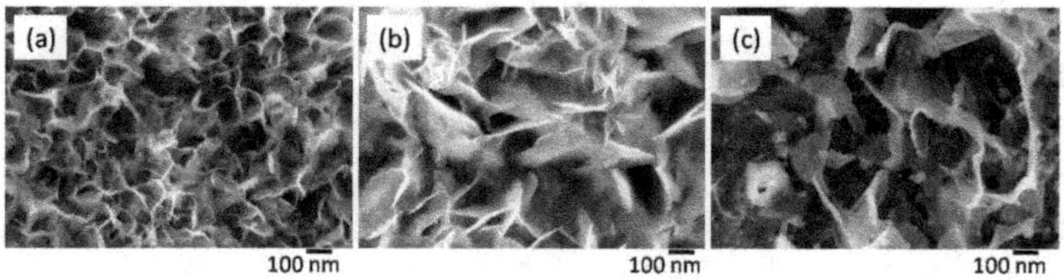

Figure 10.1: SEM images of aluminium (a) magnesium (b) alloy and stainless steel (c) after 40 min sonication at 57 W/cm² *Source: Royal Society of Chemistry*

of the containers embedded into the coating as well as to incorporate them into the coating without agglomeration. A lot of effort will be needed in the future to attain active properties to the galvanic coatings.

Simultaneously, for the majority of metals, there is native passivity which is based on the spontaneous formation of a **thin oxide layer** (the passive film), in the specific environment. This film can slow down corrosion (dissolution) reactions by many orders of magnitude. For example, the typical aluminium surface is covered by a 3 to 7 nm thick natural oxide film. However, this thin layer is not sufficient to protect against corrosion agents and does not yield good adhesion to subsequent layers of the coating [4]. Therefore, the metal surface is always pretreated before use. The most extensively used surface pretreatment procedures are based on aggressive chemicals including chromate solutions [5].One of the functions of the surface pretreatment is to produce a porous oxide layer on the metal surface with roughness sufficient for mechanical interlocking.

High intensity ultrasonic irradiation has been used for a long time in metal technology including metal crystallisation, forming and finishing [6]. The application of ultrasound to a metal melt generally leads to metals with improved grain refinement and homogeneity [7].

For example, the collapse of cavitation bubbles near an aluminium surface yields surface cleaning (destruction and removal of impurities);

- significant increase of roughness of the interfacial layer and, therefore, adhesion of further films to the surface
- formation of a thick oxide layer [8].

The SEM images (Figure 10.1a-c) of metal surfaces sonicated at 57 W/cm² indicate the formation of highly developed metal surfaces. It should also be noted that the intensity and duration of ultrasound treatment dramatically influence the development of porous metal structures. The thickness of the modified layer estimated by TEM analysis of ultramicrotomed plates is about 200 nm with a foam cell size of ~7 nm.

The scanning vibrating electrode technique (SVET) allows measuring current density maps over the selected surface of the sample thus monitoring local cathodic and anodic activity in the corrosion zones [9]. Figure 10.2 shows the monitoring of corrosion activity of a steel plate with and without ultrasound treatment, immersed into a corrosive solution (0.1 M NaCl) for 12 hours. It is observed that an intensive corrosion process results in the development of defects throughout the whole surface of the sample, finally leading to the total corruption of the unmodified steel surface (Figure 10.2a and c). Figure 10.2e shows the time monitoring of anodic and cathodic activity of a steel plate without ultrasound treatment (curve 1 and 4) immersed into a corrosive solution (0.1 M NaCl) for 12 hours. It can be seen that corrosion increases dramatically with time reaching a very high value of anodic activity ~300

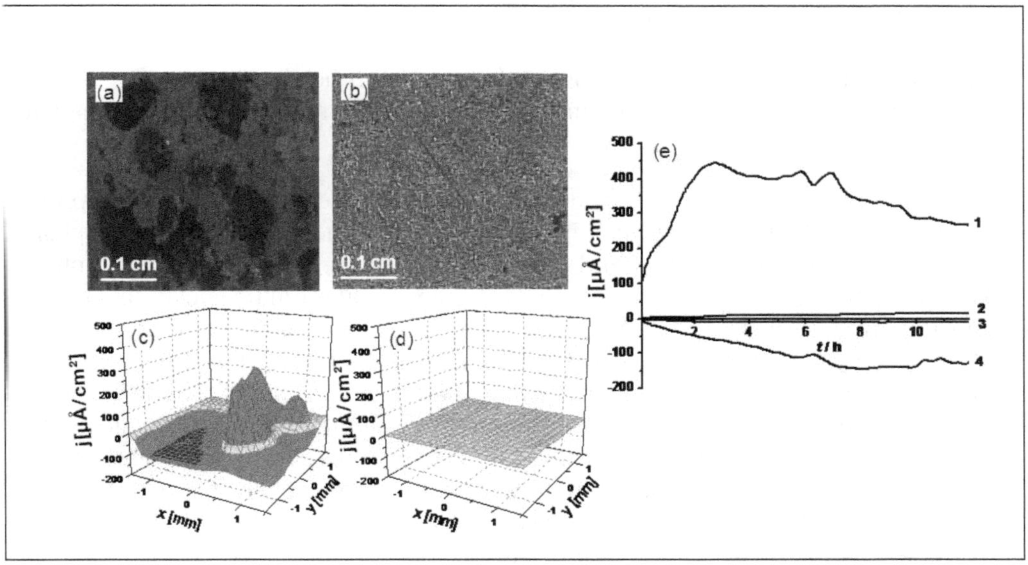

Figure 10.2: Optical microscopy images (a,b) and 3D current density maps obtained by SVET (c,d) of the scratched steel plates in a corrosion test of unmodified steel (a,c) and steel after ultrasound treatment (57 W/cm², 60 min) (b,d). (e) Time monitoring of the anodic (curves 1 and 2) and cathodic (curves 3 and 4) activity on the steel surface before sonication (curve 1 and 4) and after sonication at 57 W/cm² (curve 2 and 3). Source: Royal Society of Chemistry

to 400 μA/cm². The steel samples after high intensity ultrasonic irradiation (Figure 10.2e curve 2 and 3) exhibit dramatically different behaviour. The increase in corrosion activity is very low, after 12 hours of immersion in 0.1 M NaCl the anodic activity is ~8 μA/cm². In the long run such pretreated metal surfaces are useful for subsequent application of organic multi-layers for corrosion protection.

10.1.1 Self-assembling monolayers

The 1980s saw an explosion of interest in "self-assembled monolayers" (SAMs)-densely packed, oriented monolayer films of long-chain surfactants that spontaneously chemisorbed from solution onto solid surfaces [10]. SAMs make ideal models for the study of fundamental processes at organic surfaces, such as wetting [11], adsorption [12], adhesion [13], chemical reactivity [14] and as a result corrosion stability. They also provide a testing ground for the physical principles that govern the assembly of other supramolecular structures, such as membranes and micelles. Potential applications of SAMs are in the traditional areas of surface chemistry-adhesion [15], wetting, corrosion [16], and lubrication.

Among different approaches, the self-assembled monolayer technique has recently attracted much attention in the coating and corrosion protection research area [17]. While self-assembled monolayers on substrates such as glass and gold electrodes have been studied extensively in the last few decades [18], the application of self-assembled monolayer and multi-layer films on metal and metal alloy substrates such as aluminium, aluminium alloy and steel and other corroding metals is still relatively new. Simultaneously, they are very promising due to the formation of a unique corrosion stable layer. The self-assembled monolayers, often made from amphiphilic hydrocarbon molecules, are expected to function as a barrier to prevent the permeation of corrosion accelerants such as moisture and electrolytes into the metal substrate, therefore improving the corrosion resistance of the substrate materials. Three categories of molecules are commonly used for this purpose:

- **fatty acids** with carboxylic end groups that form electrostatic interactions with metal substrates,
- **alkyl thiols** that link to the steel substrate by iron-sulfur bonding and
- **alkyl silanes** that react with the metal oxide from the metal and metal alloy substrates.

Maege et al. [19] reported a systematic study and comparison of self-assembled monolayers on aluminium alloy substrates using amphiphilic molecules including fatty acids, phosphonic or phosphoric acids, **alkyl silanes and alkyl siloxanes.** The fatty acids and phosphonic acids or phosphoric acids tend to bond with Al_2O_3 from the aluminium substrates through electrostatic interactions, and the long alkyl chains interact together through van-der-Waals interactions to form stable monolayer films. For alkyl silanes, a two-dimensional hydrolysis-condensation reaction occurs between silane and Al_2O_3 from the aluminium substrate, leading to a polymerised monolayer film. Due to the film formation, the metal alloy surface becomes much more hydrophobic, and the corrosion resistance of the substrate was found to increase significantly. Among all these molecules being studied, it was found that the phosphonic acid monolayer gave the best corrosion protection according to humidity test. Despite these extensive studies, many issues associated with this research area remain unclear. For example, the surface chemistry of metal alloys is much more complicated compared to the well-defined and much smoother glass or gold surface. Such a complex surface chemistry is certain to bring some significant effects on the self-assembling process and the properties of the resulting thin films. The surface properties of metal alloys are often affected by the surface cleaning and treatment procedures, and therefore, so is the molecular self-assembling process on the metal alloy surface.

Four different alkylsilane compounds, which include octadecyltrichlorosilane ($C_{18}H_{37}SiCl_3$), octadecyltrimethoxysilane ($C_{18}H_{37}Si(OMe)_3$), octyltrimethoxysilane ($C_8H_{17}Si(OMe)_3$) and octadecyldimethylchlorosilane ($C_{18}H_{37}SiMe_2Cl$), were suggested in a study [20] to compare the self-assembled thin film formation from alkly silane compounds with different chemical structures and to access the corrosion resistance of thin film modified Al 2024 substrates. It is well known that with different chemical structures, the hydrolysis-condensation kinetics of sol-gel reactions vary from one type of molecule to the other, as do the self-assembled monolayer or multi-layer structures and properties. For example, some alkyl-silane compounds tend to form monolayer films with more closely packed and ordered monolayer films, while others will result in films with more defective domains. It is hypothesised that films with less defective domains will result in a better corrosion protection of the aluminium alloy substrate.

It should also be noted that the traditional means of forming an **organic monolayer film** is to spread an insoluble compound on an aqueous subphase, compress the film mechanically with a barrier until the molecules are densely packed and oriented approximately normal to the surface, and then to transfer this monolayer, if desired, to a solid substrate by "dipping" using the Langmuir-Blodgett method" [21]. Self-assembled monolayers could be formed by chemisorption providing a convenient technique to modify surfaces (Figure 10.3). The highly ordered and packed monolayer can prevent solution from transferring to the metal surface so it can effectively protect the metal from corrosion [22]. Monolayers of alkane thiols, dialkyl disulfides, and other sulphur-containing molecules prepared on copper and gold by adsorption from solution were described in numerous reports [23]. A few studies about iron corrosion were reported [24]. It was shown that the 2-mercaptobenzothiazole, an N- and S-containing heterocyclic compound, can form monolayers on iron. These compounds have been widely investigated because they can form hydrophobic complexes with many metals such as iron, copper, cobalt, nickel, etc. and therefore are used as corrosion inhibitors [25]. Moreover, based upon the spontaneous adsorption and orientation of

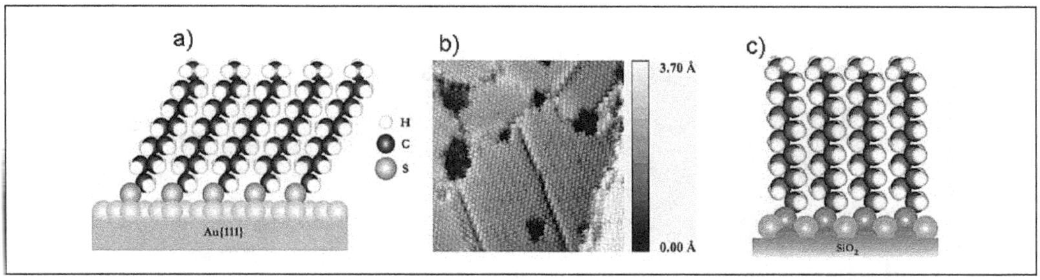

Figure 10.3: (a) Schematic of an n-dodecanethiolate monolayer self-assembled on an atomically flat gold substrate with van der Waals interactions between neighboring hydrocarbon chains.
(b) STM image of a 250 Å x 250 Å area of an n-decanethiolate self-assembled monolayer. A step edge along with several film defects can be seen, including domain boundaries (protruding and depressed ridged lines) and substrate vacancy islands (dark, circular depressions). Tunneling conditions: Vsample=-1 V, it=10 pA. *Source: Elsevier*

molecules with a reactive group bonded chemically to the substrate, 2-mercaptobenzothiazole can form self-assembled monolayers on metals. During the adsorption process, the corresponding heat changes could be detected by microcalorimetry. For copper protection self-assembled monolayers formed from alkanethiols, alkyl thiosulfates, mercaptoalkanols and Schiff bases were suggested [26]. The agent inositol hexaphosphate, also known as phytic acid and its salts have been suggested to be used as environmentally benign inhibitors [27].

However much sophisticated work has been performed using the **layer-by-layer**(LbL) **technique**. The multicomponent nano-network formed by **polyelectrolyte multi-layers** using the LbL technique [28] could offer new opportunities for anticorrosion coatings. The LbL deposition procedure involves the step-wise electrostatic assembly of oppositely charged species (e.g., polyelectrolytes and inhibitors, or others: proteins, nanoparticles) on the substrate surface with nanometre scale precision and allows the formation of a coating with multiple functionality. The coating properties can be controlled by the number of deposition cycles and the types of polyelectrolytes used. Polyelectrolytes exhibit very good adhesion to the substrate surface and are able to seal surface defects. The conformation of polyelectrolytes is mostly dependent on their nature and adsorption conditions and much less dependent on the substrate and charge density of the substrate surface. Polyelectrolyte coatings are expected to cover many kinds of surfaces including non-ionic and polar substrates. The polyelectrolyte multi-layers offer a broad range of applications in the fields of nonlinear optics, light emission, sensing, separation, bioadhesion, biocatalytic activity, drug delivery and specific bio-applications based on surface modifications [29].

In the last years, polyelectrolyte multi-layers have attracted great interest in corrosion protection [30]. Corrosion processes are accompanied by a number of reactions changing the composition and properties of both metal surface and local neighbouring environment (e.g., formation of oxides, diffusion of metal cations into the coating matrix, local changes of pH and electrochemical potential) [31]. One approach to prevent corrosion propagation on metal surfaces is the suppression of accompanying physico-chemical reactions. Polyelectrolytes are macromolecules that carry a relatively large number of functional groups that are either charged or, under suitable conditions, can become charged. Polyelectrolyte films are able to change their chemical composition with the pH value since the degree of dissociation of the polyelectrolytes is influenced by the local pH value. The films can be enriched in one polymer compared to the other by fabrication in a pH-regime where one of the polymers is weakly charged while the other is strongly charged. Active species deposited as a compo-

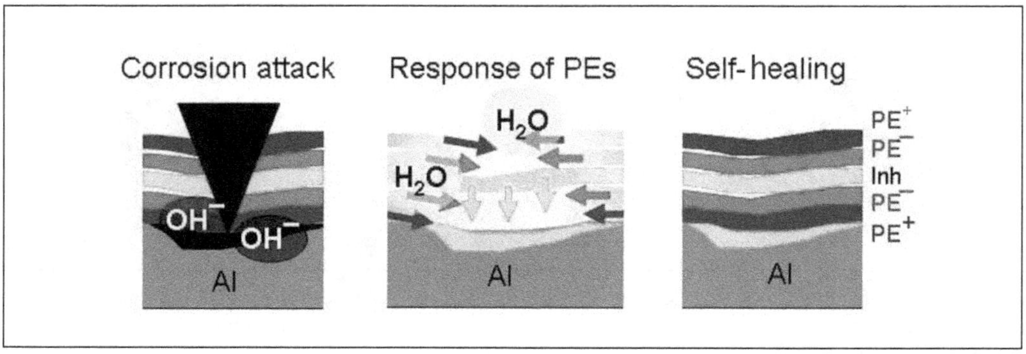

Figure 10.4 : Schematic mechanism of self-healing action of a "smart" polyelectrolyte anticorrosion coating. Corrosion attack causes pH change of the system stimulating response of the polyelectrolyte coating: pH buffering, rearrangement of polymer chains and release of corrosion inhibitor. (PE+: positively charged polyelectrolyte, PE-: negatively charged polyelectrolyte, Inh - corrosion inhibitor)

nent of the polyelectrolyte nano-network or as a dopand can be released on demand. The sensitivity of the polyelecrolyte film to a variety of physical and chemical conditions (like shift of the pH value or mechanical impact, etc.) of the surrounding media provides the ability of regulated release of the inhibitor species entrapped into multi-layers. The schematic mechanism of a smart self-healing process is shown in Figure 10.4.

The following mechanisms of the response of the polyelectrolyte multi-layer system to corrosion attack can be proposed:

- polyelectrolytes have buffering activity and can stabilise the pH value at the metal surface in corrosive media;
- inhibitors are released from polyelectrolyte multi-layers only after start of the corrosion process and directly into the rusted area, preventing corrosion propagation;
- polyelectrolytes forming the coating are relatively mobile and have the tendency to seal and to eliminate the mechanical cracks of the coating.

Thus, the polyelectrolyte nano-network is effective in both suppression of corrosion propagation and healing of first traces of corrosion microdefects. Two issues related to formation of the polyelectrolyte coatings and their anticorrosion activities must be addressed. The first

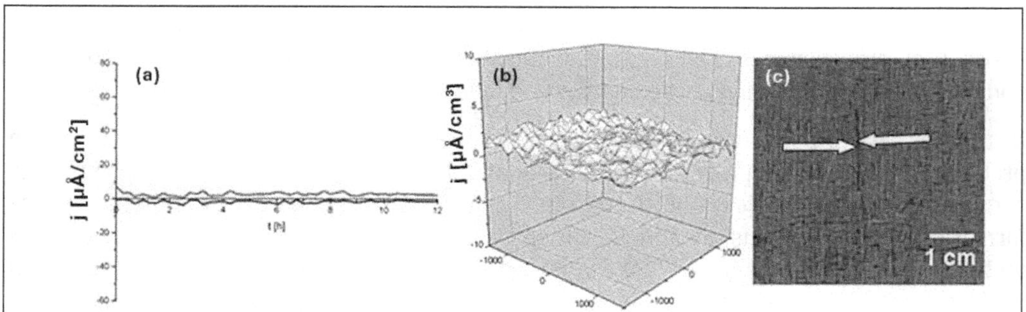

Figure 10.5 : Time dependence of the current density changes of the scratched aluminium alloy covered by the (PEI-PSS)10 coating obtained by scanning vibrating electrode technique: curve above: anodic current, curve below: cathodic current (scale units: μA cm^{-2}, spatial resolution 150 μm, solution: 0.1 M NaCl) (a); snapshot measured in 3 h of the experiment time (b); photograph of the behaviour of the scratched surface during the SVET experiments (c). The scratch is highlighted by arrows

issue is the nature of the alloy surface, polymer adhesion to the surface and coating stability in aggressive environment. Aggressive solutions as well as temperature changes or UV irradiation could alter the structural integrity of multi-layer films. The second issue is related to the anticorrosion activity of the coating. The sensitivity of the nano-network formed by different polyelectrolytes to external stimuli generated by corrosion degradation (changes of pH value, interruption of coating integrity) is most important and therefore subject to extensive studies.

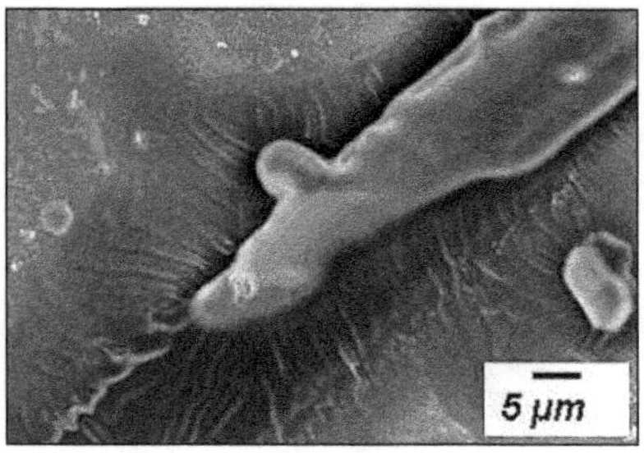

Figure 10.6: Scanning electron microscopy image of the surface of the PEI-PAA coating after 12 h immersion in 0.1M NaCl

Weak polyelectrolytes are increasingly used in multi-layer thin films because greater control over film properties can be achieved by varying the ionisation of the weakly charged groups through adjustments of the pH value. Therefore, one expects a different current density map for the substrates coated with multi-layers composed of a combination of weak and strong polyelectrolytes (see Figure 10.5). A combination coating of weak polyethylene imine (PEI) and strong polystyrene sulphonate (PSS) exhibits very high anticorrosion protection. The scratched aluminium surface is shown in Figure 10.5 (c). The scratch without any traces of corrosion degradation is highlighted by arrows. No ionic flux can be seen on the time monitoring measured by using SVET during 12 hours. The control snapshot in Figure 10.5 (b) exhibits only background noise. The differences between strong and weak polyelectrolyte pairs are presumed to reflect the relative degree of association of oppositely charged polyelectrolyte segments. Polymer pairs which form stronger complexes are expected to better associate in contact with aqueous solutions and would thus be less sensitive to pH changes in contrast to weak-strong and weak-weak polyelectrolyte pairs.

pH changes following the corrosion process cause ionisation of the functional groups of the weak polyelectrolytes which results in an increased repulsion between uncompensated charges. Small counter-ions can penetrate the layer structure to compensate these charges. The higher ionic concentration inside polyelectrolyte multi-layers as compared to the surrounding solution increases the osmotic pressure resulting in permeation of water into the polyelectrolyte multi-layers. Thus, the polyelectrolyte multi-layers start to swell and, therefore, increase their mobility. Figure 10.6 shows SEM images of the PEI-PAA covered aluminium plates after 12 hours immersion in 0.1 M NaCl. The scratched surface tends to recover in aqueous solution due to chain rearrangement in the swollen polymer multi-layers.

A corrosion inhibitor incorporated into the protective coating as a component of the LbL film is responsible for the most effective mechanism of corrosion suppression. Quinolines are environmentally friendly corrosion inhibitors attracting more and more attention as an alternative to the harmful chromates. 8-HQ was found to prevent the adsorption of chloride ions thus improving the corrosion resistance due to the formation of an insoluble chelate of aluminium which protects the metal substrate [32]. Both mechanical (scratch) and chemical (changes of local pH) rupture of the polyelectrolyte network cause the release of the encapsulated inhibitor in the damaged part of the metal surface only in response to corrosion attack, resulting in termination of the corrosion process and prolongation of corrosion protection.

10.1.2 Conducting polymers

Electrically conducting polymers have been the subject of continuous research and development due to their potential applications in many technological areas such as rechargeable batteries, sensors, electromagnetic interference shielding, electrochromic display

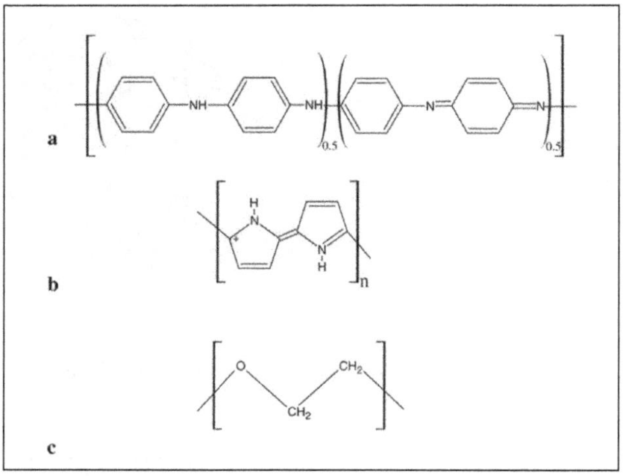

Figure 10.7: Chemical structures of selected conductive polymers: (a) poly(aniline); (b) poly(pyrrole); (c) poly(ethylene oxide)

devices, smart windows, molecular devices, energy storage systems, membrane gas separation, and so on [33]. Most recently, several studies have been carried out or are still in progress, evaluating various conducting polymers in different configurations regarding their corrosion protection towards oxidisable metals. Mengoli et al. were the first to examine the protective behaviour of polyaniline (PAni, Figure 10.7) on stainless steel, and then in 1985, DeBerry showed that the electrochemically synthesised PAni acts as corrosion protective layer on stainless steel in 1 M H_2SO_4. Since then, many research groups have systematically investigated the electrochemical synthesis of various conducting polymers on oxidisable metal surfaces for corrosion protection purposes.

Although only positive results are reported, in most publications on conducting polymers for corrosion protection, it is generally known from industrial trials of conducting polymer based coatings that the practical performance is usually by far inferior to coatings that do not contain conducting polymers (also referred to as intrinsically conducting polymers). This is due to the fact that in most published reports only simple immersion tests were performed, where there is a good chance for the conducting polymer to passivate small defects [34], while under atmospheric corrosion conditions or the presence of too large defects this principle does not work out and fast coating break-down is observed [35]. Such a sudden coating break-down is postulated to be an inherent property of all conducting polymers [36]. However some works show that by preventing formation of over-extended percolation networks of the conducting polymer in the composite coating this danger can be averted, making conducting polymers suitable for intelligent corrosion protection [37]. Thus the coatings can be designed to show efficient corrosion-triggered intelligent release of inhibitor anions [38]. Coatings which follow these design rules generally seem to perform well [39] and especially effectively inhibit fast delamination originating from external defects. However, even without defects in the coating, corrosion at the buried interface between the coating and the metal seems to be a serious problem. Williams and McMurray [40] observed the growth of a 3 μm oxide film within just 140 h at the interface between an iron sample and a composite coating of PAni dispersed in a non-conducting polymeric matrix. They explained this fast interfacial corrosion by galvanic coupling between PAni and iron, i.e. oxidation of the iron fuelled by reduction of the PAni, which in turn is re-oxidised by environmental oxygen (which is reduced in this process). In other words, the conducting polymer serves as a mediator between metal and atmospheric oxygen, thereby promoting corrosion rather than preventing it. When the same coating was applied to zinc, only a comparatively thin oxide layer of about 0.5 μm was formed in this case by interfacial corrosion. Williams et al. explain this difference with the different acidity of the corresponding metal cations: while the iron(II) (and also aluminium) cation is quite acidic and thus can effectively buffer the pH value inside the coating, for the case of zinc alkalisation of the interface occurs as a consequence of the accompanying oxygen reduction, as the lower acidity of the zinc cation does not suffice for buffering the pH value. Hence, PAni stays intact in the case of iron or aluminium as substrate while it loses its conductivity due to de-protonation caused by high pH value in the case of zinc as substrate. They also observed that the work function measured on coated iron or aluminium substrates is determined by the work function of PAni, while the work function measured on coated zinc is more or less equivalent to the one of zinc itself.

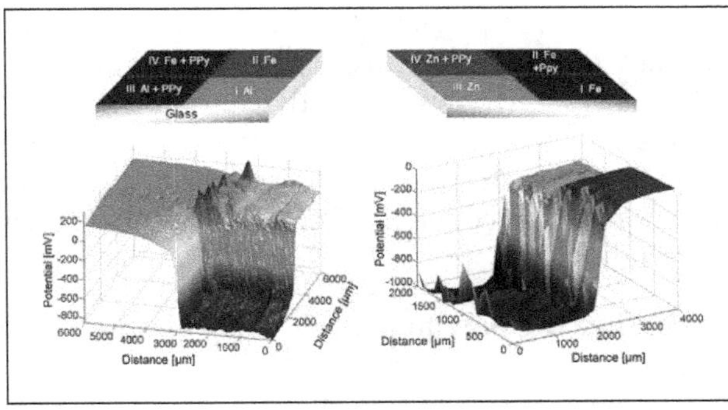

Figure 10.8: Potential maps obtained by SKP over an iron/aluminium couple and an iron/zinc couple, half covered by a layer of polypyrrole (Conquest) and half uncovered, as indicated in the drawings depicted above the corresponding potential maps. Source: Elsevier

Nazarov and Thierry [41] report a similar observation: polypyrrole (PPy) covered zinc in wet atmosphere showed a low work function (respectively potential; see for details about the correlation of potential and work function measured by Scanning Kelvin Probe [42]), however upon drying the potential went up to the level corresponding to PPy. They attribute this change to the formation of an ordered water layer at the buried interface that induces a dipole potential causing the decrease in the potential measured by Scanning Kelvin Probe. The observation presented in Figure 10.8 shows an Fe/Al couple and an Fe/Zn couple partly coated with a PPy film. The observation is that while on iron and aluminium more or less the original potential of the existing film is measured, on zinc the low potential of metallic zinc is obtained, although on the surface the existing film still seemed to be in its oxidised state, i.e. the work function on the surface of the film should be determined by non-reduced PPy.

Problems with anodic corrosion at the interface between metal and composite coating still exist. In principle the high potential of the conducting polymer should ensure that the metal in direct contact with it is kept reliably within its passive range. However, in a composite coating not all parts of the coated metal surface will be in direct contact with the conducting polymer, but just in its vicinity. Hence, over time the metal surface that is just close

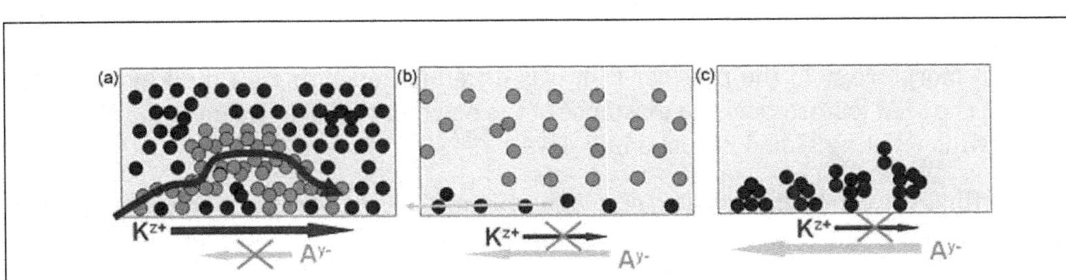

Figure 10.9: Schematic explanation of the different ion exchange behaviour of different conducting polymer based composite coatings in the delamination case. Left: composite coating with macroscopic percolation networks of conducting polymer. This causes cation permselective behaviour and hence prevents anion release. Middle: low density of conducting polymer. Only a few particles of the conducting polymer are in contact with the metal and are thus electrochemically active. No cation permselectivity is expected here and anion release may be possible, but only to the limited extent permitted by the small amount of active polymer. Right: if the conducting polymer forms microscopic networks, but not macroscopic ones, then cation permselectivity is still prevented, while allowing significant anion release. Source: Elsevier

to the conducting polymer, but not in direct contact with it, might form a weak galvanic element with the conducting polymer, due to water and ions diffused through the coating to the interface. The low conductivity at the interface then prevents the full polarisation to the potential of the conducting polymer and hence the metal may not be kept in its passive range. On the contrary it may be polarised towards increased corrosion activity. Different ion exchange behaviour of different conducting polymer based composite coatings is possible (Figure 10.9).

The **conducting polymers** could be synthesised from organic solution. The first attempts to electrodeposition of conducting polymers on active metals as substrates and investigations of their properties were undertaken in non-aqueous solutions. The most significant advantage of organic solvents in comparison to water is the better solubility of monomers, resulting in more concentrated solutions. Thus, formation of the polymer layer is much faster, and the anodic dissolution of the substrate does not proceed. In 1982, Skotheim and co-workers were among the first scientists to examine the formation of conducting polymer layers on active metals as substrates [43]. The purpose of their investigation was to explain how the substrate material influences the initiation of the polymer deposition. **Polypyrrole** (PPy) was chosen as an example of a conducting polymer. The polymer film was deposited from a monomer solution in acetonitrile with $BF^{4}-$ anion as a counter-ion. Electropolymerisation was performed on the substrates: Pt, Au, Pd, Ag, Ti, Cr, Ni, Al, In and Fe. These metals were deposited as a thin film (200 to 300 nm) on a silicon support. The PPy layer could not be grown on Al, In, Ag and Fe because the thin film of these metals dissolved before the polymerisation potential was reached. On other metals (Ni, Ti and Cr) an oxide layer was formed before the polymer deposition and blocked off the metal surface. The authors found that the polymerisation process generally proceeded similarly on different electrode materials; however, the PPy film structures were different when grown on various substrates.

The **properties of a PPy**-toluene sulphonate film electrodeposited on different electrodes have been investigated by Cheung et al [44]. As support were chosen: platinum, titanium, aluminium, mild steel and brass. Electropolymerisation was carried out under galvanostatic control from a solution of the pyrrole monomer in propylene carbonate. Characterisation of the polymer included elemental analysis, scanning electron microscopy (SEM), Raman spectroscopy and conductivity measurements. It was found that the potential of pyrrole oxidation increased when Ti, Fe and Al were used as substrate instead of Pt. The redox potential of the PPy-toluene sulphonate film depends on the substrate, while the electrical conductivities of the free-standing polymer films were independent of the substrate nature. However, morphology of the polymer film was different depending on the kind of metal support. The SEM features showed existence of the polymer region adjacent to the electrode surface which distinguished it from a bulk layer.

The PPy film has also been deposited onto Al and Ti surfaces from organic solvents by other researchers [45]. Chun et al. electrodeposited the PPy film on stainless steel from benzene solution in diethyl ether and studied the morphology, mechanical properties and conductivity of the film. They found that the PPy film was flexible and could be cut into any shape, so the film had properties very useful from a practical point of view. The influence of metal support (Pt, Ni, Ti and stainless steel) on the PPy electrochemical properties has been studied by Zalewska et al. [46]. The polymer coating was deposited from acetonitrilic as well as from aqueous solutions. It was found that the properties of the polymer film depend on the substrate material, especially when electropolymerisation was carried out onto a nickel electrode. An assumption about the formation of the PPy-Ni composite film has been formulated. The electropolymerisation of PPy on Zn-Pb-Ag (65 % Zn, 10 % Pb and 25 % Ag) alloys was studied in organic solvents (acetonitrile, nitrobenzene and propylene carbonate) [47]. It has been stated that the well adherent and homogeneous polymer layer can be obtained after pretreatment of the zinc alloy with aqueous solution of Na_2S. Bazzaoui et al. [48] synthesised well adherent PPy film on Zn from acetonitrile, nitrobenzene and propylene carbonate media in the presence of para-toluene sulphonate counter-ions.

The **formation of a PPy layer** on iron surface from 1-butyl-3-methyl imidazolium has been studied by Fenelon and Breslin [49]. Adherent and highly conducting PPy films have been obtained. Electropolymerisation of acetylene with copper catalyst on Pt and Cu from acetonitrile solution was performed [50]. The polyindole film deposited on Fe from acetonitrilic-water solution containing tetrabutylammonium tetrafluoro-borate has been examined by Sazou [51]. It has been found that increasing concentration of H_2O in the electrolyte leads to increasing concentration of Fe^{3+} ions in solution due to dissolution of the iron electrode, thereby inducing a mechanical instability of the polymer film.

Electrodeposition of polymer layers from aqueous solutions is also one of the possibilities. However the organic solvents have many disadvantages in comparison to water. They are:

- toxic,
- highly flammable and
- more expensive,

so for practical purposes aqueous solutions are preferred. Thus, electrodeposition of conducting polymers onto active metal surfaces from water solutions has become a challenge. PPy has been chosen as a subject of a great number of studies due to its relatively easy preparation from aqueous solutions. Schirmeisen and Beck [52] studied the formation of a PPy layer on inert (Pt, Au and glassy carbon) and active (Cu, Ti, Fe and stainless steel) supports in aqueous and non-aqueous solutions (acetonitrile, methanol) in the presence of various inorganic ions. They stated that electrodeposition of the PPy coating proceeds in aqueous solutions only when substrates do not dissolve simultaneously with the polymerisation process. Thus, it is very important to select anions which promote the passivation of the metal surface in contact with the electrolyte [53]. In the case of iron, nitrate ion is such an anion. Excellent and well adhering PPy coatings have been obtained on the Fe surface in KNO_3 solution. Husler and Beck studied the process of electropolymerisation of PPy on Al and Ti surfaces [54]. Formation of the strongly adhering PPy film on Al from an aqueous solution is reported [55]. The polymer film was deposited on the aluminium surface after its passivation in contact with diluted acids: oxalic, nitric and sulphuric. Lacase et al. obtained well adhering PPy coatings on Al and other active metals (Zn, Cu, Ni and Fe) from aqueous or aqueous-alcoholic solutions containing salicylate ions without any pretreatment of the surface [56]. The kinetics of the pyrrole electropolymerisation and morphology of the PPy layer on stainless steel was studied in an aqueous medium by Su and Iroh [57]. Apart from the influence of the current density, temperature and pyrrole concentration on the process of PPy polymerisation the influence of the concentration of the counter-ion benzene sulphonate has also been shown. It was established that smoother and more compact PPy coatings were formed at lower electrolyte concentration, while the deposit became less dense and porous when the concentration was higher. When polymerisation was carried out in the presence of oxalate, anion formation of a thin layer of $FeC_2O_4{\cdot}2H_2O$ on the steel surface was shown. This layer caused passivation of steel, which allowed a deposition of the PPy film.

A **single-step electropolymerisation** of pyrrole on Fe and Steel has been studied [58] in presence of malate as counter-ion in the electrolyte. The malate counter-ion slows down the iron dissolution which allows to obtain uniform and well adherent PPy coatings. The authors found that a PPy layer formed galvanostatically had a cauliflower-like structure and was thicker than a film synthesised in potentiodynamic mode. Formation of the PPy coating on steel coated with various types of zinc alloys was examined in the presence of oxalate in aqueous solution [59]. Satisfactory PPy coatings were obtained on galvanised steel only after passivation of the electrode surface in a medium-acidic solution containing high pyrrole concentration. The possibility of the PPy film formation on Zn in aqueous solutions was also examined [60]. The authors found that the polymer layer could be deposited on Zn only when the zinc surface was previously passivated by treatment in 0.2 M Na_2S solution.

The mechanism of the **two-step electropolymerisation** process and the influence of the current density have been discussed. Later the one-step electrosynthesis of PPy on Zn and mild steel from aqueous solution containing salicylate was proposed [61]. The authors found that in the presence of salicylate anions a passivation of the substrate surface took place, allowing to obtain strongly adherent PPy films with current efficiencies close to 100 %.

One important issue in electropolymerisation of conducting polymers on active metal surfaces is to obtain good adhesion of the polymer film to the metal. This problem has been carefully studied [62] for deposition of PPy on titanium. The authors employed various methods of chemical pretreatment of the titanium surface and made a recommendation how to obtain an extremely adhesive PPy film on the Ti surface. Aluminium and steel (or iron), as metals widely employed in practice, are the most often examined substrates for electrodeposition of conducting polymer coatings. Usually the polymer film is deposited on Al after passivation of its surface by aqueous solutions of acids: oxalic, nitric or sulphuric. Deposition from an aqueous oxalic acid solution of the PPy layer on the Al surface and modification of the PPy film by cementation of copper were performed [63]. The high activity of the PPy layer on Al in the presence of NO^{3-} ions in alkaline solutions is connected with the fact that this anion limits anodic dissolution of Al [64]. Formation of the combined Al_2O_3/PPy layer in aqueous solution containing sulphonate-based anions was examined [65]. Earlier it has been established that, the PPy film has a tendency to overoxidise when electropolymerisation is accompanied by growth of thickness of the Al_2O_3 layer, so that the conductivity of the polymer dramatically decreases. To avoid this effect, it has been proposed to use Tiron (4,5-dihydroxy-1,3-benzenedisulphonic acid di sodium salt) as an electron transfer mediator in order to catalyse formation of the PPy layer by reduction of the polymer deposition potential by nearly 500 mV [66]. The authors noted that the polymer film obtained on the surface of AA 2024 T3 had high conductivity and good adhesion to the aluminium alloy in the presence of Tiron. The authors suggest that salicylate may also act as an electron mediator in the formation of the PPy film.

PPy coatings have also been deposited on zinc and copper surfaces [67]. Prissanaroon et al. obtained well adherent PPy coatings on an electropolished Cu electrode from aqueous dodecylbenzene sulphonic acid solution. The authors showed that a thin Cu_2O layer was formed during transfer processes allowing a PPy coating to be formed. Electropolishing of Cu leads to partial passivation of the Cu surface and prevents copper dissolution while the electropolymerisation process takes place without hindrance. The redox behaviour of PPy layers on Cu was similar to the observed when PPy coatings formed on noble metals. The synthesis of PPy on copper surfaces could also be carried out aqueous electrolyte containing sodium salicylate. The PPy coatings were uniform and well adherent to Cu surface. Formation of the thin intermediate layer was observed during oxidation of copper in sodium salicylate solution. The existence of that layer allows to avoid copper dissolution. Processes of PPy electropolymerisation on a steel substrate from aqueous solutions of oxalic acid have also been studied.

Furthermore, the process of PAni layer formation on active metals has been often investigated. Abalayeva and co-workers studied the formation of the **polyaniline film** on Ti, Ta, Pb, Ni, Ti and Al from aqueous solutions containing 0.1 M H_2SO_4 [68]. In these solutions, the oxide film on Ti, Ta and Pb (formed during contact of the metals with air) remained untouched, but the aluminium oxide dissolved, thereby necessitating a special procedure of passivation before the electropolymerisation. Such a procedure is recommended in ref. [69] and as a result PAni layer on the Al surface could be grown. The electrochemical properties of the PAni film in dependence of the way of electropolymerisation were examined, too. Formation of the PAni film on the iron surface from an aqueous solution has been investigated. The electrosynthesis and the degradation of PAni films on iron from oxalic acid electrolyte have been studied. The characteristics of PAni coatings prepared on stainless steel electrodes by various methods galvanostatic, potentiostatic and potentiodynamic have been given [70]. Electropolymerisation of aniline on stainless steel, Mg, Ni, Ti, Al and Pb has been studied in aqueous solutions of Na_2ClO_4, $H_2C_2O_4$ and H_2SO_4 [71].

Some **other polymer coatings** or their combination with PPy and PAni or doping with conducting polymers were examined on active metal surfaces. In literature [72] the polymers employed as anticorrosive additives are PAni emeraldine salt and base, a PAni emeraldine salt composite with carbon black, a PPy composite with carbon black and poly(3,4-ethylenedioxythiophene) doped with poly(styrene sulphonate). Initially, the structural, thermal and mechanical properties of the unmodified epoxy paint as well as the modified coatings, obtained by the addition of 0.3 % of conducting polymers, have been characterised. After this, controlled accelerated corrosion assays in an aggressive medium were developed using coated steel panels. Results indicate that the protection against corrosion imparted by the formulations modified by the addition of poly(3,4-ethylenedioxythiophene) doped with poly(styrene sulphonate), PAni emeraldine salt and, especially, PAni emeraldine base is significantly higher than that of unmodified paint. In contrast, the use of a conducting polymer composite with carbon black reduces the efficacy of the coating. Results indicate that some conducting polymer compositions should be considered as a suitable alternative to replace inorganic anticorrosive pigments currently used in paint formulations. Pisarevskaya et al. studied electropolymerisation of the polyparaphenylene films onto the surface of Au, Pt, glassy carbon, Ni and Ti. Later, Pisarevskaya et al. applied other metals like Cr, Nb, Ni-Ru, Cu and Ru for the electropolymerisation of the polyparaphenylene from a solution containing sulphuric acid and benzene as an emulsion [73]. The polymer films were obtained by the cyclo-voltammetry method, and the main purpose of this work was to elucidate the influence of substrate morphology on polymer film behaviour. It has been found that the polymer film morphology, as well as its adhesion, strongly depends on the substrate morphology. The electrochemical and other properties of poly(3-methylthiophene) coating on type 430 stainless steel have been examined by ref. [74]. The effects of the polymer preparation conditions on the behaviour of poly(3-methylthiophene) films have been studied. Smooth and well adhering coatings were obtained in oxalic acid solutions on an iron oxide passive film. The electropolymerisation of 3-methylthiophene on the Mo, Ni and Fe surfaces from a lithium perchlorate solution was also studied [75]. The authors show that cations of the metal-support electrode may be incorporated into a polymer matrix, and in result, they have an impact on the electroactivity of the polymer film. Influence of electrosynthesis conditions (temperature and current density) on the morphology of poly(o-anisidine) coatings deposited from an aqueous solution on low carbon steel was discussed in ref. [76]. Recently, electrochemical synthesis of poly(N-methylpyrrole) films on copper electrodes from diluted oxalic acid has been achieved [77]. The corrosion inhibiting properties of this coating on copper were investigated in aqueous 0.1 M NaCl solution using potentiodynamic polarisation. They were observed to be comparable to those of PPy films. A physical barrier effect is the most likely protection mechanism. Electrosynthesis of the poly(N-methylpyrrole) on steel surfaces was studied by Su and Iroh and the existence of the interphase Fe(II)-oxalic thin layer has also been established. The growth rate of poly(o-phenylenediamine) (o-PD) coatings on type 304

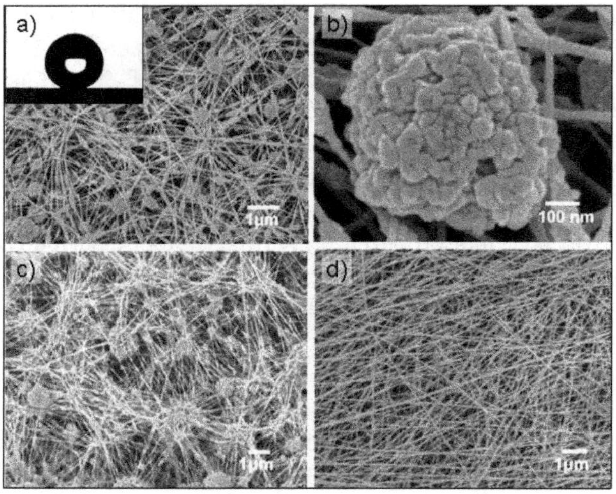

Figure 10.10: (a) SEM image of an electrospun PAni/ polystyrene (PS) composite film with lotus-leaf-like structure prepared from a 3.72 wt.-% PS:ABSA/DMF solution (inset: water droplet); (b) Magnified view of a single sub-microsphere from c,d) SEM images of PAni/PS composite films prepared from solutions with different concentrations of PS [wt.-%]: (c) 5.26 wt.-%; (d) 7.10 wt.-%.

Source: American Chemical Society

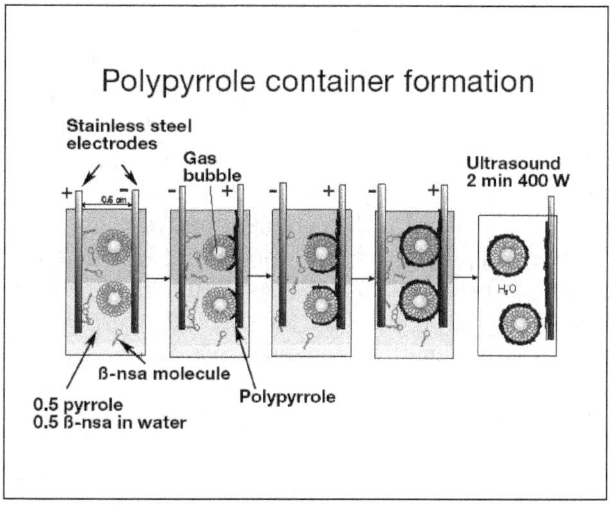

Figure 10.11 : Scheme of polypyrrole container formation: (a) gas bubble formation, (b-d) polypyrrole film growth, and (e) separation of polypyrrole containers by ultrasound. The redox reaction may lead to a PPy shell on the bubbles. Varying O_2 content, surfactants, and parameters of the redox scan, one can obtain and manipulate PPy containers and electrodes of different roughness.

Source: American Chemical Society

stainless steel in diluted phosphoric acid and the structure of the coating were investigated, too [78]. It was found that the growth rate of the o-PD film on stainless steel is much lower than the same process on a Pt electrode, and under potentiostatic conditions the polymer film was obtained only after prolonged polarisation time. The electroformation of a poly (3,4-ethylenedioxythiophene) layer on a Ni electrode from an aqueous solution of $LiClO_4$ and sodium dodecylsulphate has been carried out and the mechanism of the PEDT film growth is analysed [79]. A poly (2,5-dimethoxyaniline) coating on low carbon steel has been grown and examined [80]. Soluble conducting poly(aniline-co-ortho-toluidine) copolymer was used as corrosion inhibitor for carbon steel [81]. These results show an increase in inhibition efficiency as a function of copolymer concentration to attain an optimum (70 %) for a concentration close to 100 ppm. This is due to the higher coverage of the carbon steel electrode surface as shown by the Temkin adsorption isotherm. The **potentiodynamic polarisation** studies reveal that this copolymer acts as a mixed-type inhibitor. As mentioned above the conducting polymer coatings may be used as blends with conventional coatings (non-conducting polymers, paints) or as primer coatings. Electrosynthesis of **polybithiophene** and its bilayers with PAni coatings on stainless steel in diluted oxalic acid containing monomer and sodium dodecyl sulfate was carried out by the potentiodynamic synthesis technique [82]. Smooth and adherent films were obtained on the steel surfaces. Tests for corrosion protection of polymer coated and uncoated substrates were performed in highly aggressive media, revealing that among the protective coatings obtained, polybithiophene as homopolymer and

PAni/polybithiophene as bilayer exhibited the most effective anticorrosive properties. Some interesting results were obtained in the area of designing stable, superhydrophobic, and conductive PAni/polystyrene films for corrosive environments (Figure 10.10) [83].

A PAni/polystyrene composite film with a lotus-leaf-like structure is prepared via a simple electrospinning method. The film shows stable superhydrophobicity and conductivity, even in many corrosive solutions, such as acidic or basic solutions over a wide pH range, and also in oxidising solutions. The special surface composition and morphology are the two important aspects that induce such unusual properties. The polystyrene content can strongly influence the morphology of composite films, which thus display different superhydrophobicities and conductivities.

How corrosion protection by conducting polymers might work is actually subject of intense research [84]. The redox activity offered by conducting polymers together with positive equilibrium potentials relative to those of Fe and Al, as is the potential of chromate, suggest that the mechanism is likely to involve anodic protection. The most popular mechanism is the so-called "ennobling" that is based on the assumption that a conducting polymer could provide anodic galvanic protection keeping the metal passive in small defects, or may act as oxidiser, improving the oxide layer at the conducting polymer/metal interface. The possible reaction between the oxidised form of the polymer (conducting polymer $(CP)^{m+}$) and the metal (Me) is:

$$1/n\ Me + 1/m\ CP^{m+} + y/n\ H_2O \rightarrow 1/n\ Me_y^{(n-y)+} + 1/m\ CP^0 + y/n\ H^+$$

the conducting polymer being oxidised by atmospheric or dissolved oxygen:

$$m/4\ O_2 + m/2\ H_2O + CP^0 \rightarrow CP^{m+} + mOH^-$$

In addition, the electrons produced during metal oxidation at the defect can move into the polymer and thus dislocate the oxygen reduction process from the metal/CP interface to the CP/solution interface. This alleviates the local pH increase at the former and in turn the cathodic disbonding of the coating. All of the processes mentioned above may simultaneously contribute to substrate protection. It was also suggested in our group that conducting polymers scan be used as a reservoir of corrosion inhibitors that could be released as a result of the reduction of the conducting polymer [85]. PPy microcontainers were successfully prepared by electrochemical polymerisation of pyrrole on the surface of stainless steel electrodes (Figure 10.11). The size of the nano-containers and thickness of the PPy shell (up to complete filling of the container with grown polypyrrole) can be varied by changing both the scan speed of the electrode potential and potential range. The PPy shell of the microcontainers exhibits very strong barrier properties in acidic media at $2 < pH < 7$ and a high permeability at $pH > 7$, providing effective encapsulation of low molecular weight species at low pH values.

10.1.3 Biopolymers

Biological polymers (biopolymers) offer a degree of functionality not available in synthetic conducting polymers. Carbohydrate polymers (polysaccharides) are very frequently produced in nature [86]. Starch, cellulose, and chitin are some of the most abundant natural polymers on earth. Biopolymers are a renewable resource and have a wide range of uses, functioning as energy storage, transport, signaling, and structural components. Agriculture-based biopolymers could have an economic advantage over synthetic, petroleum-based products in the future.

Starch is composed of a mixture of **linear and branched polysaccharides** (Figure 10.12a). Amylose is a linear polymer of α(1,4)-linked anhydroglucose units. Amylopectin is a highly branched polysaccharide composed of linear α(1,4)-anhydroglucose unit chains with branch-

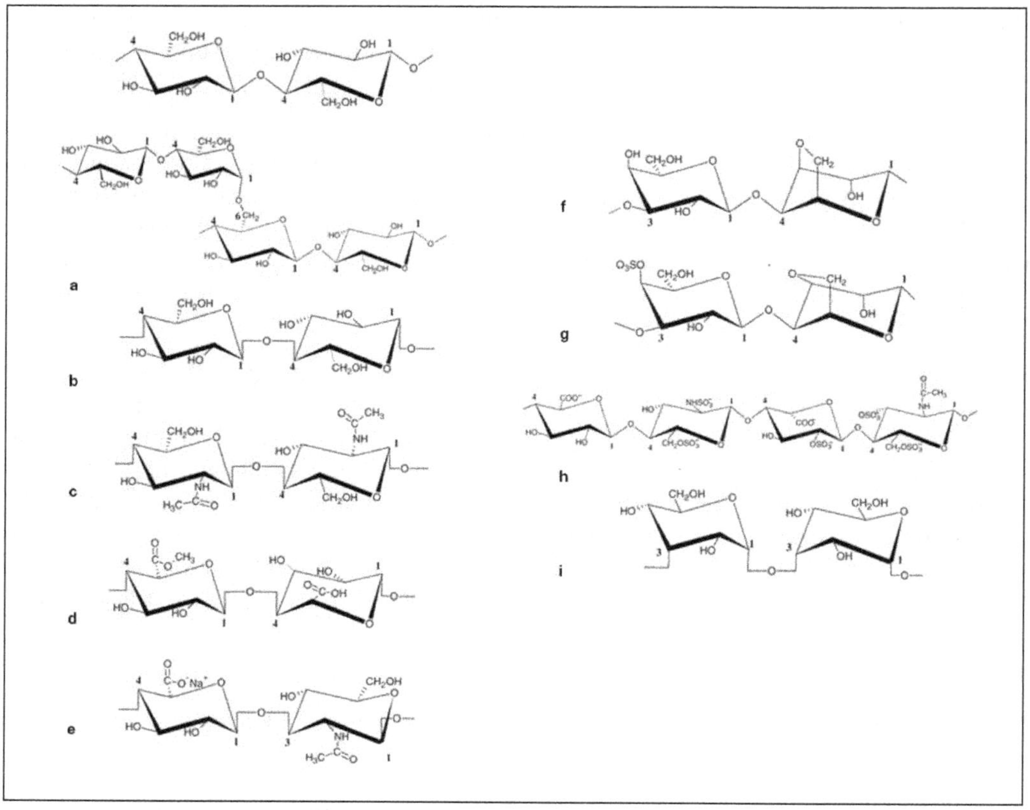

Figure 10.12: structures of polysaccharides: (a) amylose and amylopectin, (b) cellulose, (c) chitin, (d) pectin, (e) hyaluronic acid, (f) agarose, (g) carrageenan, (h) heparin, (i) pullulan

points being α(1,6)-linkages. While starch is packed into granules in its native state, the intrinsic properties of starch materials are exhibited when the granular structure is broken down by mechanical and thermal means. Thermoplastic starch, produced by reactive extrusion, can be doped with metal halides [87]. The amount of moisture in the system affects both the mobility of polymer and ion within the system and the quality of the physical properties of thermoplastic starch. Starch is naturally insulating and its proton mobility is low [88].

Cellulose is a β(1,4)-linked D-glucan (Figure 10.12b). Cellulose xanthate was dissolved in water, mixed with polyethylene glycol, and precipitated with ethanol [89]. The electrical conductance of the polyethylene glycol-plasticised cellulose xanthate depended on the presence of free ions, since the electron conduction was essentially zero. Hydroxyethyl cellulose was blended with poly(ethylene oxide) to form good films, which is essential for some applications [90]. Hydroxyethyl cellulose was also grafted with propylene oxide and then blended with poly(ethylene oxide) (Figure 10.12c) to form crosslinked polyether networks. Cellulose serves as a flexible substrate for coating initial conducting polymers [91].

Chitosan is the N-deacetylated form of chitin with the repeat unit β(1,4)-linked 2-acetoamido-2-deoxy-D-glucose (Figure 10.12c). Because of its functional groups, chitosan has many potential applications. In recent work, chitosan was plasticised [92] and doped with sodium perchlorate ($NaClO_4$) [93], sodium iodide (NaI), lithium trifluoromethanesulphonate ($LiCF_3SO_3$) [94], and lithium acetate ($LiCH_3CO_2$) [95]. The crystallinity of the material increased with deacetylation, preventing mobile species from migrating, due to rigid polymer chains. Such chitosan films could serve as membranes for alkaline polymer electrolyte batteries and fuel cells as well as in corrosion protective films. Pore resistance is a convenient way to represent corrosion protection. Pore resistance indicates the level of ionic conductivity from the electrolyte penetration into the coating, i.e. as pore resistance increases, ionic conductivity decreases. In the case of corrosion protection, penetration of corrosives (salts, moisture, etc.) is undesirable, so the conductivity imparted by ionic species (a measurement of the ionic mobility) within the matrix should be low. Recently the group of Zhitomirsky [96] has made progress in constructing effective chitosan composite coatings. They use the electrophoretic deposition method for the fabrication of chitosan carbon **nanotube-hydroxyapatite** and hydroxyapatite-$CaSiO_3$-chitosan composite coatings (Figure 10.13).

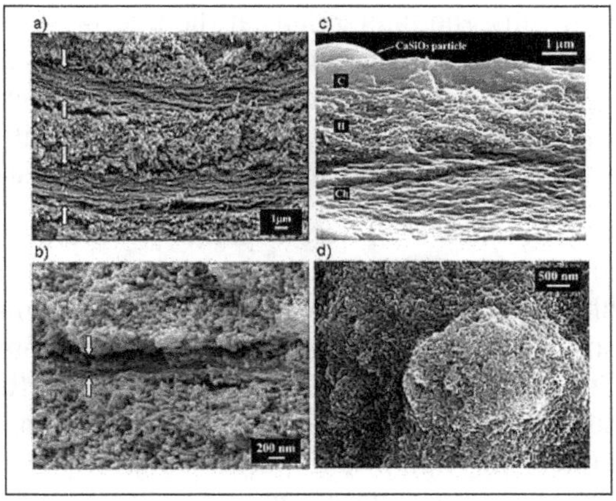

Figure 10.13 : (a, b) SEM images of laminates of chitosan-hydroxyapatite layers, prepared from 0.5 g/l chitosan solutions containing 1 g/l hydroxyapatite, and chitosan-multiwalled carbon nanotubes layers prepared from 0.5 g/l chitosan solutions containing 0.1 g/l multiwalled carbon nanotubes, arrows show chitosan-multiwalled carbon nanotubes layers, deposition voltage 20 V. (c) SEM image of the cross-section of a CaSiO3-chitosan(C)/hydroxyapatite-chitosan(H)/chitosan(Ch) three-layer coating on a graphite substrate, prepared from the 0.5 g/l chitosan suspensions containing 0.3 g/l $CaSiO_3$ (C) or 1 g/l hydroxyapatite (H); (d) SEM top view of coating (c) on stainless steel substrate. Source: Elsevier

The use of chitosan enabled the co-deposition of **hydroxyapatite and carbon nanotubes** or $CaSiO_3$ particles and offered the advantage of room temperature processing of composite materials. The thickness of the individual layers was varied in the range of 0.1 to 20 µm. Composite chitosan multi-walled carbon nanotube layers were in the range of 0.5 to 10 µm which was controlled by variation of the deposition voltage in the range of 20 to 50 V and deposition time in the range of 1 to 10 min. It was shown that biopolymer multi-walled carbon nanotube layers for hydroxyapatite films can be prepared as monolayer nanocomposites containing multi-walled carbon nanotube layers and hydroxyapatite in a biopolymer matrix or laminates, containing **biopolymer multi-walled carbon nanotube layers** separated by biopolymer-hydroxyapatite layers. The thickness of individual layers can be varied in the range of 0.2 to 5 µm by variation in deposition time. Obtained films provided corrosion protection of NiTi shape memory alloys in Ringer's physiological solution. The use of poly(itaconic acid)-modified chitosan coatings for mitigating corrosion of aluminium substrates was reported [97]. The synthetic poly(itaconic acid)-polymer was added to graft it onto the liner chitosan chains and to crosslink between the chitosan chains. The grafting and cross-linking reactions were generated by a strong chemical affinity between NH_2 groups in chitosan and COOH groups in poly(itaconic acid), leading to the formation of amide bonds. When this polymer was applied as environmentally benign waterbased coating to an aluminium substrate by a simple dip-withdrawing method, the following three factors played an essential role in mitigating the rate of corrosion of the Al:

- firstly, the polymer conformation, containing hydrophobic amide bonds, together with a minimal amount of hydrophilic unreacted COOH in poly(itaconic acid) and NH_2 in chitosan;
- secondly, the lesser susceptibility of the coating films to moisture due to the enhanced degree of grafting and cross-linking;

- thirdly, the interactions at the interfaces between the grafted chitosan and the Al substrate, developing the formation of the -COO-Al linkage. The poly(itaconic acid) -grafted and -cross-linked chitosan polymer coating films generated from the proper proportions of poly(itaconic acid) to chitosan included all these factors and displayed low ionic conductivity, thereby imparting a salt-spray resistance of 694 h for coated Al panels.

Pectin and starch were separately grafted with polyorganosiloxane and blended with an antimicrobial agent [98]. Pectin is a polygalacturonic acid methyl ester with varying degrees of methylation (Figure 10.12d). The copolymer mixture was then applied to aluminium substrates and evaluated for resistance to water, microbial attack, and corrosion protection. Pectin polyorganosiloxane composite was thermally more stable than pectin, was less susceptible to moisture due to a loss of hydrophilic groups, and imparted greater corrosion protection. Aluminium coated with pectin alone had a slightly higher pore resistance than the bare aluminium surface. However, aluminium coated with pectin polyorganosiloxane or starch polyorganosiloxane showed an increase of at least one order of magnitude over bare aluminium. In addition, the incorporation of an antimicrobial agent into the graft polymer played a role in the overall susceptibility of the coating to degradation by water or bacteria. There is higher corrosion protection due to both favorable interactions of polyorganosiloxane with aluminium oxide layers on the metal surface and the reduced hydrophilic nature of the polysaccharide matrix.

Collier et al. [99] coupled the biocompatibility of **hyaluronic acid** (Figure 10.12e) with the electroactive properties of PPy for corrosion protection (see Chapter 10.1.2). Hyaluronic acid is a negatively charged glycosaminoglycan and was chosen for its biocompatibility and known activity in wound healing and tissue regeneration. Hyaluronic acid was blended with PPy in a single thin layer **(PPy hyaluronic acid)** or deposited on a layer of PPy-poly(styrenesulphonate). The PPy hyaluronic acid/PPy poly(styrenesulphonate) laminate exceeded the biological responses (such as morphology, hyaluronic acid activity, cell compatibility, nontoxicity, increased vascularisation) of the control sample of PPy-poly(styrenesulphonate).

Heparin was incorporated into PPy [100]. Heparin is a linear polysaccharide of glucuronic acid and glucosamine containing sulfates (Figure 10.12h). Heparin is widely known as an anticoagulant and has specific affinity for thrombin. It was shown that polyelectrolytes (such as heparin) can be incorporated in initial conducting polymers during synthesis. Cyclic voltammetry indicated that the **heparin-PPy** was electroactive upon oxidation/reduction in a reversible process. The primary mode of conduction was cation exchange, with the cations moving in and out of the heparin-PPy matrix to maintain charge balances during redox reactions. Cation conduction could be attributed to the large anion (i.e. heparin) immobilised in the PPy matrix. The cation conductivity of heparin-PPy was significantly higher than that of doped PPy. The heparin retained its biological activity on the surface of the heparin-PPy blend. In a separate work, heparin and dextran sulphate were blended with PPy to create hydrophilic polyelectrolytes that were able to store and release proteins [101]. This process could be controlled by varying the potential or the charge.

While currently most biopolymers have not achieved the level of conductance of some synthetic conducting polymers, it is only a matter of time for them to be be engineered and developed. The information presented here demonstrates the viability of using naturally occurring **polysaccharides** for biobased conducting polymers in applications such as anticorrosion technology. Other areas could also be affected: batteries, tissue engineering, actuators, electrochromic paper, molecular wires, and biosensors. There are major opportunities for the development of biotechnology using integrated biological, chemical, and engineering approaches, with bioreactors, fermentations, and organisms specifically designed to produce electroactive biopolymers. Electroactive biopolymers have better biocompatibility and they are more environmentally friendly than synthetic conducting polymers. Sustainability could be achieved by tailored biosynthesis of conductive biopolymers, using plants and microbes. The use of renewable resources to produce nontraditional biobased products will affect the worldwide dependence on petroleum-based and synthetic products and feed stocks.

10.2 Nanomaterials

New modern technologies require new materials. During the past decade, the movement towards nanodimensions in many areas of technology raised a huge interest in nano-structured materials. Many new corrosion protection treatments based on nano-structured materials were developed over the last decade. These include the synthesis of different **nanocomposites** and their individual use, or, for example, in combination with electrically conducting polymers, nanocrystalline deposits, nanowires, nanotubes, polyelectrolytes or polymer nanolayers. The addition of nanomaterials, even in small quantities, into conventional coatings improves the properties of the coatings and, whatis more important, makes these coatings multi-functional thereby, increasing their durability. Despite the fact that nanocomposite coatings are already employed in maritime, automobile and aerospace industries, the main drawback is still the costs of the nanomaterials, especially with functionality, added to the coating. Further sustained development of coating components will provide efficient solutions in the future for new generations of coatings where the "passive" component will be added to the "active" one.

10.2.1 Nanocomposites

Layered materials such as anionic clays (e.g., layered double hydroxides) [102] and cationic clays (e.g., montmorillonite) [103] have been widely investigated as additives in organic anti-corrosion coatings or as **polymer-clay** nanocomposite corrosion-resistant coatings. Moreover, zeolites [104] have been explored as corrosion-resistant coating materials. Hydrophobic layers [105] of surfactant molecules on the surface have recently been proposed as corrosion inhibitors but suffer from the drawbacks that the layers have limited stability and molecule-sized defects which allow water to reach the underlying surface. These problems should be mitigated if the surfactant could be incorporated in an inorganic host matrix, a thin film of which has been previously strongly bonded to a metal, for example aluminium, surface.

Layered double hydroxides (LDHs) are one such potential inorganic host. They can be expressed by the general formula

$$[M^{2+}_{1-x}M^{3+}_{x}(OH)_2]A^{n-}_{x/n} \cdot mH_2O,$$

where the cations $M^{2+}(Mg^{2+}, Zn^{2+}, Fe^{2+}, Co^{2+}, Cu^{2+}$, and others) and

M^{3+} ($Al^{3+}, Cr^{3+}, Fe^{3+}, Ga^{3+}$, and others)

occupy the octahedral holes in a brucite-like layer and the anion An- is located in the hydrated **interlayer galleries** [106]. The ability to vary the composition over a wide range allows materials with a wide variety of properties of the preparation. The LDHs have been applied as catalysts and catalyst supports [107], polymer stabilisers [108], and traps for anionic pollutants [109]. In addition, they have been considered for the delivery of drugs at cellular level, with several studies showing their low toxicity and high biocompatibility, together with the ability to control release of active species under certain circumstances [110]. Therefore, the nano-structured systems used for drug-delivery purposes are prospective systems for the incorporation/encapsulation of corrosion inhibitors.

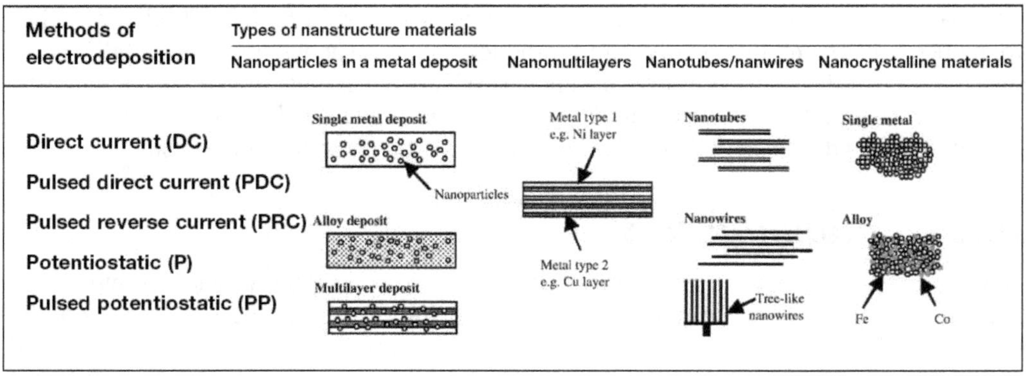

Figure 10.14 : Various types of nanostructured materials which may be produced by electrodeposition technique. *Source: Elsevier*

Application of **hydrotalcite-like compounds** in corrosion science has covered different aspects. In some studies, LDHs have been produced in-situ, on the top of metallic substrates as protective films [111].

Hydrotalcite-based conversion films have demonstrated good corrosion protection, and some research groups have been trying to improve the interaction between these conversion films and organic coatings [112]. A different perspective is to use these anionic clays as containers for corrosion inhibitors and incorporate them into the organic coatings. In this case, the target is two-fold, i.e. not only to release the species that impart active protection but also to trap corrosive agents (Cl^-, SO_4^{2-}). Regarding this topic, several works by Buchheit et al. [113], Williams and McMurray [114] and others [115] can be found in the literature. Within the LDH family, a class of materials with emerging importance is that constituted by the LDHs loaded with organic anions [116]. Two short studies on these materials are available in the literature. Williams and McMurray prepared LDHs with different organic species (benzotriazole, ethyl xanthate, and oxalate) by re-hydration of commercial hydro-talcite [$Mg_6Al_2(OH)_{16}\cdot CO_3\cdot 4H_2O$]. The resulting layered systems were inserted into a poly(vinylbutyral) coating, prepared by a bar cast on top of AA 2024-T3. In another report, Kendig and Hon [117] prepared LDHs intercalated with 2,5-dimercapto-1,3,4-thiadiazolate in a similar way and studied the inhibiting properties of this anion with respect to the oxygen reduction on copper. Another recently developed possibility [118] is the use of Zn-Al and Mg-Al LDHs loaded with quinaldate and 2-mercaptobenzothiazolate whose anions were synthesised via anion-exchange reaction. Spectrophotometric measurements demonstrated that the release of organic anions from these LDHs into the bulk solution is triggered by the presence of chloride anions, evidencing the anion-exchange nature of this process. The anticorrosion capabilities of LDHs loaded with organic inhibitors toward the AA2024 aluminium alloy were analysed by electrochemical impedance spectroscopy. A significant reduction of the corrosion rate was observed when the LDH nanopigments are present in the corrosive media. The mechanism by which the inhibiting anions can be released from the LDHs underlines the versatility of these environmentally friendly structures and their potential application as **nanocontainers** in self-healing coatings (see Chapter 10.3.2).

Simultaneously, conductive polymers discussed in Chapter 10.1.2 could serve both ways, as blocks in nanocomposite construction i.e. as **nanoparticular dispersions** (for example PAni, PPy [119]) in paints at low concentrations to enhance anticorrosion properties; or as matrix for the incorporation of inorganic materials, for example LDHs (montmorillonite clay, Fe_3O_4, ZnO) [120]. In all cases model coatings showed high corrosion protection in laboratory scale for aluminium and steel.

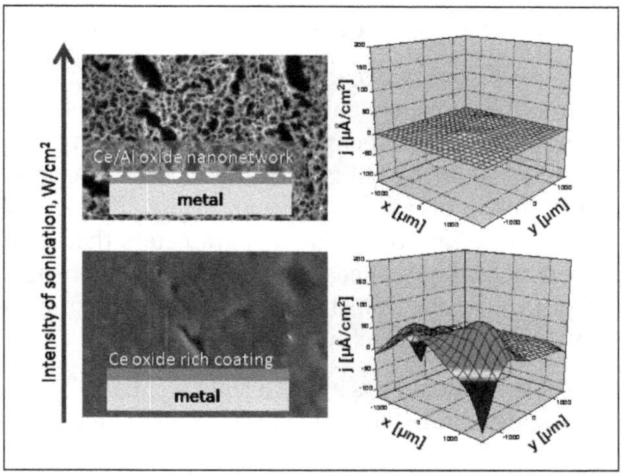

Figure 10.15 : SEM and SVET study of the aluminium alloy AA2024 covered by a cerium/aluminium nanonetwork (up) and a dense cerium oxide layer (down)

Moreover, recent literature on the electrodeposition of metallic coatings containing nanosized particles was surveyed. Nanosized particles, suspended in electrolyte by agitation and/or use of surfactants, can be codeposited with metal. The inclusion of nanosized particles can give

- an increased microhardness and corrosion resistance,
- modified growth to form a nanocrystalline metal deposit and
- a shift in the reduction potential of a metal ion.

Many operating parameters influence the quantity of incorporated particles, including current density, bath agitation (or movement of work piece) and electrolyte composition. High incorporation rates of the dispersed particles have been achieved using

- a high nanoparticle concentration in the electrolyte solution,
- smaller sized nanoparticles;
- a low concentration of electroactive species,
- ultrasonication during deposition and
- pulsed current techniques.

Compositional gradient coatings are possible, with a controlled distribution of particles in the metal deposit, and the theoretical models used to describe the phenomenon of particle co-deposition within a metal deposit are critically considered [121]. Figure 10.14 shows the categories of nano-structured materials along with various electrodeposition manufacturing techniques. Readers are directed to a number of earlier reviews [122] for a detailed background to composite electrodeposition of larger particles which typically have sizes in the range of 1 to 80 µm.

Formation of **cerium/aluminium oxide anticorrosion** protective layers on metal surfaces has been done by ultrasonic irradiation of high intensity [123]. The mechanism of metal surface modification in the presence of cerium(III) aqueous solution results from two factors, i.e. activation of the metal surface by localised heating and activation of cerium ions which are diffused within liquid jets shot at the metal surface with high velocity. The ultrasonically increased reactivity of cerium and the engineered metal surface resulted in a novel type of cerium enriched protective layer, a nanonetwork, where cerium and aluminium oxides are interlaced in a mixed layer strongly connected to the metal surface. The anticorrosion activity of the novel cerium/aluminium oxide system was demonstrated by using the scanning vibrating electrode technique (Figure 10.15). At high intensity of sonication (30 min at 57 W/cm^2) a 20 nm cerium/aluminium nanonetwork deposited on the engineered metal surface can be formed. The cerium/aluminium nanonetwork exhibits very good adhesion to both metal surface and components of the anticorrosion system (here, silica/zirconia based sol-gel film). Corrosion resistance of the metal surface covered by the cerium/aluminium nanonetwork is significantly enhanced. Release of cerium ions incorporated into the interfacial layer is observed after mechanical rupture (cracks) of the protective coating.

10.2.2 *Sol-gel derived ceramic and hybrid coatings*

One of the prospective candidates for the substitution of chromate pretreatments are sol-gel derived thin films. Sol-gel films have a good adhesion to both metallic substrates and organic top coats. Sol-gel technology can offer various ways to prepare functional coatings with different properties. Sol-gel coatings can be prepared using hydrolytic and non-hydrolytic procedures in organic or aqueous media [124]. The preparation of sol-gel materials can be controlled by variation of many parameters that influence the processes, and in turn, the properties of the final coatings. The sol-gel procedure is a method that allows the deposition of a thin oxide film at temperatures near to room temperature in contrast to traditional ceramic processing technologies.

The **sol-gel process** has an excellent potential to be used for the preparation of **inorganic or organically modified** protective coatings. Sol-gel processes are mainly based on hydrolysis and condensation reactions of metal alkoxides (M(OR)n). The resulting oxide materials present structures varying in range from nanoparticulate sols to continuous polymer gels depending on the rate of each of the forming reactions and subsequent drying and processing steps. The close control of parameters of sol-gel reactions leads to the design of new advanced materials with interesting properties for many applications. Materials with different compositions can be easily obtained in the form of glasses, fibres, ceramic powders and thin films. Silicon-organic based sol-gel chemistry is more developed nowadays when compared to other metal-organic based precursors. The main aspects of sol-gel chemistry and application of sol-gel derived materials have already been described in detail in a number of reviews and books [125].

The first sol-gel processes were mainly based on simple alkoxide precursors giving access to ceramic or glass-like inorganic products. However, the low reaction temperature also allows the introduction of organic groups in inorganic materials leading to a novel class of materials composed of both inorganic and organic moieties [126]. The inorganic component confers enhanced mechanical properties while the organic component leads to increased flexibility [127] and functional compatibility with organic paint systems. The classical approach to join the properties of different materials led to the development of composite materials, where two different phases with complementary physical and chemical properties were mixed. However, the properties of a hybrid composite material do not only depend on properties of individual components, but also on synergies between them, providing better properties than any of the two constituents.

Hybrid inorganic-organic sol-gel composites can be prepared in two ways, i.e. the organic and inorganic groups can be linked by stable chemical bonds, or the organic component can be merely embedded in an inorganic material [128]. Another important factor is the size of organic-inorganic domains in the final composite. Improved physical properties can be achieved by the reduction of the domain size to nanoscale, increasing the interfacial surface area. Further reduction in domain size leads to the formation of molecular composites [129]. This may be accomplished by incorporating the polymerisable inorganic group and an organic constituent in the same molecule when the organic group is bound to oxide-forming elements in a hydrolytically stable way.

Nanocomposites with a well defined structure can be obtained using preformed nano-building blocks, such as nanoclusters, nanoparticles or nanolayered materials. The nano-building blocks strategy confers several benefits providing a monodispersed, well defined nanostructure of the final hybrid material. The incorporation of an organic component into the inorganic materials opens a wide range of possibilities for functionalising and tailoring of final materials. The addition of chromophore molecules or luminescent groups leads to hybrids with advanced optical properties [130].

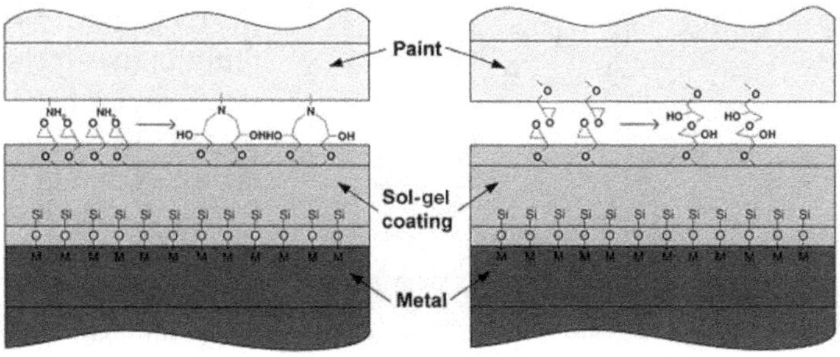

Figure 10.16 : Schematic representation of enhanced compatibility of different paint systems with an epoxy-functional hybrid sol-gel coating. Source: Royal Society of Chemistry

Another key property, which can be tuned using modification of the organic functionality, is **compatibility** and **adhesion** of thin protective hybrid sol-gel films to organic paint systems applied on top. **Chemical bonding** is possible between the sol-gel film and the top coat conferring enhanced adhesion in comparison to chromate pretreatments where adhesion is mainly attributed to mechanical interlocking, dispersion forces and hydrogen bonds. Figure 10.16 depicts possible ways to obtain an enhanced chemical compatibility of an epoxy-based or amino-containing paint system to a sol-gel pretreatment by functionalising the organic component of the hybrid film with epoxy groups. Other functional groups were used for pre-treatment functionalisation in the case of poly(urethane) and thioester paint systems [131].

A number of **waterbased stable sol-gel** systems have been developed. Functional groups varied from amino, epoxy, vinyl, and allyl groups can be incorporated into the sol-gel network to interact with organic polymer resins [132]. A strong motivation for the study of inorganic/ organic hybrid coatings, especially when a significant portion of inorganic network components is involved, is their potential to ensure an extended service life as compared to traditional organic-polymer-based coatings. Current waterbased epoxy resins (organic polymer) tend to degrade within 3 years of weathering exposure, making them imcompatible with outdoor use as topcoats. Inorganic networks are resistant to degradation by aging under UV light. A hybrid coating containing significant portions of inorganic network integrated with the epoxy polymers may last much longer in a UV environment, moreover the healing of a sol-gel matrix could be achieved under UV [133]. Figure 10.17 illustrates an inorganic/ organic hybrid coating structure: (top) reaction and curing of sol-gel network with colloidal particles and silanes with functional groups; (centre) reaction and curing of water-reducible epoxy based organic primer; (bottom) reaction and curing of a hybrid coating system. The flexible long strands are epoxy resin polymers. The short zigzag strands are amine-based curing agents.

One more advantage is an increased of a mechanical strength (such as abrasion resistance) of the coating. A better incorporation of inorganic salt corrosion inhibitors and fillers is possible in this case, too. The possibility to develop a true single step coating combing the functions of the conversion coating, primer, and topcoat becomes more realistic based on sol-gel hybrid systems. The coating surfaces of the hybrids are glossy and have shown water repellency. The coatings are also thicker than normal primer coatings (2 mm or thicker) so that various pigments may be incorporated in the system without sacrificing the performance of the coating. In a sol-gel/epoxy resin hybrid coating, the combination of weathering stability and structural strength of a sol-gel network with the flexibility and room temperature cure of an epoxy resin will hopefully be possible. Figure 10.17 shows the curing process and final structure depicted in a simplified manner.

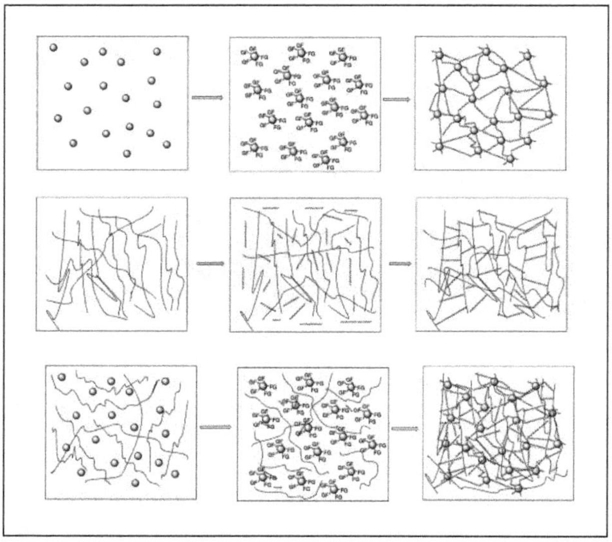

Figure 10.17 : Illustration of inorganic/organic hybrid coating structure: (top) reaction and curing of sol-gel network with colloidal particles and silanes with functional groups; (middle) reaction and curing of water-reducible epoxy-resin-based organic primer system; (bottom) reaction and curing of a hybrid coating system. Note: FG: functional groups such as amino, epoxy groups. The spheres are sol-gel nanoparticles. The flexible long strands are epoxy resin polymers. The short zigzag strands are amine-based curing agents. *Source: Elsevier*

In the work [134] inorganic sol-gel **pretreatments** show good coupling between the metallic substrate and organic coatings. The high adhesion of sol-gel films to metals is due to the formation of strong covalent Si–O–Me bonds. This generates corrosion stable metal/sol-gel film interfaces with noticeable self-repair ability. The addition of an organic component to the sol-gel systems leads to the formation of thicker and more flexible films with better compatibility with organic topcoats. Hybrid films have gained interest because they combine the properties of organic polymer materials and ceramics. Hybrid coatings can be prepared over a continuous compositional range from almost organic to almost inorganic. An advantage of the organically modified composites is the possibility to prepare thick, crack-free coatings that can provide an effective barrier to corrosive agents. Moreover, the hybrid films can be reinforced by doping with nanoparticles that improve corrosion protection properties. This improvement results from the reduced porosity and cracking potential of such films.

In the long run, sol-gel derived ceramic and hybrid coatings exhibit a high potential as substitutes for the environmentally unfriendly chromate surface pretreatment methods. Inorganic sol-gel derived films offer good adhesion between metals and organic paint. However, they cannot provide adequate corrosion protection due to their high crack-forming potential. The introduction of an organic component in an inorganic sol-gel system leads to the formation of thicker, more flexible and functionalised films with enhanced compatibility to different organic topcoats. The incorporation of nanoparticles in the hybrid sol-gel systems increases the corrosion protection properties due to lower porosity and lower cracking potential along with the enhancement of mechanical properties. Furthermore, the incorporation of inorganic nanoparticles can be a way to insert corrosion inhibitors, preparing inhibitor nanoreservoirs for "self-repairing" pretreatments with controlled release properties (see Chapter 10.3.2).

10.3 Self-healing coatings

Protective systems should not only react adequately on external impacts, but also respond to changes in their internal structure, and combine different damage prevention and repair mechanisms in the same system. Metallic or composite joints bonded by adhesives containing these types of intelligent release systems also exhibit improved durability against environmental degradation. The multi-level self-healing concept is based on gradual active feedback of the protective systems to the environmental conditions. The protection could be provided by incorporation of nanotraps (inorganic nanoparticles able to absorb aggressive/corrosive species if their level exceeds a critical value, see also Chapter 10.2.1). Moreover it could be based on the use of water displacing compounds, which are released from nanocontainers as soon as the first microdefects appear in the polymer matrix. Further growth of the defects triggers the release of polymerisable precursors entrapped in other nanocapsules (Figure 10.18a). Then a new thin polymer film is formed, to cover the damaged area and repair the layer, preventing crack propagation. The encapsulation of organic and inorganic corrosion inhibitors in 50 to 500 nm containers suppresses corrosion and delamination processes occurring in open defects or at cut edges (Figure 10.18b).

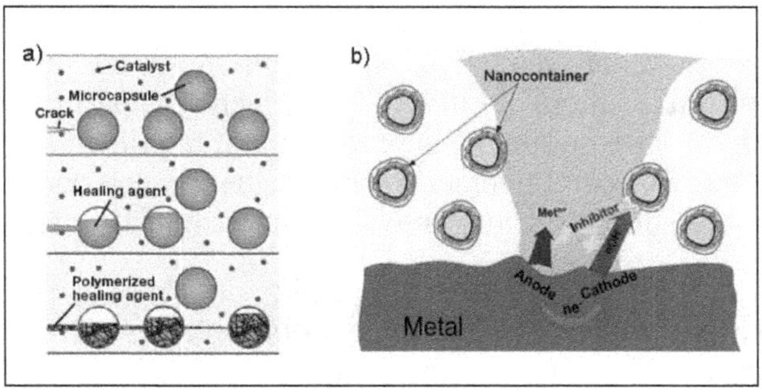

Figure 10.18 : Schematic presentation of the protection. Source (a) the Nature Publishing Group and (b) the American Chemical Society

In general the term **"self-healing"** in material science means **self-recovery** of the initial properties of the material after destructive action of the external environment. The same definition can be applied to functional coatings. However, a partial recovery of the main functionality of the material can also be considered as a self-healing ability. Thus, in the case of corrosion protective coatings, the term "self-healing" can be interpreted in different ways [135]. The classical understanding of self-healing is based on the complete recovery of the functionalities of the coating due to a real healing of the defect based on the recovery of the coating integrity. However, the main function of anticorrosion coatings is the protection of an underlying metallic substrate against an environment-induced corrosion attack. Thus, it is not obligatory to recuperate all the properties of the film in this case. The hindering of the corrosion activity in the defect by any mechanisms employed by the coating itself can already be considered as self-healing, because the corrosion protective system recovers its main function, namely the corrosion protection, after being damaged.

The development of an **active healing mechanism** for the polymer coatings and adhesives becomes an urgent issue for the respective industrial applications. Recently, several attempts were made in this field using different protection/healing mechanisms for protective coatings and adhesives. Therefore an active protection based on "self-healing" of defects in coatings is necessary to provide a long-term effect. A multi-level self-healing approach as described combines - within one system - several damage prevention and repair mechanisms, which are activated depending on type and intensity of the environmental impact (pH, irradiation, potential change, etc.).

10.3.1 Self-repairing polymer films

The development of new materials with stimuli-responsive physical properties is a rapidly expanding area of research [136]. Notable recent developments in this field include self-healing polymers [137] which, when fractured, can regain the physical properties of the original polymer either autonomically [138], or in response to an external stimulus such as heat [139] or pressure [140]. Healable polymeric systems may for example contain encapsulated monomers and polymerisation catalysts, or latent functionalities which are able to participate in thermally-reversible [141], covalent bond-forming reactions [142], It has also been shown that non-covalent interactions, specifically hydrogen bonds [143], may be used to effect healing within a supramolecular polymer blend (albeit in the presence of a plasticising solvent). In the latter system, it is proposed that fracturing propagates via dissociation of the weak supramolecular interactions rather than by scission of covalent bonds, so that re-assembly of the supramolecular network restores the original physical properties of the material.

Thus, due to the self-healing ability of polymers, protective coatings doped with **organically modified hollow glass microspheres** exhibited an enhanced resistance to erosive wear contributing to more intense corrosive wear [144]. The large specific surface area of glass microspheres coated with a specially developed polymer finish, ensures robust and tight binding between capsules and epoxide matrix. Polydispersity as well as shape irregularity enable effective filling of structural voids in the epoxide plastic, reducing the coating porosity and improving its mechanical properties. Sometimes the active agent (inhibitor, sealant etc.) to be encapsulated is either only water-soluble or has an essentially high solubility in aqueous media for incorporation into the waterborne polymeric coating. In this case, **micro- or nanocapsules** with aqueous cores and charged polymeric shells which are dispersible in aqueous solutions have to be synthesised. An interesting possible route for preparing such a container via w/oil (w/o)/w-emulsion and solvent evaporation was shown in [145].

Two types of polystyrene microcapsules with aqueous interior and sizes of 6 and 26 μm, respectively, were formed depending on the chemistry of the stabilising surfactants and their concentrations. The electrodeposition of microcapsules on a metallic coating carried out in the next step was only satisfactory when their size was essentially smaller than the coating thickness (60 μm), stabilising their integrity. A similar way to produce a composite metal coating with organic liquid-containing microcapsules by electroplating of metal substrate was chosen in order to achieve the sustained release of a liquid lubricant, lowering the friction between repeatedly contacting surfaces (self-lubricating coating) [146]. Microcapsules with liquid organo-silica resin as the core and polyvinyl alcohol as the shell material were prepared via o/w emulsification using an alkyl polyoxyethylene polyether type surfactant as a stabiliser. Subsequent electroplating from the mixture of electrolyte solution and capsule dispersion completes the formation of the composite capsule-contained metal coating. When the samples were subjected to a scratch or wear test, surface wetting in the immediate vicinity of the damaged site was observed, confirming the release of liquid from the broken microcapsules and the ability of the coating to provide self-lubrication.

For effective performance of self-healing composite coatings, the ability to **recover cracks** generated during service of a polymer is essential and may be attained by incorporation of microcontainers with encapsulated liquid sealant. Urea-formaldehyde microcapsules filled with drying linseed oil were used for the healing of cracks in an epoxy coating [147]. Microcapsules were synthesised by in-situ polymerisation in an o/w-emulsion. Initially fully water-compatible urea and formaldehyde react in a continuous aqueous medium to form poly(urea-formaldehyde). As the molecular weight of this polymer increases, the fraction of polar groups gradually decreases until the polymer molecules become hydrophobic and are deposited on the surface of o/w emulsion droplets. The obtained microcapsules hereby obtained were then incorporated into an epoxy coating. Since the outer shell surface of the microcapsules was very rough, a good binding with the coating matrix was provided. Upon cracking of the coating the encapsulated linseed oil is released and fills the crack in the matrix. Finally, oxidation of linseed oil by atmospheric oxygen led to the formation of a continuous film inside the crack. Superior corrosion resistance of such a self-healing coating was demonstrated by comparative exposure of two specimens in a salt-spray cabinet.

When scratched, the sample coated by film containing capsules remained free from corrosion after up to 72 h of exposure. In contrast, the control specimen suffered from corrosion after 48 h.

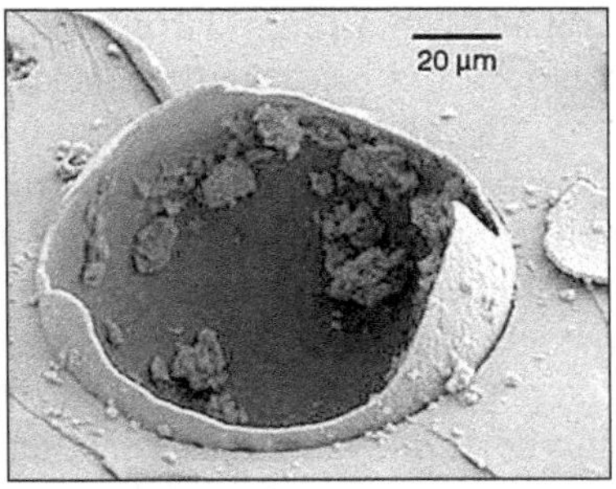

Figure 10.19 : SEM image of the fracture plane of a self-healing material with a ruptured urea-formaldehyde microcapsule in a thermosetting matrix. Source: Elsevier

Significant progress in the preparation of protective composite coatings with incorporated organic **micro- and nanocontainers** was achieved in the past decade in the group of S. R. White [148]. In the most well-known paper, published in 2001, they showed the first working prototype of polymeric coatings with the ability to "autonomically heal cracks" in the structural polymers. The proposed self-healing coating contained microcapsules (50 to 200 mm) with poly(urea-formaldehyde) shells filled by the healing agent - dicyclopentadiene monomer. These capsules were then mixed with epoxy resin and Ru-based Grubbs' catalyst in order to prepare a self-healing epoxy specimen. A developing crack ruptures the embedded microcapsules and releases healing agent into the crack channel due to capillary forces (see Figure 10.19).

Polymerisation of the healing agent was triggered by bonding the crack faces. Healing efficiency was about 60 % (see Figure 10.20). The proposed self-repairing approach had, however, some essential drawbacks for application of the protective self-healing coatings:

- The concentration of Grubbs' catalyst has to be quite high increasing the price of the coating, especially in the case of large surfaces to be protected;
- The relatively large size of capsules (50 to 200 µm) inhibits their application to coatings with thickness less than 0.5 mm because of integrity failure.

On the other hand, lowering the microcapsule size without reducing the self-healing efficiency requires much higher concentration in the coating matrix. Samples with smaller microcapsules must exhibit a very good adhesion to the epoxy matrix to ensure capsule rupture under stress. Much smaller capsules (nanocapsules) filled by dicyclopentadiene as a **self-healing agent** were synthesised very recently using ultrasonic treatment for the preparation of initial o/w-emulsions. Up to 2 % v/v of these capsules can be dispersed in an epoxy matrix leading to a slight decrease of tensile strength accompanied by a significant increase in fracture toughness. An increase of fracture toughness up to 59 % was found for a capsule volume fraction of 0.015. Because of their size, smaller capsules were found to be suitable for applications in self-healing thin films and coatings.

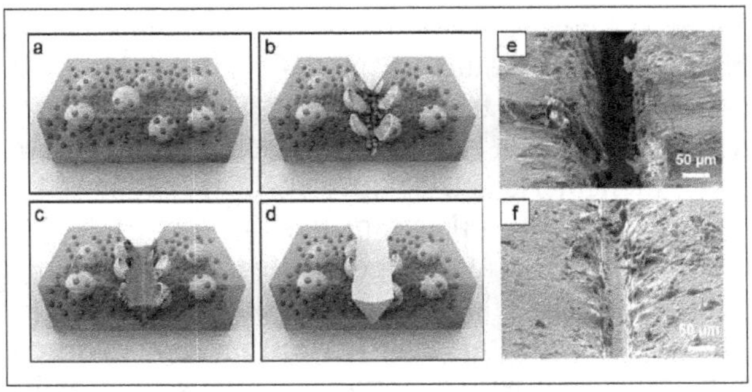

Figure 10.20 : Self-healing process. (a) Self-healing coating containing microencapsulated catalyst and phase-separated healing agent droplets (small) in a matrix on a metallic substrate. (b) Damage to the coating layer releases catalyst and healing agent. (c) Mixing of healing agent and catalyst in the damaged region. (d) Damage healed by cross-linked polymer, protecting the substrate from the environment. (c,d) SEM images of the scribed region of (e) the control coating and (f) the self-healing coating after healing (d) Source: Wiley

A less-expensive and chemically more stable self-healing composition was based on the direct incorporation of phase separated droplets containing hydroxyl end-functionalised polydimethyl-siloxane and polydiethoxysiloxane in the coating matrix. Only the catalyst, di-n-butyltin dilaurate or dimethyl dineodecanoate tin, was embedded in the polymer matrix in the microencapsulated state, using polyurethane microcapsules of a size up to 450 µm. The advantages of this self-repairing system were, however, considerably reduced by the large size of the microcapsules and poor adhesion of the healing agent (polysiloxanes) and matrix, leading to the addition of adhesion promoter, and the extreme toxicity of the employed catalyst. The last point could be especially critical, taking into account the details of preparation of the polyurethane shell. The reaction between polyurethane and ethylene glycol was done in an aqueous medium and droplets of o/w-emulsion were stabilised by 15 % solution of gum arabic. These factors can significantly contribute to the side reaction of polyurea formation, which is accompanied by the evolvement of CO_2 and causes porosity of the capsule shell. Attempts to use various microencapsulated solvents embedded in a polymer matrix as a self-healing system were less successful. The encapsulation of five highly efficient solvents with a high degree of polarity failed, and among the polar solvents used only chlorobenzene emerged as a potential candidate for self-healing. Nevertheless, experiments with poly(ureaformaldehyde) microcapsules (160 µm) filled with chlorobenzene showed sufficient healing efficiency for the epoxy matrix. The idea of microencapsulation of solvents as active agents for self-healing coatings was further developed by including a reactive epoxy resin component in the capsule core. New, less toxic solvents as compared with previously used chlorobenzene were employed in the **self-healing core mixture.** The use of ultrasonication and a co-stabiliser for the emulsion preparation step yielded much smaller capsules up to submicrometre size. The resin-solvent capsules of micrometre size were successfully incorporated in the epoxy matrix and ruptured by a propagating crack. An epoxy film was then observed to be accumulated on the crack surface evidencing that a reactive epoxy resin was delivered to the damaged coating site. Almost the same approach was utilised when a self-repairing coating for corrosion protection of aluminium alloys was developed using cage-like microcapsules with an oil core and a silica gel shell [149]. Two types of these microcapsules were prepared by interfacial polycondensation on the surface of Pickering o/w emulsion droplets composed of methyl methacrylate/tetraethyl orthosilicate mixture stabilised by interfacially-active polystyrene nanoparticles. The application of polystyrene nanoparticles with two different sulphonation times, 60 and 9 h, yields microcapsules with closed and open state of surface pores, respectively. Closed microcapsules were then used as a source of self-healing polymer methacrylate and open highly-porous capsules of smaller size were impregnated with a solution of a polymerisation initiator. Both types of microcapsules were deposited on a surface of aluminium alloy AA 2024 and then coated with a sol-gel film. Scratching led to the break-up of capsules and a monomeric self-healing agent was released, forming a polymeric film. However, pictures showing microcapsules of both kinds demonstrate a certain, relatively high porosity. This was also demonstrated for "closed" microcapsules containing a monomeric sealant, which could open the possibility of spontaneous polymerisation especially when polymerisation catalyst diffuses from the open microcapsules. Scratch dimensions are huge compared with the microcapsule size of both types. Therefore, the scratch has to destroy not only methacrylate-filled microcapsules but also neighbouring catalyst-loaded containers causing release and immediate polymerisation of methacrylate.

The next promising step is now to proceed to constructing the response in a way that it will, in turn, affect the stimulus. This feedback loop has to be used to anneal defects, e.g. to construct a self-repairing coating. It is shown that there are conceptual solutions but still many technical and economical issues must be solved to arrive at practical solutions. The stimuli predominantly have been of mechanical and chemical nature but in the future one may envision more of these, e.g. friction, biocide. One should, however, also be aware that feedbackloops are ubiquitous in nature and technology, and mostly, they are not used to remove a stimulus but to keep it at a defined level, e.g. concentration, pH value, potential. Mimicking this, however, is a more demanding task which should be tackled in the future. On the other hand one can envision much potential in constructing these more sophisticated regulatory loops, e.g. maintaining a fixed humidity, potential or enzyme concentration. There is a high demand for such feedback active systems in the most urgent areas of society, energy, health and raw materials, and we look forward to interesting solutions in these areas in the near or far future.

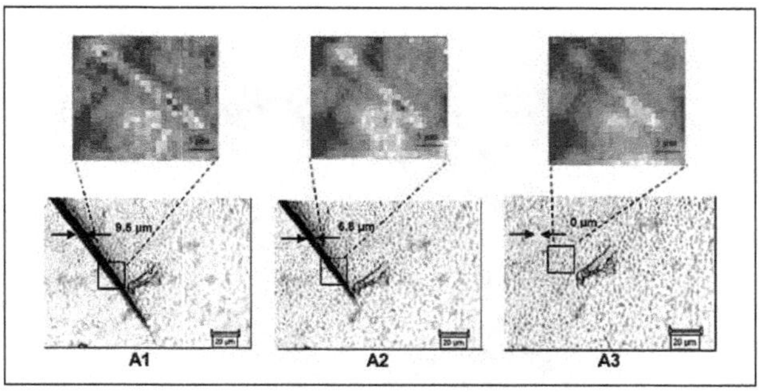

Figure 10.21 : IR (upper) and optical (lower) images of self-repairing oxetane-substituted chitosan polyurethane networks recorded as a UV exposure time. A1, 0 min; A2, 15 min; A3, 30 min. (SOM provides details regarding spectroscopic changes detected by IR imaging.)

Source: Science Publishing Group

Also, the development of a new generation of thermosetting polymers that are of considerable technical and commercial importance [150] is extremely interesting (see Figure 10.21). The use of the UV portion of the electromagnetic radiation for repairing mechanical damages in coatings offers an ambient temperature approach to self-healing . This approach might avoid the placement of other, often elaborate, network components that is controlled by the chemistries and morphologies of polymer networks in a number of critical applications and technologies.

10.3.2 Inhibitor release

The use of different types of **micro- and nanocontainers for encapsulation**, controlled delivery and release of active materials has become more and more popular nowadays. The containers can have an inorganic, organic or composite origin, and can be applied for encapsulation of drugs, vitamins, quantum dots, bactericides, oils, biopolymers, corrosion inhibitors, low molecular weight compounds and many other active molecules. For example, micrometre sized capsules with a layer-by-layer assembled polyelectrolyte shell are used for the encapsulation and release of drugs, DNA, dendrimers and enzymes [151]; inorganic halloysite nanotubes were demonstrated to be suitable for loading of ferments and inorganic nanoparticles [152].

Hydrogels were used for the encapsulation of phospholipids, drugs, as liposome reactors and plant growth media [153]. Micelles and microemulsions are probably the most investigated systems for the encapsulation of active biomaterials. There are numerous publications on this topic and we address several recent reviews to provide better understanding of the current achievements in the use of micelles and microemulsions for encapsulation [154].

The most interesting feature of containers is their ability to control release/upload. The **controlled release**/upload is, in general, dependent on the interaction between the material of shell or stopper and the environment surrounding the container. The shell/stopper of containers should be sensitive to external impact or changes in the local environment; moreover, it should selectively react only to one or two triggers changing its permeability while others should leave the shell intact. The **permeability of the container shell** is

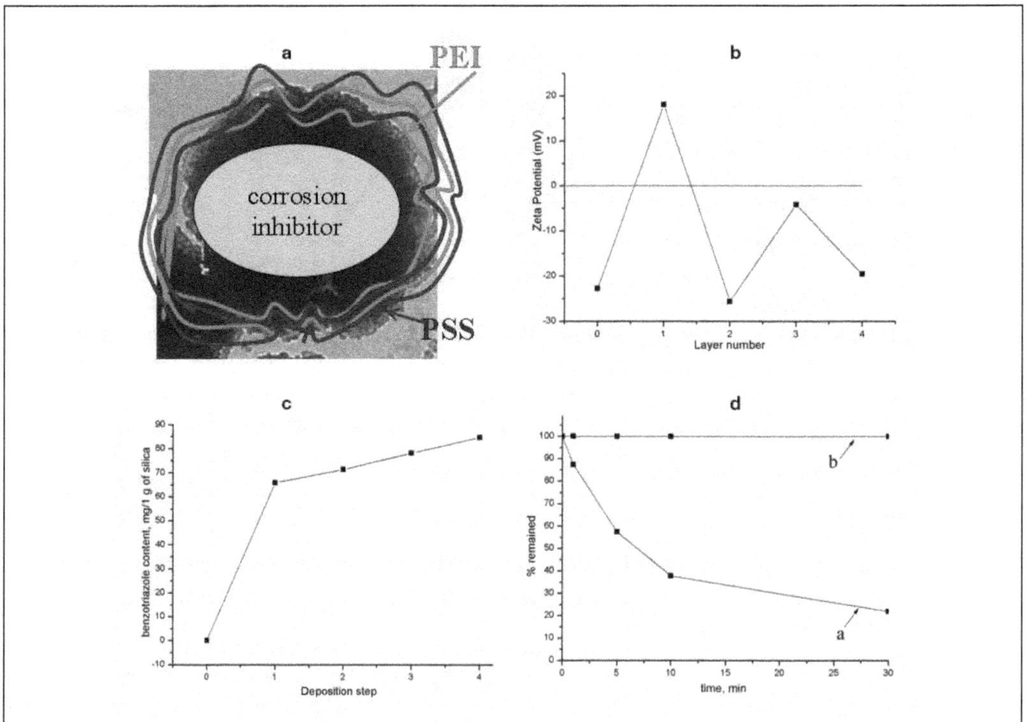

Figure 10.22: Container loading with corrosion inhibitor and modification by polyelectrolyte shell: (a) scheme; (b) changes of zeta potential during the procedure of polyelectrolyte shell formation; (c) loading of the interior of SiO_2 containers with 2-(benzothiazol-2-ylsulfanyl)-succinic acid under vacuum; (d) the release of inhibitor from nanocontainers at (a) pH.10.1 and (b) neutral pH. Source: Wiley

determined by balance of electrostatic interactions of surface charges that tend to expand the surface and, therefore, produce pores in the shell, and hydrophobic interaction contributing to surface energy that tends to reduce the surface [155]. Establishing the equilibrium between these forces requires work against film elasticity [156]. The following **triggers** are utilised for opening/closing of the container shell/stopper:

- local pH changes
- temperature changes
- electromagnetic irradiation
- mechanical pressure (also ultrasonic treatment)
- humidity
- electric (electrochemical) potential
- ionic strength and
- dielectric permeability of the solvent.

The simplest trigger for opening/closing of the container shell is the pH shift in the local environment. Polyelectrolyte capsules, hydrogels and emulsions with weak acidic or basic functional groups in the shell are sensitive to it, displaying reversible and/or irreversible changes of the shell permeability in a wide pH range (e.g. at low pH < 4 or high pH > 9) [157]. The other important triggers are external electromagnetic irradiation and mechanical impact. For the opening by electromagnetic irradiation, a container shell should have sensitive components, for example, like metal (silver) nanoparticles, for IR [158], dyes for visible light [159] and semiconductors (TiO_2 particles) for UV [160]. The mechanical impact requires a certain level of rigidity or brittleness of the shell, because the elastic shell can undergo

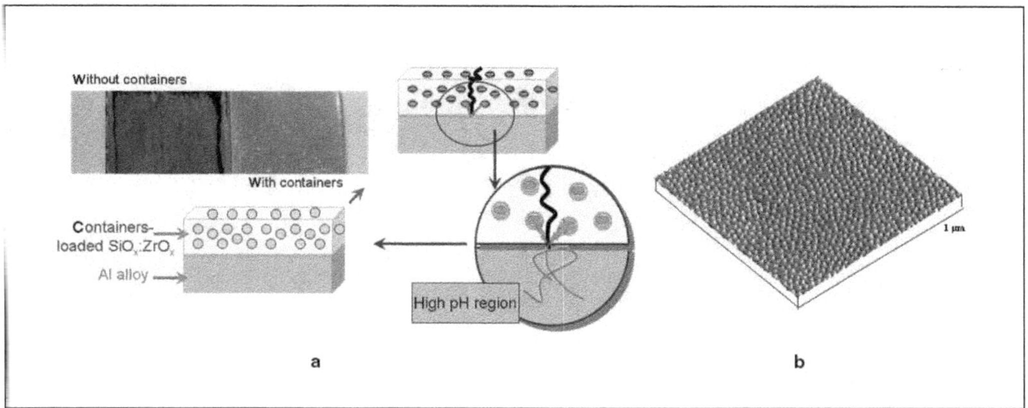

Figure 10.23 : (a) Scheme of the self-healing effect of anticorrosion coatings and (b) AFM images of container-loaded SiOx:ZrOx film. Source: Wiley

deformation under pressure instead of rupture [161]; containers with diameters <100 nm are hardly destroyed with reasonable mechanical force because they tend to escape from the force direction. On the contrary, ultrasonic treatment can be applied to irreversibly open containers of any size [162]. Other triggers for capsule opening were demonstrated for hydrogels (temperature, electric potential, high ionic strength) and polyelectrolyte capsules (temperature, high ionic strength, dielectric permeability of the solvent) [163] These triggers are usually involved in feedback active systems of specific functionality (e.g. coatings with electrochemically reversible permeability) [164].

The strategy of preparation of nanocontainers based on mesoporous oxide core is presented in [165]. The fabrication of a polyelectrolyte shell around a container can possibly be done by layer-by-layer assembly (Figure 10.22a) of oppositely charged species, thereby allowing to prevent the spontaneous release of a loaded corrosion inhibitor. The precision of one adsorbed layer thickness is about 2 nm. Polyelectrolyte nanocontainers completely repeat the shape of the templating colloids. The polyelectrolyte shell lends controlled release properties to the nanocontainers. The opening of the shell can be induced only by changing the surrounding pH value to the acidic or alkali region (which happens when the corrosion starts) [166], while in neutral pH the polyelectrolyte shell remains intact preventing undesirable leakage of the entrapped inhibitor. The formation of oppositely charged layers was determined by zeta-potential measurements (Figure 10.22b). The inhibitor content in the nanocontainers is equal to 85 mg/g of the initial SiO_2 particles (Figure 10.22c). The release kinetics of the entrapped inhibitor into water at alkaline pH is represented in Figure 10.22d.

As can be seen, the release of the **corrosion inhibitor** (2-(benzothiazol-2-ylsulfanyl)-succinic acid) from containers with a polyelectrolyte shell is triggered at pH 10.1, similar to the pH value in the cathodic zone during the corrosion process of aluminium [167]. Thus, modification of containers with polyelectrolyte shell is an important issue to achieve controlled release of the encapsulated inhibitor as well as to prevent its undesirable leakage from the coating [168]. The next step in the anticorrosion self-healing coating formation is the incorporation of the containers in the coating. It is important to achieve their uniform distribution (Figure 10.23). One of the important characteristics of corrosion is the local change of the pH value in the corrosive area. Depending on the corrosion mechanism, intermetallic inclusions, and surface properties of the metal substrate, the local pH shift could occur both to the acidic and alkaline region [169]. This local pH change influences the polyelectrolyte shell, in particular, its opening and inhibitor release and, consequently, the healing of the corrosion area that can be achieved. This effect is self-regulated and an inhibitor release

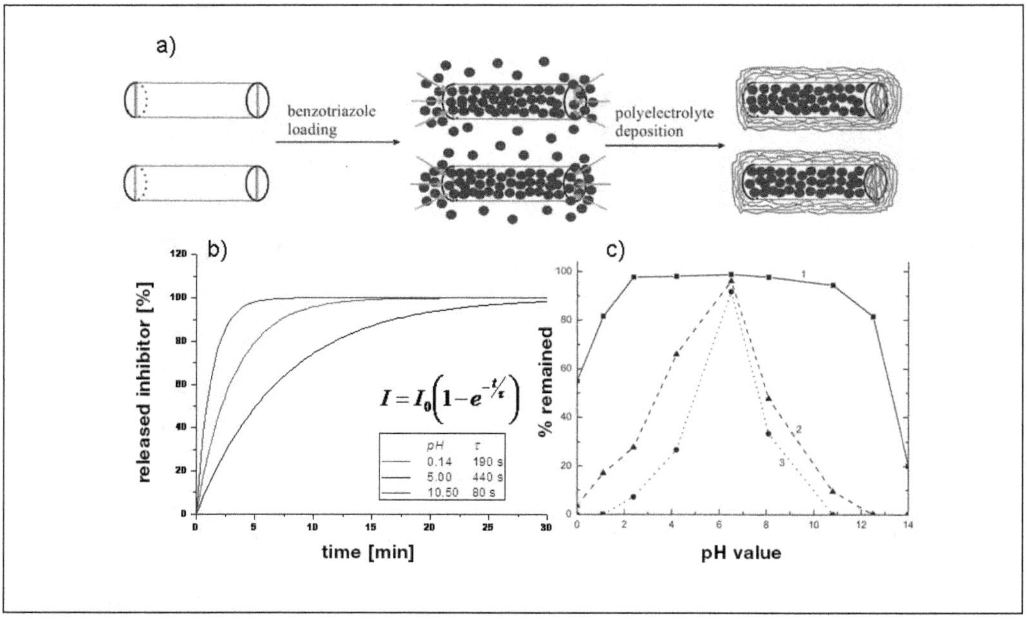

Figure 10.24 : a) Schematic illustration of the procedure for benzotriazole loading for halloysite G nanotubes. (b, c) Release of the benzotriazole at different pH values from a) uncovered halloysite G nanotubes and c) from 1) halloysite G nanotubes modified by PDADMAC/PSS layers, 2) halloysite G nanotubes modified by PAH/PSS layers, and 3) halloysite G nanotubes modified by PAH/PMA layers (after 30 min of incubation) a, c). *Source: Wiley*

takes place just in the region of the pH-change (Figure 10.23a) trigging the self-healing process. When the pH returns to the initial value, the shell is closed and the release of the inhibitor stops. The SVET maps have shown that the highest protection was provided by the container-loaded sol-gel film. The effect of very low corrosion in the case of container-loaded films could be explained through the additional and very important characteristics of self-healing. During the corrosion process the pH shifts to the alkaline region. It was shown in Figure 10.22d that in alkaline pH the release of a loaded inhibitor takes place and, consequently, the inhibitor heals damaged areas, pH returns to neutral, and the polyelectrolyte shell closes until a new corrosion attack.

The **photocatalytic activity** inherent in some semiconductors which could be a core of light sensitive containers (TiO_2, TiO_2:In_2O_3 composite, etc.) opens up fresh opportunities to exert remote control over corrosion protection efficiency under ultraviolet (UV) irradiation [170]. The local change in pH at the **semiconductor surface** under UV irradiation leads to conformational transitions in the polyelectrolyte shells that switch them into the opened state. The sensitisation of containers to near-IR LASER irridation can be attained with the use of metal and semiconductor particles absorbing at long wavelengths. In this case the container opening will be due to the conformational transitions of thermal nature occurring in the polyelectrolyte shell under irradiation. These approaches permit to exert effective remote control over release properties of immobilised nanocontainers and to ensure rapid release (much faster than in response to the chemical stimuli) of the chemical species encapsulated within the containers.

Halloysite is an economically viable raw material that can be mined from the corresponding deposit as a raw mineral [171]. As for most natural materials, the size of halloysite particle varies within 1 to 15 μm of length and 10 to 100 nm of inner diameter depending on the deposits. Embedding of the corrosion inhibitor (for example, benzotriazole) inside the inner

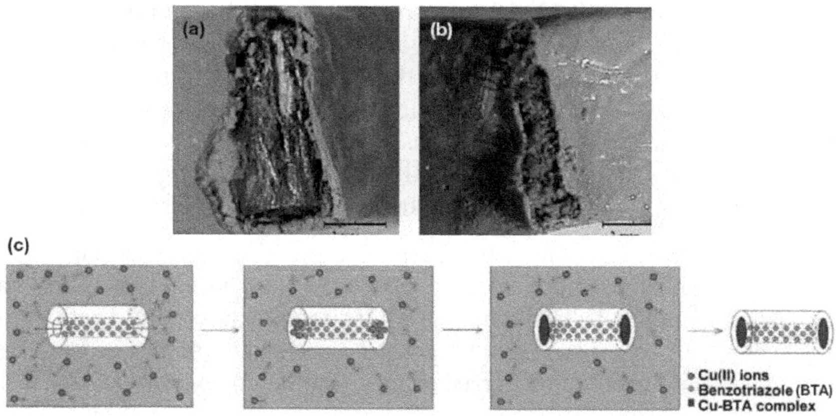

Figure 10.25 : Images of scratched copper strips painted with oil-based blue paint (a) containing halloysite nanotubes loaded with benzotriazole and (b) without these tubes, after exposure to a corrosive environment for 10 days. c) Illustration of stopper formation at halloysite tube endings by interaction of leaking benzotriazole with Cu(II) ions. Source: American Chemical Society

volume of the halloysite G nanotubes was performed according to the adapted procedure described by Price et al. [172]. Benzotriazole molecules have good solubility in water, so pure benzotriazole-filled halloysite can be applied directly in self-healing anticorrosion coatings, however spontaneous leakage and uncontrolled release of embedded benzotriazole is possible (Figure 10.24). To attain controlled release properties in the halloysite nanotubes, their surface could be modified by LbL deposition of polyelectrolyte bilayers [173] (Figure 10.24), so the openings at the edges became blocked with polyelectrolytes. In this work the use of different polyelectrolyte composition, was suggested, see [174]: benzotriazole-loaded halloysite G coated with (poly(diallyl dimethylammonium chloride)/(PDADMAC)/poly(styrene sulphonate)/(PSS))$_4$ layers, benzotriazole-loaded halloysite G coated with ((poly(allylamine hydrochloride) (PAH)/PSS)$_4$ layers, and benzotriazole-loaded halloysite G coated with (PAH/poly(methacrylic acid) (PMA))$_4$ layers. The pH-dependency of benzotriazole release from halloysite nanocontainers is depicted in Figure 10.24. The increase of the release rate both at low and high pH value is observed for all polyelectrolyte shells. Controllable release of benzotriazole from halloysite nanotubes could also be achieved by the formation of metal-benzotriazole complex caps (stopper) at halloysite tube endings by the interaction of leaking benzotriazole and metal ions from the bulk solution [175].

This suggested method requires only a short rinsing of benzotriazole-loaded halloysite nanotubes with an aqueous solution containing metal ions. In Figure 10.25, the formation of stoppers at **halloysite tube** openings is demonstrated (using Cu(II) ions). The complex of copper ions and benzotriazole has been studied thoroughly by various authors and described in the review by Cease et al. [176]. Benzotriazole forms stable 2D complexes with most of the transition metals, and Cu(II), Fe(II), Fe(III), and Co(II) salts were used for stopper formation. Benzotriazole release characteristics were studied in detail, with the formation of stoppers with Cu(II) ions providing the best release control. The decrease in the benzotriazole release rate depends on a number of parameters, such as the chemistry and morphology of halloysite samples, the concentration and type of the metal ion, and the concentration of benzotriazole available at tube openings. One of the important parameters influencing benzotriazole release rates is the concentrations metal ions that are used for the formation of metal-benzotriazole stoppers. This is an indication that the efficiency of the encapsulation strongly depends on the concentration of the metal ions used for the formation of stoppers. Another important factor that influences the benzotriazole release rate is the concentration of benzotriazole at the tube endings available for the formation of metal-benzotriazole films. This can be controlled by washing loaded nanotubes with deionised water after loading them with benzotriazole and prior to processing them with metal ions. The corrosion inhibition efficiency of halloysite nanocontainers for aluminium and copper samples was demonstrated by monitoring the localised corrosion current density on scratches; it could also be determined visually by the exposure of a scratched metal sample to a highly corrosive environment (Figure 10.25) which showed a significant reduction in the metal corrosion rate.

Thus, to gain a solid fundamental knowledge on these “smart” release systems and inherent prompt feedback effects is an important task. One important concept is to utilise nanoscale reservoirs allowing a uniform distribution of the inhibitor across the coating. For this reason, the nano-entrapment of active species is an indispensable issue. Moreover the release of the chemical agents from the containers can be trigged employing different external stimuli including:

- pH changes due to corrosion reaction and reactions of hydrolysis
- pH changes and changes in the ionic strength resulting from treatment of the coating with corresponding solutions
- electrochemical switching (when the conducting polymer/polyelectrolyte composite is used as the host for capsules);
- photocatalytic degradation of capsule shells under irridation at corresponding wavelength.
- For the construction of containers the following tasks should always be taken into account:
- entrapment of inhibitors with different nano-containers (polyelectrolyte, nanotubes, etc.)
- tailoring of nanocontainers for specific triggers
- uniform incorporation of nanoreservoirs into coatings
- study of controllable release of active species encapsulated inside nanocontainers by changing pH, ionic strength, compound modeling agents (chemical and inhibitors contaminants), electrochemical potential of the coating
- optimisation of corrosion protection and self-healing properties of novel coatings;
- investigation of container-loaded coatings to envisage their important practical application for determination and destruction of chemical compounds in water and in air.

10.4 Conclusions

This review comes timely since there have been many developments in recent years to fabricate new types of **nanocomposite** coatings and thin films for corrosion protection that are responsive to **environmental** stimuli. The future step is to proceed towards a higher level of intelligence to construct the response in a way that it affects the stimulus in turn. While conceptual solutions do exist, still many technical and economical issues must be solved to arrive at a practical level. Predominantly, the stimuli have been of mechanical and electrochemical nature, but in future one may envision other stimuli, e.g. friction. The release of active materials (e.g., corrosion inhibitors) only occurs when triggered by environmental (corrosion) processes or defects in the coating integrity, which prevents a leakage of the active component from the coating and increases the coating durability. Moreover, the active coating might comprise several active functionalities (e.g., antibacterial, anticorrosion and antistatic) when several types of nanocomposite materials are incorporated simultaneously into a coating matrix. This will surely be a matter of intense future research, which, as a result, may lead to highly sophisticated surfaces. However, one should also be aware that feedback loops are ubiquitous in nature and technology, and mostly they are not used to remove a stimulus but to keep it at a defined level, e.g. concentration, pH, potential. Mimicking this, however, is a more demanding task which should be tackled in the future. This requires additional efforts to investigate kinetic and structural properties of the nanocomposite coatings. However, there is no doubt that nanomaterials in coatings will create countless new opportunities in research and also in technologies for the fabrication of a new generation of active composite films.

10.5 Literature

[1] World Corrosion Organization (WCO): Global Needs for Knowledge Dissemination, Research, and Development in Materials Deterioration and Corrosion Control. 2009

[2] (a) Directive 2000/53/EC of the European Parliament and of the Council of 18 September 2000 on end-of life vehicles; (b) EU-Directive 67/548/EWG

[3] A. D. Jones, Principles & Prevention of Corrosion, 2nd Ed., Prentice Hall, NJ (1996); (b) R. S. Munn, The Modeling of Galvanic Corrosion Systems Using Numerical Methods with Particular Attention to Boundary Conditions of Nonlinear Polarisation", Ph.D. Thesis, University of Connecticut, 1986

[4] P. Schmuki, Solid State Electrochem. 6 (2002) 145

[5] (a) P. Berthier, Annales de chimie et de physique 17 (1821) 55; (b) W. Vernon, J. Inst. Metals 48 (1932) 121; (c) D. Medous, The limits of Groth, A. Signet. Book from New American Library, N.Y. (1972)

[6] (a) Chemistry with ultrasound. Ed. T. J. Mason. Elsiver Sci. Pub.: 1990, Vol. 28, 195 p.(b) B. Langenecker, C. W. Fountain, O. V. Janes, Metal Prog. 85 (1964) 97; (c) B. A. Shenoi, K. S. Indira, R. Subramanian, Metal Finish. 77 (1979) 41

[7] O.V. Abramov, Ultrasonics. 25 (1987) 73

[8] (a) D. V. Andreeva, D. Fix, H. Möhwald, D. G. Shchukin, Adv. Mater. 20 (2008) 2789; (b) E. V. Skorb, D. G. Shchukin, H. Möhwald, D. V. Andreeva, Nanoscale. (2010) DOI:10.1039/C0NR00074D

[9] (a) D. He, V. J. Gelling, D. E. Tallman, G. P. Bierwagen, G. G. Wallacec, J. Electrochem. Soc. 147(2000) 3667; (b) D. He, V. J. Gelling, D. E. Tallman, G. P. Bierwagen, J. Electrochem. Soc. 147 (2000) 3661

[10] For a general review, see A. Ulman: An Introduction to Ultrathin Organic Films, Academic Press, Boston 1991

[11] (a) C. D. Bain, G. M. Whitesides, JACS 110 (1988) 5897; (b) T. H. Ong, R. N. Ward, P. B. Davies, C. D. Bain, JACS 114 (1992) 6243; (c) C. Redon, F. Brochard-Wyatt, E Rondelez, Phys. Rev. Lett. 66 (1991) 715

[12] K. L. Prime, G. M. Whitesides, Science 252 (1991) 1164

[13] M. K. Chdudhury, G. M. Whitesides, Science 225 (1992) 1230

[14] N. Balachander, C. N. Sukenik, Langmuir 6 (1990) 1621

[15] E. P. Plueddemann: Silane Coupling Agents, 2nd Ed. Plenum Press, New York

[16] S. L. Michelhaugh et al., Corrosion 47 (1991) 322

[17] (a) G.A. Bommarito, A.V. Pocius, Thin Solid Films 327 (1998) 481; (b) S. Meth, C.N. Sukenik, Thin Solid Films 425 (2003) 49; (c) I. Maege, E. Jaehne, A. Henke, H.-J.P. Adler, C. Bram, C. Jung, M. Stratmann, Prog. Org. Coat. 34 (1998) 1; (d) C.M. Ruan, T. Bayer, S. Meth, C.N. Sukenik, Thin Solid Films 419 (2002) 95; (e) K. O. Berg, P. Persson, A. Shchularev, B. Eliasson, Thin Solid Films 397 (2001) 102

[18] (a) A. Ulman, An Introduction to Ultrathin Organic Films from Langmuir-Blodgett to Self-Assembly, Academic Press, San Diego, CA, 1991; (b) A. Ulman, Chem. Rev. 96 (1996) 1533; (c) M. Giza, P. Thissen, G. Grundmeier, Langmuir 24 (2008) 8688; (d) K. Wapner, M. Stratmann, G. Grundmeier, Inter. J. Adh. Adh. 28 (2008) 59

[19] I. Maege, E. Jaehne, A. Henke, H.-J.P. Adler, C. Bram, C. Jung, M. Stratmann, Prog. Org. Coat. 34 (1998) 1

[20] D. Wang, Y. Nia, Q. Huoa, D. E. Tallman Thin Solid Films 471 (2005) 177

[21] R. K. Smith, P. A. Lewis, P. S. Weiss, Prog. Surf. Sci. 75 (2004) 1

[22] (a) G.K. Jennings, T.-H. Yong, J.C. Munro, P.E. Laibinis, J. Am. Chem. Soc. 125 (2003) 2950; (b) H.Y. Ma, C. Yang, B.S. Yin, G.Y. Li, S.H. Chen, J.L. Luo, Appl. Surf. Sci. 218 (2003) 143

[23] (a) A. Ulman, Chem. Rev. 96 (1996) 1533; (b) K. Nakano, T. Sato, M. Tazaki, M. Takagi, Langmuir 16 (2000) 2225; (c) J.C. Love, L. A. Estroff, J. K. Kriebel, R. G. Nuzzo, G. M. Whitesides, Chem. Rev. 105 (2005) 1103

[24] (a) Y. Feng, S. Chen, H. Zhang, P. Li, L. Wu, W. Guo, Appl. Surf. Sci. 253 (2006) 2812; (b) S. Zor, B. Yazıcı, M. Erbil, Corros. Sci. 47 (2005) 2700; (c) K. Nozawa, K. Aramaki, Corros. Sci. 41 (1999) 57; (d) G. Paliwoda-Porebska, M. Stratmann, M. Rohwerder, K. Potje-Kamloth, Y. Lu, A.Z. Pich, H.-J. Adler, Corros. Sci. 47 (2005) 3216; (e) D.-S. Kong, S.-H. Chen, L.-J. Wan, M.-J. Han, Langmuir 19 (2003) 1954

[25] (a) G. Contini, V. Di Castro, S. Stranges, R. Richter, M. Alagia, J. Phys. Chem. A 106 (2002) 2833; (b) L.Y. Goh, Z.Weng,W.K. Leong, J.J. Vittal, J. Am. Chem. Soc. 124 (2002) 8804; (c) C.M. Whelan, M.R. Smyth, C.J. Barnes, J. Electroanal. Chem. 441 (1998) 109

[26] (a) K. G. Jennings, P. E. Laibinis, Colloids Surf. A: Physicochem. Eng. Aspects 116 (1996) 105; (b) M. Epple, M.A. Bittner, K. Kuhnke, K. Kern, W.Q. Zheng, A. Tadjeddine, Langmuir 18 (2002) 773; (c) F. Li, Y. Lu, G. Xue, Q. Cao, Chem. Phys. Lett. 264 (1997) 376 ; (d) Z. Quan, S. Chen, Y. Li, X. Cui, Corros. Sci. 44 (2002) 703

[27] (a) Y. Shinohara, Y. Ota, H. Yamamoto, H. Kunieda, Japanese Patent, Japanese Kokai Tokkyo Koho JP 61 030,685 [86 30 685] (1986); (b) H. Baba, T. Kodama, T. Fujii, Boshoku Gijutsu 34 (1985) 10; (c) H. Yang, Y. Yanga, Y. Yang, H. Liu, Z. Zhang, G. Shen, R. Yu, Anal. Chim. Acta 458 (2005) 159.

[28] (a) G. Decher, Science 277 (1997) 1232; (b) M. A. C. Stuart, W. T. S. Huck, J. Genzer, M. Muller, C. Ober, M. Stamm, G. B. Sukhorukov, I. Szleifer, V. V. Tsukruk, M. Urban, F. Winnik, S. Zauscher, I. Luzinov, S. Minko, Nature Mater. 9 (2009) 101; (c) C. D. Bucur, Z. Sui, J. B. Schlenoff, JACS 128 (2006) 13690; (d) M. Agarwal, Y. Lvov, K. Varahramyan, Nanotechnology 9 (2006) 5319; (e) M. Michel, A. Taylor, R. Sekol, P. Podsiadlo, P. Ho, N. Kotov, L. Thompson, Adv. Mater. 19 (2007) 3859; (f) M. Agarwal, Q. Xing, B. S. Shim, N. Kotov, K. Varahramyan, Y. Lvov, Nanotechnology 20 (2009) 215602; (g) S. Vozar, Y.-C. Poh, T. Serbowicz, M. Bachner, P. Podsiadlo, M. Qin, E. Verploegen, N. Kotov, A. J. Hart, Rev. Sci. Instrum. 80 (2009) 023903

[29] (a) D. M. DeLongchamp, M. Kastantin, P. T. Hammond, Chem. Mater. 15 (2003) 1575; (b) S. S. Shiratori, M. F. Rubner, Macromolecules 33 (2000) 4213; (c) A. Voigt, E. Donath, H. Moehwald, Macromol. Mat. Eng. 282 (2000) 13; (d) T. R. Farhat, J. B. Schlenoff, Langmuir 17 (2001) 1184; (e) E. Tjipto, J. F. Quinn, F. Caruso, J. Polymer Sci.: Part A: Polym. Chem. 45 (2007) 4341; (f) L. Vagharchakian, B. Desbat, S. Henon, Macromolecules 37 (2004) 8715; (g) P. R. Roberge, Corrosion basics. 2nd Edition. NACE Press Book, 2005; p. 364

[30] O. Kachurina, E. Knobbe, T. L. Metroke, J. W. Ostrander, N. A. Kotov, Inter. J. Nanotechnol. 1 (2004) 347.

[31] (a) G. S. Frankel, D. Landolt, in Encyclopedia of Electrochemistry, Stratmann, M.; Bard, A.J., Eds.; Wiley-VCH: Weinheim, 2003; (b) H. S. Nalwa, Handbook of polyelectrolytes and their applications, Japan, 2002

[32] (a) L. Garriguess, N. Pebere, F. Dabosi, Electrochimica Acta 41 (1996) 1209; (b) G. P. Cicileo, B. M. Rosales, F. E. Farela, J. R. Vilche, Corrosion Science 40 (1998) 1915; (c) S. Szunerits, D. R. Walt, Anal. Chem. 74 (2002) 886

[33] (a) J. Heinze, Topics in Current Chemistry, vol. 152, Springer-Verlag, Berlin, 1990; (b) J. Simonet, J. Rault-Berthelot, Prog. Solid State Chem. 21 (1991) 1; (c). Rubinson, H.B. Mark Jr., in: A. Wieckowski (Ed.), Interfacial Electrochemistry, Theory, Experiment and Applications, Marcel Dekker, 1999; (d) M.E.G. Lyons, in: I. Prigogine, A. Stuart, A. Rice (Eds.), Adv. In Chem. Physics, vol. XCIV, John Wiley & Sons Inc., 1996, p. p. 299; (e) H. S. Nalwa, in: H. S. Nalwa (Ed.), Handbook of Organic Conductive Molecules and Polymers, vol. 2, John Willey & Sons, New York, 1997. (f) H. K. Song, G. Palmore, R. Tayhas, Adv. Mater., 18 (2006) 1764

[34] T.D. Nguyen, M. Keddam, H. Takenouti, Electrochem. Solid State Lett. 6 (2003) B25. A. Michalik, M. Rohwerder, Z. Phys. Chem. 219 (2005) 1547

[35] G. Paliwoda-Porebska, M. Stratmann, M. Rohwerder, U. Rammelt, L. Minh Duc, W. Plieth, J. Solid State Electrochem. 10 (2006) 730

[36] M. Rohwerder, A. Michalik, Electrochim. Acta 53 (2007) 1300

[37] G. Paliwoda, M. Stratmann, M. Rohwerder, K. Potje-Kamloth, Y. Lu, A.Z. Pich, H.-J. Adler, Corros. Sci. 47 (2005) 3216

[38] P. J. Kinlen, V. Menon, Y. Ding, J. Electrochem. Soc. 146 (1999) 3690. M. Kendig, M. Hon, L.Warren, Prog. Org. Coat. 47 (2003) 183

[39] G. Williams, A. Gabriel, A. Cook, H. N. McMurray, J. Electrochem. Soc. 153 (2006) B425. R. J. Holness, G. Williams, D. A.Worsley, H. N. McMurray, J. Electrochem. Soc. 152 (2005) B73

[40] G. Williams, H. N. McMurray, Electrochem. Solid State Lett. 8 (2005) B42

[41] A. Nazarov, D. Thierry, Protect. Met. 41 (2005) 105

[42] M. Rohwerder, F. Turcu, Electrochim. Acta 53 (2007) 290. R. Hausbrand, M. Stratmann, M. Rohwerder, J. Electrochem. Soc. 155 (2008) C369

[43] T. A. Skotheim (Ed.), Handbook of Conducting Polymers, Marcel Dekker, Heidelberg, New York, 1986

[44] K. M. Cheung, D. Bloor, G. C. Stevens, Polymer 29 (1988) 1709

[45] (a) F. Beck, P. Hulser, J. Electroanal. Chem. 280 (1990) 159; (b) J. Rodrigues, T.F. Otero, H. Grande, J. P. Moliton, A. Moliton, T. Trigaud, Synth. Met. 76 (1996) 301

[46] (a) T. Zalewska, Ph. D. Thesis, Gdansk University of Technology, 2000; (b) T. Zalewska, A. Lisowska-Oleksiak, S. Biallozor, V. Jasulaitiene, Electrochim. Acta 45 (2000) 4031

[47] M. Bazzaoui, E. A. Bazzaoui, L. Martins, J.I. Martins, Synth. Met. 130 (2002) 73

[48] M. Bazzaoui, E.A. Bazzaoui, L. Martins, J.I. Martins, Synth. Met. 128 (2002) 103

[49] A. M. Fenelon, C. B. Breslin, J. Electrochem. Soc. 152 (2005) D6

[50] Y. Sahin, K. Pekmaz, A. Yildiz, Synth. Met. 129 (2002) 117

[51] D. Sazou, Synth. Met. 130 (2002) 45

[52] M. Schirmeisen, F. Beck, J. Appl. Electrochem. 19 (1989) 401

[53] S. Biallozor, A. Kupniewska, Synthetic Metals 155 (2005) 443

[54] (a) P. Hulser, F. Beck, J. Appl. Electrochem. 20 (1990) 596; (b) F. Beck, P. H.ulser, R. Michaelis, Bull. Electrochim. 8 (1992) 35
[55] P. Hulser, F. Beck, J. Appl. Electrochem. 20 (1990) 596
[56] P. C. Lacase, C. A. Ferreira, S. Aeiyach, J. J. Aaron, French Patent PSACitroen no. 9214091-24/11 (1992)
[57] (a) W. Su, J. O. Iroh, Electrochim. Acta 42 (1997) 2685; (b) W. Su, J. O. Iroh, Electrochim. Acta 46 (2000) 1
[58] J. I. Martins, N. Bazzaoui, T. C. Reis, E. A. Bazzaoui, L. Martins, Synth. Met. 129 (2002) 221
[59] A. De Bruyne, J. L. Delplancke, R. Winand, J. Appl. Electrochem. 27 (1997) 867
[60] B. Zaid, S. Aeiyach, P. C. Lacaze, H. Takenouti, Electrochim. Acta 43(1998) 2331
[61] J. Petitjean, S. Aeiyach, J. C. Lacroix, P. C. Lacaze, J. Electroanal. Chem. 478 (1999) 92
[62] (a) K. Idla, O. Inganas, M. Strandberg, Electrochim. Acta 45 (2000) 2121; (b) S.T. Earley, D.P. Dowling, J. P. Lowry, C. B. Breslin, Synth. Met. 148 (2005) 111
[63] L. Makhloufi, H. Hammache, B. Saidani, Electrochem. Commun. 2 (2000) 552
[64] S. B. Saidman, O. V. Quinzani, Electrochim. Acta 50 (2004) 127
[65] K. Naoi, M. Takeda, H. Kanno, M. Sakakura, A. Shimada, Electrochem. Commun. 2 (2000) 552
[66] (a) P. H.ulser, F. Beck, J. Appl. Electrochem. 20 (1990) 596; (b) D.E. Tallman, C. Vang, G.G. Wallace, G. P. Biergwagen, J. Electrochem. Soc. 149 (2002) C173
[67] (a) W. Prissanaroon, N. Brack, P.J. Pigram, J. Liesegang, Synth. Met. 142 (2004) 25; (b) A.C. Cascalheira, S. Aeiyach, P. C. Lacase, L. M. Abrantes, Electrochim. Acta 48 (2003) 2523; (c) W.J. Hamer, L. Koene, J. H. W. de Wit, Mater. Corros. 55 (2004) 653
[68] (a) V.V. Abalayeva, I. I. Kogan, Synth. Met. 63 (1994) 109; (b) I. I. Kogan, V. V. Abalayeva, Synth. Met. 63 (1994) 153; (c) V. V. Abalayeva, O. N. Jefimov, Elektrokhim. (Rus.) 31 (1995) 598; (d) V.V. Abalayeva, O. N. Jefimov, Elektrokhim. (Rus.) 32 (1996) 728; (e) M. M. Gvozdenovi , B. N. Grgur, Prog. Organ. Coat. 65 (2009) 401; (f) D.O. Flamini, S. B. Saidman, Corros. Sci. 52 (2010) 229
[69] (a) D. Sazou, Ch. Georgolios, J. Electroanal. Chem. 429 (1997) 81; (b) A. Eftekhari, Synth. Met. 125 (2001) 295; (c) A. L. Correa-Borroel. S. Gutierrez, E. Arce . R. Cabrera-Sierra . P. Herrasti, J Appl Electrochem. 39 (2009) 2385
[70] (a) A.M. Fenelon, C.B. Breslin, Synth. Met. 144 (2004) 125 ; (b) S.K. Mondal, K.R. Prasad, N. Munichandraiah, Synth. Met. 148 (2005) 275
[71] (a) K.R. Prasad, N. Munichandraiah, Synth. Met. 123 (2001) 459; (b) K.R. Prasad, N. Munichandraiah, Synth. Met. 130 (2002) 17; (c) X.W. Guo, V.F. Jing, C.Q. Zhai, C. Lu, W.J. Ding, Synth. Met. 135 (2003) 169; (d) A. Olad, H. Rasouli, J. Appl. Pol. Sci. 115 (2010) 2221
[72] E. Armelin, A. Meneguzzi, C. A. Ferreira, C. Aleman, Surf. Coat. Technol. 203 (2009) 3763
[73] (a) M.D. Levi, E.Yu. Pisarevskya, E.B. Molodkina, A.I. Danilov, Synth. Met. 54 (1993) 210; (b) E.Yu. Pisarevskya, S.S. Sedova, A.L. Ignatov, N.A. Tomilova, Electrokhim. (Rus.) 30 (1994) 918
[74] S. Ren, D. Barkey, J. Electrochem. Soc. 139 (1992) 1021
[75] I. Marawia, A. Khaskelisa, A. Galalb, J. F. Rubinsona, R. P. Popatc, F. J. Boerioc, H. B. Mark Jr., J. Electroanal. Chem. 434 (1997) 61
[76] P. M. Koinkara, M. G. Wankhedea, M. A. Morea, P. P. Patil, S. A. Gangalb, Synth. Met. 130 (2002) 193
[77] M. I. Redondo, E. Sanchez de la Blanca, M.V. Garcia, M.J. Gonzalez-Tejera, Prog. Organ. Coat. 65 (2009) 386
[78] (a) W. Su, J.O. Iroh, Electrochim. Acta 44 (1999) 3321; (b) L.F. D'Elia, R.L. Ortiz, O.P. Marquez, J. Marquez, Y. Martinez, J. Electrochem. Soc. 148 (2001) C297
[79] Z. Mokhalif, F. Plumior, J. Delhalle, Appl. Surf. Sci. 212–213 (2003) 172
[80] V. Patil, S. R. Sainkar, P. P. Patil, Synth. Met. 140 (2004) 57
[81] A. Benchikh, R. Aitout, L. Makhloufi , L. Benhaddad, B. Saidani, Desalination 249 (2009) 466
[82] N. O. Pekmez, E. Abacı, K. Cınkıllı, A. Yagan, Prog. Organ. Coat. 65 (2009) 462
[83] Y. Zhu, J. Zhang, Y. Zheng, Z. Huang, L. Feng, L. Jiang, Adv. Funct. Mater. 16 (2006) 568
[84] (a) J. He, D. E. Tallman, G. P. Bierwagen, J. Electrochem. Soc. 151 (2004) B644; (b) G. Paliwoda-Porebska, M. Rohwerder, M. Stratmann, U. Rammelt, L. M. Duc, W. Plieth, J. Solid State Electrochem. 10 (2006) 730; (c) G. Paliwoda-Porebska, M. Stratmann, M. Rohwerder, K. Potje-Kamloth, Y. Lu, A. Z. Pich, H. J. Adler, Corros. Sci. 47 (2005) 3216; (d) L. Cecchetto, D. Delabouglise, J. P. Petit, Electrochim. Acta 52 (2007) 3485; (e) K. Kamaraj, S. Sathiyanarayanan, G. Venkatachari, Prog. Org. Coat. 64 (2009) 67; (f) M. Trueba, S. P. Trasatti, Prog. Org. Coat. 66 (2009) 265
[85] B. Parakhonskiy, D. Andreeva, H. Moehwald, D. G. Shchukin, Langmuir 25 (2009) 4780
[86] V. L. Finkenstadt, Appl. Microbiol. Biotechnol. 67 (2005) 735
[87] (a) V. L. Finkenstadt, J. L, Willett, J Polym. Environ. 12 (2004) 43; (b) V. L. Finkenstadt, J. L, Willett, Carbohydr. Polym. 55 (2004) 149
[88] V. L. S. Lopes, D. C. Dragunski, A. Pawlicka, J. P. Donoso, Electrochim Acta 48 (2003) 2021
[89] A. M. A. Nada, M. Dawy, A. H. Salama, Mater. Chem. Phys. 84 (2004) 205
[90] (a) A. M. Regiani, G. O. Machado, J. LeNest, A. Gandini, A. Pawlicka, Macromol. Symp. 175 (2001) 45; (b) C. Schoenenberger, J. LeNest, A. Gandini, Electrochim. Acta 40 (1995) 2281
[91] H. Weiss, O. Pfefferkorn, G. Kotora, B. D. Humphrey, J Electrochem. Soc. 136 (1989) 3711

[92] Y. Wan, K. A. M. Creber, B. Peppley, V. T. Bui, Polymer 44 (2003) 1057
[93] (a) R. H. Y. Subban, A. K. Arof, Phys. Scr. 53 (1996) 382; (b) R. H. Y. Subban, A. K. Arof, S. Radhakrishna, Mater. Sci. Eng. B. B38 (1996)156
[94] Z. Osman, A. K. Ibrahim, A. K. Arof, Carbohydr. Polym. 44 (2001) 167
[95] H. Yoshida, F. Takei, N. Sawatari, Fujitsu Sci. Tech. J 38 (2002) 39
[96] (a) K. Grandfield, F. Sun, M. FitzPatrick, M. Cheong, I. Zhitomirsky, Surf. Coat. Technol. 203 (2009) 1481; (b) X. Pang, T. Casagrande, I. Zhitomirsky, Journal of Colloid and Interface Science 330 (2009) 323
[97] T. Sugama, M. Cook, Progr. Organ. Coat. 38 (2000) 79
[98] (a) T. Sugama, Mater. Lett. 25 (1995) 291; (b) T. Sugama, J. E. DuVall, Thin Solid Films 289 (1996) 39
[99] J. H. Collier, J. P. Camp, T. W. Hudson, C. E. Schmidt, J Biomed. Mater. Res. 50 (2000) 574
[100] D. Zhou, C. O. Too, G.G. Wallace, React Funct Polym 39 (1999) 19
[101] A. J. Hodgson, M. J. John, T. Campbell, A. Georgevich, S. Woodhouse, T. Aoki, N. Ogata, G. G. Wallace, Proc SPIE 2716 (1996) 164
[102] a) G. Williams, H. N. McMurray, Electrochem. Solid-State Lett. 6 (2003) B9; b) G.Williams, H. N. McMurray, Electrochem. Solid-State Lett. 7 (2004) B13; c) R. G. Buchheit, H. Guan, S. Mahajanam, F. Wong, Prog. Org. Coat. 47 (2003) 174; d) R. G. Buchheit, S. B. Mamidipally, P. Schmutz, H. Guan, Corrosion 58 (2002) 3; e) W. Zhang, R. G. Buchheit, Corrosion 58 (2002) 591
[103] a) J. M. Yeh, C. L. Chen, Y. C. Chen, C. Y. Ma, K. R. Lee, Y. Wei, S. Li, Polymer 43 (2002) 2729; b) J. M. Yeh, S. J. Liou, C. Y. Lai, P. C. Wu, T. Y. Tsai, Chem. Mater. 13 (2001) 1131
[104] a) A. Mitra, Z. B. Wang, T. G. Cao, H. T. Wang, L. M. Huang, Y. S. Yan, J. Electrochem. Soc. 149 (2002) B472; b) D. E. Beving, A. M. P. McDonnell, W. S. Yang, Y. S. Yan, J. Electrochem. Soc. 153 (2006) B325; c) X. L. Cheng, Z. B. Wang, Y. S. Yan, Electrochem. Solid-State Lett. 4 (2001) B23; d) J. K. Choi, Z. P. Lai, S. Ghosh, D. K. Beving, Y. S. Yan, M. Tsapatsis, Ind. Eng. Chem. Res. 46 (2007) 7096
[105] a) L. N. Mitchon, J. M. White, Langmuir 22 (2006) 6549; b) G. HNhner, R. Hofer, I. Klingenfuss, Langmuir 17 (2001) 7047; c) H. Gao, N. N. Gosvami, J. Deng, L. Tan, M. S. Sander, Langmuir 22(2006) 8078; d) P. E. Hintze, L. M. Calle, Electrochim. Acta 51 (2006) 1761; e) I. L. Liakos, R. C. Newman, E. McAlpine, M. R. Alexander, Langmuir 23 (2007) 995
[106] a) D. G. Evans, X. Duan, Chem. Commun. (2006) 485; b) G. R. Williams, D. OOHare, J. Mater. Chem. 16 (2006) 3065; c) D. G. Evans, R. C. T. Slade, Struct. Bonding 119 (2006) 1; d) J. He, M. Wei, B. Li, Y. Kang, D. G. Evans, X. Duan, Struct. Bonding 119 (2006) 89
[107] S. Albertazzi, F. Basile, A. Vaccari, In Clay Surfaces: Fundamentals and Applications; Wypych, F., Satyanarayana, K. G., Eds.; Elsevier: Amsterdam, The Netherlands, 2004; p 497
[108] (a) A. Sorrentino, G. Gorrasi, M. Tortora, V. Vittoria, U. Constantino, F. Marmottini, F. Padella, Polymer 46 (2005) 1601; (b) F. Leroux, J.-P. Besse, Chem. Mater. 13 (2001) 3507
[109] S. J. Palmer, R. L. Frost, T. Nguyen, Coord. Chem. Rev. 253 (2009) 250
[110] (a) S.-Y. Kwak, W. M. Kriven, M. A. Wallig, J.-H. Choy, Biomaterials 25 (2004) 5995; (b) J.-H. Choi, S.-Y. Kwak, Y.-J. Jeong, J.-S. Park, Angew. Chem., Int. Ed. 39 (2000) 4041; (c) Z. P. Xu, Q. H. Zeng, G. Q. Lu, A. B. Yu, Chem. Eng. Sci. 61 (2006) 1027
[111] (a) R. G. Buchheit, S. B. Mamidipally, P. Schmutz, H. Guan, Corrosion 58 (2002) 3; (b) J.-Y. Uan, B.-L. Yu, X.-L. Pan, Metall. Mater. Trans. A 39A (2008) 3233; (c) F. Zhang, M. Sun, S. Xu, L. Zhao, B. Zhang, Chem. Eng. J. 141 (2008) 362; (d) J.-l. Yi, X.-M. Zhang, M.-A. Chen, R. Gu, Scr. Mater. 59 (2008) 955; (d) F. Zhang, L. Zhao, H. Chen, S. Xu, D. G. Evans, X. Dan, Angew. vChem., Int. Ed. 47 (2008) 2466; (e) W. Zhang, R. G. Buchheit, Corrosion 58 (2002) 591; (f) R. G. Buchheit, H. Guan, J. Coat. Technol. Res. 1 (2004) 277
[112] (a) R. B. Leggat, W. Zhang, R. G. Buchheit, S. R. Taylor, Corrosion 58 (2002) 322; (b) R. B. Leggat, S. A. Taylor, S. R. Taylor, Colloids Surf. A 210 (2002) 83
[113] (a) R. G. Buchheit, H. Guan, S. Mahajanam, F. Wong, Prog. Org. Coat. 47 (2003) 174; (b) S. P. V. Mahajanarn, R. G. Buchheit,R. G. Corrosion 64 (2008) 230
[114] (a) G. Williams, H. N. McMurray, Electrochem. Solid-State Lett. 7 (2004) B13; (b) G. Williams, H. N. McMurray, Electrochem. Solid-State Lett. 6 (2003) B9; (c) H. N. Murray, G. Williams, Corrosion 60 (2004) 219
[115] (a) B. Chico, J. Simancas, J. M. Veja, N. Granizo, I. Diaz, D. Fuente, M. Morcillo, Prog. Org. Coat. 61 (2008) 283; (b) X. Yu, J. Wang, M. Zhang, P. Yang, L. Yang, D. Cao, J. Li, Solid State Sci. 11 (2009) 376
[116] (a) B. K. G. Theng, In The Chemistry of Clay-Organic Reactions; Theng, B. K. G., Ed.; Wiley: New York, 1974; (b) F. Leroux, J.-P. Besse, In Clay Surfaces: Fundamentals and Applications; Wypych, F., Satyanarayana, K. G., Eds.; Elsevier: London, 2004
[117] M. Kendig, M. Hon, Electrochem. Solid-State Lett. 8 (2005) B10
[118] (a) S. K. Poznyak, J. Tedim, L. M. Rodrigues, A. N. Salak, M. L. Zheludkevich, L. F. P. Dick, M. G. S. Ferreira Appl. Mater. Interfaces 1(2009) 2353; (b) M. L. Zheludkevich, S. K. Poznyak , L. M. Rodrigues, D. Raps, T. Hack, L. F. Dick, T. Nunes, M. G. S. Ferreira, Corr. Sci. 52 (2010) 602
[119] (a) B. Wessling, J. Posdorfer, Synth. Met. 102 (1999) 1400

[120] (a) J. M. Yeh, C. L. Chen, Y. C. Chen, C. Y. Ma, K. R. Lee, Y. Wie, S. X. Li, Polymer 43 (2002) 2729; (b) J. M. Yeh, C. P. Chin, J. Appl. Polym. Sci. 88 (2003) 1072; (c) M.G. Hosseinia, M. Raghibi-Boroujenia, I. Ahadzadeha, R. Najjarb, M.S. Seyed Dorrajic, Prog. Organ. Coat. 66 (2009) 321. (d) J. M. Yeh, S. J. Liou, C. Y. Lai, P. C. Wu, Chem. Mater.13 (2001) 1131; (e) A. Malinauskas, J. Malinauskiene, A Ramanavicius, Nanotechnology 16 (2005) R51; (f) H.-W. Chen, T.-P. Lin, F.-C. Chang, Polymer 43 (2002) 5281;(g) B. Garcia, A. Lamzoudi, F. Pillier, H. N. T. Le, C. Deslouis, J. Electrochem. Soc. 149 (2002) B560; (h) S. H. Sonawane, B. M. Teo, A. Brotchie, F. Grieser, M. Ashokkumar Ind. Eng. Chem. Res. 49 (2010) 2200

[121] C. T. J. Low, R. G. A. Wills, F. C. Walsh, Surf. Coat. Technol. 201 (2006) 371

[122] (a) M. Musiani, Electrochim. Acta 45 (2000) 3397; (b) A. Hovestad, L. J. J. Janssen, J. Appl. Electrochem. 25 (1995) 519; (c) D. Clark, D. Wood, U. Erb, Nanostruct. Mater. 9 (1997) 755; (d) C. Kerr, D. Barker, F. C. Walsh, J. Archer, Trans. Inst. Met. Finish. 78 (2000) 171; (e) J. L. Stojak, J. Fransaer, J. B. Talbot, Adv. Electrochem. Sci. Eng. 7 (2001) 193; (f) W. Gao, Z. Li, Mater. Res. 7 (2004) 175; (g) C. T. J. Low, C. Ponce-de-leon, F.C. Walsh, Aust. J. Chem. 58 (2005) 246

[123] E. V. Skorb, D. G. Shchukin, H. Moehwald, D. V. Andreeva, Langmuir (2010) in press

[124] (a) A. Atkinson, D. L. Segal, J. Sol-Gel Sci. Technol. 13 (1998) 133; (b) C. J. Brinker, G. Scherrer, Sol-Gel Science: The Physics and Chemistry of Sol-Gel Processing, Academic Press, San Diego, CA, 1990

[125] (a) J. Livage, M. Henry, C. Sanchez, Prog. Solid State Chem. 288 (1988) 259; (b) L. L. Hench, J. K. West, Chem. Rev. 90 (1990) 33; (c) L. C. Klein, Sol-Gel Technology for Thin Films, Fibers, Preforms, Electronics and Specialty Shapes, Noyes Publications, Park Ridge, NJ, 1988

[126] (a) P.Gomez-Romero, C. Sanchez, Functional Hybrid Materials, Wiley-VCH, 2003; (b) Y. Chujo, T. Saegusa, Adv. Polym. Sci. 100 (1992) 11; (c) P. Judeinstein, C. Sanchez, J. Mater. Chem. 6 (1996) 511; (d) A. Morikawa, Y. Iyoku, M. Kakimoto, Y. Imai, J. Mater. Chem. 2 (1992) 679; (e) C. Sanchez, F. Ribot, New J. Chem. 18 (1994) 1007; (f) R. J. P. Corriu, Polyhedron 17 (1998) 925; (g) P. V. Braun, P. Osenar, V. Tohver, S. B. Kennedy, S. I. Stupp, J. Am. Chem. Soc. 121 (1999) 7302; (h) S. I. Stupp, P. V. Braun, Science 277 (1999) 1242

[127] A. J. Atanacio, B. A. Latella, C. J. Barbe, M. V. Swain, Surf. Coat. Technol. 192 (2005) 354

[128] (a) C. Sanchez, G. J. A. A. Soler-Illia, F. Ribot, C. R. Mayer, V. Cabuil, T. Lalot, Chem. Mater. 13 (2001) 3061; (b) B. M. Novak, Adv. Mater. 5 (1993) 422; (c) U. Schubert, N. Husing, A. Lorenz, Chem. Mater. 7 (1995) 2010

[129] D. A. Loy, K. J. Shea, Chem. Rev. 95 (1995) 1431

[130] C. Sanchez, B. Lebeau, F. Chaput, J-P Boilot, Adv. Mater. 15 (2003) 1969

[131] (a) J. H. Osborne, K. Y. Blohowiak, S. R. Taylor, C. Hunter, G. Bierwagen, B. Carlson, D. Bernard, M. S. Donley, Prog. Org. Coat. 41 (2001) 217 ; (b) T. L. Metroke, O. Kachurina, E. T. Knobbe, Prog. Org. Coat. 44 (2002) 185

[132] Y. J. Du, M. Damron, G. Tanga, H. Zheng, C.-J. Chu, J. H. Osborne, Progress in Organic Coatings 41 (2001) 226

[133] P. Kiruthika, R. Subbari, A. Jyothirmayi, K. Sarvani, N. Y. Hebalkar, Surfac. Coat. Technol. 204 (2010) 1270

[134] M. L. Zheludkevich, I. Miranda Salvadob, M. G. S. Ferreira, J. Mater. Chem. 15 (2005) 5099

[135] W. Feng, S. H. Patel, M.-Y. Young, J. L. Zunino, M. Xanthos, Adv. Polym. Technol. 26 (2007) 1

[136] (a) S. J. Rowan, J. P. Capadona, K. Shanmuganathan, D. J. Tyler, C. Weder, Science, 319 (2008) 1370; (b) R. Abbel, C. Grenier, M. J. Pouderoijen, J. W. Stouwdam, P. E. L. G. Lecle` re, R. P. Sijbesma, E. W. Meijer, A. P. H. J. Schenning, JACS, 131 (2009) 833; (c) S. Burattini, H. M. Colquhoun, J. D. Fox, D. Friedmann, B. W. Greenland, P. J. F. Harris, W. Hayes, M. E. Mackayc, S. J. Rowan, Chem. Commun. (2009) 6717

[137] (a) S. D. Bergman and F. J. Wudl, J. Mater. Chem., 18 (2008) 41; (b) D. Y. Wu, S. Meure and D. Solomon, Prog. Polym. Sci., 33 (2008) 479; (c) R. P. Wool, Soft Matter, 4 (2008) 400

[138] S. R. White, N. R. Sottos, P. H. Geubelle, J. S. Moore, M. R. Kessler, S. R. Sriram, E. N. Brown, S. Viswanathan, Nature, 409 (2001) 794

[139] X. Chen, M. A. Dam, K. Ono, A. Mal, H. Shen, S. R. Nutt, K. Sheran and F. Wudl, Science, 295 (2002) 1698

[140] P. Cordier, F. Tournilhac, C. Soulie.-Ziakovic, L. Leibler, Nature, 451 (2008) 977

[141] (a) S. H. Cho, H. M. Andersson, S. R. White, N. R. Sottos, P. V. Braun, Adv. Mater., 18 (2006) 997; (b) S. M. Bleay, C. B. Loader, V. J. Hawyes, L. Humberstone, P. T. Curtis, Composites, Part A, 32 (2001) 1767; (c) M. M. Caruso, D. A. Delafuente, V. Ho, N. R. Sottos, J. S. Moore, S. R. White, Macromolecules, 40 (2007) 8830; (d) J. W. C. Pang, I. P. Bond, Compos. Sci. Technol.,65 (2005) 1791; (e) J. W. C. Pang and I. P. Bond, Composites, Part A, 36 (2005) 183

[142] (a) X. Chen, A. Mal, H. Shen, S. R. Nutt and F. Wudl, Macromolecules, 36 (2003) 1802; (b) B. J. Adzima, H. Aguirre, C. J. Kloxin, T. F. Scott and C. N. Bowman, Macromolecules, 41 (2008) 9112; (c) V. S. Khire, Y. Yi, C. N. Bowman, Macromolecules, 41 (2008) 7440

[143] (a) C. Fouquey, J.-M. Lehn, A.-M. Levelut, Adv. Mater., 2(1990) 254; (b) S. H. M. Sontjens, R. P. Sijbesma, M. H. P. van Genderen, E. W. Meijer, Macromolecules, 34 (2001) 3815; (c) H. Kautz, D. J. M. van Beek, R. P. Sijbesma, E. W. Meijer, Macromolecules, 39 (2006), 4265; (d) S. Sivakova, D. A. Bohnsack, M. E. Mackay, P. Suwanmala, S. J. Rowan, JACS, 127 (2005) 18202

[144] D. Kotnarowska, Prog. Org. Coat. 31 (1997) 325

[145] A. Kentepozidou, C. Kiparissides, F. Kotzia, C. Kollia, N. Spyrellis, J. Mater. Sci. 31 (1996) 1175
[146] Z. Liqun, Z. Wei, L. Feng, Y. He, J. Mater. Sci. 39 (2004) 495
[147] C. Suryanarayana, K. Chowdoji Rao, D. Kumar, Prog. Org. Coat. 63(2008) 72
[148] (a) S. R. White, N. R. Sottos, P. H. Geubelle, J. S. Moore,M. R. Kessler, S. R. Sriram, E. N. Brown, S. Viswanathan, Nature 409 (2001) 794; (b) E. N. Brown, S. R. White and N. R. Sottos, J. Mater. Sci. 39 (2004) 1703; (c) B. J. Blaiszik, N. R. Sottos and S. R. White, Compos. Sci. Technol. 68 (2008) 978; (d) S. H. Cho, H. M. Andersson, S. R. White, N. R. Sottos, P. V. Braun, Adv. Mater. 18 (2006) 997; (e) S. H. Cho, S. R. White and P. V. Braun, Adv. Mater. 21 (2009) 645; (f) M. M. Caruso, D. A. Delafuente, V. Ho, N. R. Sottos, J. S. Moore, S. R. White, Macromolecules 40 (2007) 8830; (h) M. M. Caruso, B. J. Blaiszik, S. R. White, N. R. Sottos, J. S. Moore, Adv. Funct. Mater., 18 (2008) 1898; (g) B. J. Blaiszik, M. M. Caruso, D. A. McIlroy, J. S. Moore, S. R. White, N. R. Sottos, Polymer, 50 (2009) 990
[149] X. He and X. Shi, Prog. Org. Coat. 65(2009) 37
[150] Biswajit Ghosh and Marek W. Urban Science 323 (2009) 1458
[151] (a) T. Shutava, S. S. Balkundi, Y. M. Lvov, J. Colloid Interface Sci. 330 (2009) 276; (b) A. D. Price, A. N. Zelikin, Y. J. Wang , F. Caruso, Angew. Chem. Int. Ed. 48 (2009) 329; (c) G. B. Sukhorukov, H. Moehwald, Trends Biotechnol. 25 (2007) 93
[152] N. G. Veerabadran, R. R. Price, Y. M. Lvov, Nano 2 (2007) 115
[153] (a) N. Dong, A. K. Agarwal, D. J. Beebe, H. R. Jiang, Nature 442 (2006) 551; (b) T. Konno, K. Ishihara, Biomaterials, 28 (2007) 1770
[154] (a) S. Gupta, S. P. Moulik, J. Pharm. Sci. 97 (2008) 22; (b) A. S. Narang, D. Delmarre and D. Gao, Int. J. Pharm. 9 (2007) 9
[155] (a) K. Glinel, G. B. Sukhorukov, H. Mohwald, V. Khrenov, K. Tauer, Macromol. Chem. Phys. 2042 (2003) 1784; (b) B. S. Kim, O. V. Lebedeva, K. Koynov, H. Gong, G. Glasser, I. Lieberwith, O. I. Vinogradova, Macromolecules, 38 (2005) 5214
[156] T. Mauser, C. Dejugnat, G. B. Sukhorukov, J. Phys. Chem. B 110 (2006) 20246
[157] (a) C. C. Lin, A. T. Metters, Adv. Drug Delivery Rev. 58 (2006) 1379; (b) D. G. Shchukin, G. B. Sukhorukov, Adv. Mater. 16 (2004) 671; (c) S. K. Mehta, G. Kaur, K. K. Bhasin, Pharm. Res. 25 (2008) 227
[158] A. M. Javier, P. de Pino, M. F. Bedard, D. Ho, A. G. Skirtach, G. B. Sukhorukov, C. Planck, W. J. Parak, Langmuir 24 (2008) 12517
[159] X. Tao, J. B. Li, H. Moehwald, Chem.-Eur. J. 10 (2004) 3397
[160] D. G. Shchukin, D. V. Sviridov, J. Photochem. Photobiol. C. 7 (2006) 23
[161] A. Fery, R. Weinkamer, Polymer 48 (2007) 7221
[162] D. G. Shchukin, D. A. Gorin, H. Moehwald, Langmuir 22 (2006) 7400
[163] (a) R. V. Kulkarni, S. Biswanath, J. Appl. Biomater. Biomech. 5(2007) 125; (b) A. A. Antipov, G. B. Sukhorukov, Adv. Colloid Interface Sci. 111 (2004) 49
[164] K. Koehler, D. G. Shchukin, H. Moehwald, G. B. Sukhorukov, J. Phys. Chem. B 109 (2005) 18250
[165] E. V. Skorb, D. Fix, D. V. Andreeva, H. Moehwald, D. G. Shchukin, Adv. Funct. Mater. 19 (2009) 2373
[166] (a) G. B. Sukhorukov, E. Donath, S. Moya, A. S. Susha, A. Voigt, J. Hartmann, H. Mo.hwald, J. Microencapsulation 17 (2000) 177; (b) G. B. Sukhorukov, M. Brumen, E. Donath, H. Mo.hwald, J. Phys. Chem. B 103 (1999) 6434
[167] K. A. Yasakau, M. L. Zheludkevich, S. V. Lamaka, M. G. S. Ferreira, J. Phys. Chem. B 110 (2006) 5515
[168] (a) D. G. Shchukin, H. Mo.hwald, Small 3 (2007) 926; (b) D. G. Shchukin, H. Mo.hwald, Adv. Funct. Mater. 17 (2007) 1451
[169] (a) W. L. Zhang, G. S. Frankel, J. Electrochem. Soc. 149 (2002) B510; (b) P. Schmutz, G. S. Frankel, J. Electrochem. Soc. 145 (1998) 2295; (c) W. L. Zhang, G. S. Frankel, Electrochim. Acta 28 (2002) 1193
[170] (a) E. V. Skorb, D. G. Shchukin, H. Möhwald, D. V. Sviridov, J. of Mater. Chem. 19 (2009) 4931; (b) E. V. Skorb, D. V. Sviridov, H. Möhwald, D. G. Shchukin, Chem. Comm. (2009). 6041; c) E. V. Skorb, A. Skirtach, D. V. Sviridov, D. G. Shchukin, H. Möhwald, ACS Nano 3 (2009) 1753
[171] a) B. Singh, R. Gilkes, Clays Clay Miner. 40 (1992) 212; b) Y. Lvov, R. Price, B. Gaber, I. Ichinose, Colloids Surf. A 198 (2002) 375
[172] R. Price, B. Gaber, Y. Lvov, J. Microencapsulation 18 (2001) 713
[173] (a) Y. Lvov, D. Shchukin, H. Moehwald, P. Price, ACS Nano 2 (2008) 81;(b) E. Abdullayev, D. Shchukin, Y. Lvov, Polym. Mater. Sci. Eng. 99 (2008) 331
[174] D. G. Shchukin, H. Mohwald, Adv. Funct. Mater. 17 (2007) 1451
[175] E. Abdullayev, R. Price, D. Shchukin, Y. Lvov, Appl. Mater. Interfaces 7 (2009) 1437
[176] C. Sease, Stud. Conserv. 23 (1978) 76

11 Standards and guidelines

Jörg Sander

Like every industrial activity, the coating of metallic work pieces is subject to numerous standards, norms, guidelines and regulations. These have been primarily developed to ensure a common agreement on reliable quality of operational practice and performance. However, gaining importance with the awareness of the protection of resources and welfare, standards are also intended to create a binding framework for sustainable and responsible manufacturing.

Standards and regulations that are relevant to quality and environment while treating and coating metals in order to prevent corrosive decay, have been mentioned throughout the book where appropriate. The following list provides a collection of these norms. It is not intended to be complete. The reader is referred to standard literature, in particular to the Handbook of Coating Technology by A. Goldschmidt and H.-J. Streitberger [1], as well as the book on Coil Coating by B. Meuthen and A.-S. Jandel [2].

11.1 General information

- American Society for Testing and Materials (ASTM) International, W. Conshohocken PA, USA (www.astm.org)
- Dept. of Defense, Defense Logistics Agency, Document Automation & Production Service, Philadelphia PA, USA (www.dodssp.daps.dla.mil/...)
- European Aluminium Association (EAA), Brussels (www.eaa.net)
- European Coil Coating Association (ECCA), Brussels (www.prepaintedmetal.eu)
- European Committee for Iron and Steel Standardization (ECISS), Brussels (www.cen.eu/cen/Sectors/Sectors/Materials/Pages/ECISS.aspx)
- European Committee for Standardization (CEN), Brussels (www.cen.eu)
- Deutsches Institut für Normung (DIN), Berlin (www.din.de)
- Gütegemeinschaft für die Beschichtung von Bauteilen (GSB International), Schwäb. Gmünd (www.gsb-international.de)
- International Organization for Standardization (ISO), Geneva (www.iso.org)

11.2 General norms

Colour	DIN EN 13523-3
Corrosivity classification	ISO 12944-2
Designation of intensity, quantity, and size of defects	EN ISO 4628
Patch test to assess coating compatibility	ASTM 5064-95

Jörg Sander et al.: Anticorrosive Coatings

ISBN 978-3-86630-911-1

11.2.1 Norms on mechanical testing of organically coated metallic work pieces

Bending; coating elasticity, crack resistance (T-bend)	DIN EN 13523-7 ASTM D-4145-83
Adhesive tape; substrate adhesion	ditto, ASTM 3359-95
Coating thickness, magnetic and Eddy current methods	DIN EN 13523-1
continuous measurement by β-back-scattering	DIN EN ISO 3543
continuous measurement by thermography	DIN EN 15042-2
Cross-hatch	DIN EN ISO 2409
Erichsen dome test (Erichsen cupping)	DIN EN 13523-6
Foil adhesion, protective foils	DIN EN 13523-17
Hardness, pencil; scratch resistance	DIN EN 13523-4
Buchholz; resistance to cuts	DIN EN ISO 2815
Heat resistance; mechanical effect of ageing	DIN EN 13523-13
Impact; cracking resistance to fast deformation	DIN EN 13523-5 DIN EN ISO 6272
MEK Test; curing (cross-linking) and adhesion	DIN EN 13523-11
Peel Tests, T Peel test	ASTM D-1876-08
Pull-off Test	DIN EN ISO 4624

11.2.2 Norms on corrosion testing of organically coated metallic work pieces

Constant climate tests	
Boiling, water soak	EN 13523-9
Condensation	EN ISO 6270-1 and 2, pr EN 13523-26
Filiform	EN ISO 4623-2, EN ISO 4628-10, EN 3665
Humidity	DIN 50017 KK, pr EN 13523-25
Salt spray - neutral	EN ISO 9227 NSS (DIN 50021) ASTM B117, EN 13523-8
Salt spray - acetic acid	EN ISO 9227 AASS
Salt spray - copper acetate	EN ISO 9227 CASS
Cyclic climate tests	
Cyclic immersion test	ISO 15710, EN 3212

Cyclic salt spray/humidity tests	
- Automotive industry	VDA 621-415 Renault ECC1 D172028 VW PV1210 Volvo VCS1027,149
- Prohesion test	ASTM G85,A5
Humidity/temperature cycle tests	DIN 50017 KFW pr EN 13523-25
Humidity in SO_2 atmosphere	DIN EN 13523-13 (DIN 50018)
Weathering tests	VDA 612-430
Outdoor exposure	
Outdoor exposure testing	EN 13523-19

11.3 Selected European legislation on environmental protection

ELV

Directive 2000/53/EC, End of Life Vehicles [3]

WEEE

Directive 2002/96/EC, Waste Electrical and Electronic Equipment [4]

COM/2008/0809 final - COD 2008/0240 [5]

COM/2008/0810 final - COD 2008/0241 [6]

RoHS

Directive 2002/95/EC, Restriction of Hazardous Substances [7]

REACH

Regulation (EC) No 1907/2006, Registration, Evaluation, Authorisation and Restriction of Chemicals [8]

Surfactants, etc.

Regulation (EC) No 1272/2008, Classification, labelling and packaging of substances [9]

Commission Decision 2007/506/EC, Eco-label – Aerobic Biodegradability of Surfactants [10]

Volatile organic compounds, etc.

Directive 2008/1/EC, Integrated pollution prevention and control (IPPC) [11]

Council Directive 1999/13/EC, Volatile Organic Compounds [12]

11.4 Literature

[1] Goldschmidt, A., Streitberger, H.-J., Basic of Coating Technology, 2nd edition, BASF Coatings AG, Münster 2007, pp. 751 ff

[2] Meuthen, B., Jandel, A.-S., Coil Coating, 2nd ed., Vieweg, Wiesbaden 2008, p. 307 ff

[3] anon., Directive 2000/53/EC of the European Parliament and of the Council of 18 September 2000 on end-of life vehicles, Official Journal of the European Union, 2000, L 269, pp. 34 ff; eur-lex.europa.eu/...

[4] anon., Directive 2002/96/EC of the European Parliament and of the Council of 27 January 2003 on waste electrical and electronic equipment (WEEE) - Joint declaration of the European Parliament, the Council and the Commission relating to Article 9, Official Journal of the European Union, 2003, L 37, pp. 24 ff; eur-lex. europa.eu/...

[5] anon., COM/2008/0809 final - COD 2008/0240, Proposal for a Directive of the European Parliament and of the Council on the restriction of the use of certain hazardous substances in electrical and electronic equipment (recast) {SEC(2008) 2930} {SEC(2008) 2931}; eur-lex.europa.eu/...

[6] anon., COM/2008/0810 final - COD 2008/0241, Proposal for a DIRECTIVE OF THE EUROPEAN PARLIAMENT AND OF THE COUNCIL on waste electrical and electronic equipment (WEEE) (Recast){SEC(2008) 2933} {SEC(2008) 2934}; eur-lex.europa.eu/...

[7] anon., Directive 2002/95/EC of the European Parliament and of the Council of 27 January 2003 on the restriction of the use of certain hazardous substances in electrical and electronic equipment, Official Journal of the European Union, 2003, L 37, pp. 19 ff; eur-lex.europa.eu/...

[8] anon., Regulation (EC) No 1907/2006 of the European Parliament and of the Council of 18 December 2006 concerning the Registration, Evaluation, Authorisation and Restriction of Chemicals (REACH), establishing a European Chemicals Agency, amending Directive 1999/45/EC and repealing Council Regulation (EEC) No 793/93 and Commission Regulation (EC) No 1488/94 as well as Council Directive 76/769/EEC and Commission Directives 91/155/EEC, 93/67/EEC, 93/105/EC and 2000/21/EC, Official Journal of the European Union 49, 2006, L 396; eur-lex.europa.eu/...

[9] anon., Regulation (EC) No 1272/2008 of the European Parliament and of the Council of 16 December 2008 on classification, labelling and packaging of substances and mixtures, amending and repealing Directives 67/548/EEC and 1999/45/EC, and amending Regulation (EC) No 1907/2006, Chapter 4.1.2.9. Rapid degradability of organic substances, Official Journal of the European Union L 353, 2008, pp. 1 ff; eur-lex. europa.eu/...

[10] anon., 2007/506/EC: Commission Decision of 21 June 2007 establishing the ecological criteria for the award of the Community eco-label to soaps, shampoos and hair conditioners (notified under document number C(2007) 3127, Appendix "Environmental Criteria", Part 3.a) Aerobic Biodegradability of Surfactants, Official Journal of the European Union L 186 , 2007, pp. 36 ff; eur-lex.europa.eu/...

[11] anon., Directive 2008/1/EC of the European Parliament and of the Council of 15 January 2008 concerning integrated pollution prevention and control (Codified version) (1), Official Journal of the European Union 51, 2008, L 24; eur-lex.europa.eu/...

[12] anon., Council Directive 1999/13/EC of 11 March 1999 on the limitation of emissions of volatile organic compounds due to the use of organic solvents in certain activities and installations, Official Journal of the European Union 42, 1999, L 85; eur-lex.europa.eu/...

Authors

Dr. Jörg Sander, born 1958, obtained his Doctor's degree in chemistry at the University of Münster, Germany. 1987 he joined the Henkel Group, where he dealt with process chemicals for the surface technology of metals from both a technical and a commercial perspective for more than 20 years. Serving the target markets of steel and aluminium strip making and finishing, the production of beverage cans and the wire and extrusion industry, he managed teams in the research, product development, sales, and marketing functions. Between 1996 and 2009 he was Official Delegate for the Henkel Group to the European Coil Coating Association (ECCA), where he chaired the Appliance Task Group and related marketing project teams. Currently, he is working as industry consultant and free scientific writer.

Dr. Lars Kirmaier was born in Berlin in 1970. He studied Chemistry in Oldenburg/Germany and completed his dissertation about highly-reactive silicon intermediates in 1998. After further education in marketing and management he joined the R&D department of a leading German paint company and gained hands-on experience in solvent-free and high-solids polyurethane and epoxy resin applications. He is now with Heubach GmbH since January 2004 as laboratory manager for anticorrosive pigments. Apart from his research work he lectures on corrosion and corrosion prevention technology.

Mircea Manea was born in Bucharest, Romania. He graduated from the Polytechnic Institute of Bucharest 1976 in the field of Chemistry and Technology of Monomers and Polymers. He occupied several positions in the R&D department for coating binders, and later was appointed as resin production manager. In 1991 he entered the R&D of Beckers Acroma AB, Sweden until 1995 and later as laboratory and technical manager for Wedevåg Färg AB, Sweden and Wedevåg-Hesse Coatings AB. In 1998, he moved to Perstorp Specialty Chemicals AB, Sweden as Senior R&D Associate until 2007. He is presently acting as consultant for the PropanRaya Group, Indonesia and South East Asia. Mircea Manea has been employed as Associate Professor at the Institute of Chemistry in Bucharest between 1978 and 1987, lecturing in the field of general industrial chemistry. Since 2003 he is member of the EU expert panel for research activities and growth program (Six and Seventh Framework Program).

Dr. Dmitry G. Shchukin is a group leader at the Max Planck Institute of Colloids and Interfaces, Potsdam, Germany. He obtained his PhD in Physical chemistry in 2002. He was an Alexander von Humboldt fellow (2004 to 2005). From 2005 to 2007 he joined the Max Planck Institute of Colloids and Interfaces. He was awarded the NanoFutur Price of the German Ministry of Education and Research in 2007. His main scientific interests concern the fabrication of hollow nanocontainers and the development of feedback active coatings, interfacial sonochemistry.

Dr. Ekaterina Skorb obtained her PhD in Physical chemistry in 2008. She was LG Chem fellow (2006) at Belarusian State University and German Academic Exchange Service (DAAD) until 2008. Currently, she works at Alexander von Humboldt University at the Max Planck Institute of Colloids and Interfaces. Her main scientific interests concern the new generation of active anticorrosion and antibacterial coatings with feedback properties as well as the formation of metal films under ultrasonic excitation in solution.

Index

A

accelerated corrosion test 113, 124, 127, 130, 134, 144
acetic acid 127
acetic acid salt spray 119
acid rain 119, 120, 129
acrylate 21, 121
acrylic 113
acrylic melamine 113
acrylic resin 53, 118
active anion 18
active pigment 73, 119
addition of halogens 45
adhesion 18, 19, 21, 123, 127, 129, 133, 139, 140, 147, 155
adhesive bonding 117
advancement processes 49
AFM, atomic force microscopy 112
aged coating 155
aircraft 80, 111, 128, 129, 143, 146
alcohol 121
aldehyde 123
alkali 17, 18, 132, 139, 140
alkaline cleaner, alkaline cleaning 117, 124, 146, 147, 157
alkaline passivation 140, 141
alkalinity 17, 18, 147
alkali silicates 63
alkoxysilanes 66
alkyd 39, 51, 155
alloy, alloying 16, 18, 120, 123, 124, 132, 137-139, 141, 145, 147
alumina, aluminium oxide 122, 138, 143
aluminate 122
aluminium, Al 16-18, 113, 117, 119, 123, 124, 127, 129, 132, 138, 139, 141-150, 156
aluminium flakes 86
aluminium hydroxide 119, 138
aluminium tripolyphosphate 80
amino 100, 111
aminopropyl triethoxy silane, APS 146
ammonium hydrogen fluoride 142
ammonium zirconyl carbonate 99
anion, anionic 59, 120
annealing 124, 137
anodic 123, 124
anodising 124, 143, 147
anticorrosion properties 159
appliance 19, 110, 127, 137, 141
aqueous cleaner 157
aqueous dispersions 99
aqueous pretreatment 144, 146, 150
architecture 17, 19, 113, 118, 123, 143
aromatic polyester polyol 123
atom 15, 16, 121
atomic force microscopy, AFM 112, 122
automobile, automotive 116, 117, 127, 129, 135, 146, 156

B

backbone chemistry 104
backing coat 110
band model 16
barium metaborate 83
barrier effect 16, 18
barrier pigments 73, 85
barrier properties 47
basecoat 20
basic catalysis 50
bauxite 138
bending 110, 129, 137
benefits 55
benzyl alcohol 157
beverage can 111
binder 38, 129, 133, 155, 156
bioacrylic monomers 53
biopolymers 159, 173
bisphenol A 48, 50
bis-(triethoxysilyl) ethane, BTSE 146
black pigments 88
blister 127, 128, 129, 133, 134
blocking agents 100
blue pigments 88
Bode plot 120, 132
boiling 119, 127, 129
box section 18
brass 147
bridge engineering 17
brittle, brittleness 101, 104, 111, 112, 143
brush and scratch resistance 106

C

calcium aluminate 149
calcium borosilicate 83
calcium-modified silicate 20
calcium phosphate 141
canning 127, 142
capacitance 113, 120, 131, 132
carbamides 121

car body 18, 110, 116, 117, 141, 149
carbonate 17, 89, 137, 138
carbon dioxide 17, 19, 121
carbon steel 137
carbonyl compound 121, 123
carboxylate 100
carcinogenic 148
car component 156
car manufacture 20, 137, 141
casting 137, 139
cathode, cathodic 18, 115, 122, 132, 133
cathodic electro-dip primer 117, 122
cathodic protection 17, 18, 156
cationic 59
cationic processes 104
cationic UV coating 105
caustic soda 140
cavity sealing 116
cementation 74, 87
cerate 146
chalking 120, 127
chemical cross-linking 100
chemical resistance 19, 21, 99, 100, 127, 155
chemical vapour deposition, CVD 146
chlorate 141
chloride 17, 20, 73, 128
chlorinated rubber 46
chromate, chromating 18, 20, 99, 113, 114, 118, 119, 124, 140-144, 147-149, 156
chromate pigments 74
chromic acid 142, 143
chromiting 143
chromium chromate 142
chromium-free 20, 113, 114, 116, 118, 124, 140-149
chromium phosphate 124
chromous 142
chromous nitrate 143
chromous oxide 147
chromous oxy-hydroxide 118
chromous phosphate 142, 150
chromous sulphate 143
cladding 135, 137, 143, 144
clay 19, 138
cleaner, cleaning 111, 139, 141, 147, 149, 157
clearcoat 20, 113, 114, 115
cobalt amine 146
coil coating 19, 80, 101, 105-107, 110, 113, 118, 140, 143-147, 150
coin scratch 111
cold-rolled steel 117, 119, 137, 140-145
colorant 127
complexant 140, 143
complex fluoride 142
condensation 40, 118, 123, 127, 129, 155
conducting band 16
conducting polymer, conductive polymer 20, 159, 166
conductive pigment 116
constant climate 119, 127, 128, 129
continuous processes 66
conversion coating 16, 20, 21, 118, 140-144, 156
copolymerisation 46
copper sulphate 127
coprecipitate 119
corrosion inhibitor 162, 165, 172, 189, 190
corrosion protection primer, CPP 18, 110, 116, 156, 157
corrosion resistance 16, 18, 21, 106, 127, 128, 138-144
CPP, corrosion protection primer 18, 116-118, 157
creepage 114, 127, 128, 140
cross hatch 111, 112
cross-linking 19, 21, 101, 121, 146, 156
crown cork 110
cryolite 138
cupping 118
curing of alkyds by condensation 43
curing of alkyl silicates 68
current 130-132, 143
current density 17, 132
cut edge 17, 127, 132
cutting 110
cyclic carbonate groups 60
cyclic climate 127, 129
cyclic corrosion, cyclic corrosion test 114, 117
cyclic terpenes 44
cyclisation 40
cycloaddition 57
cyclorubber 45
cyclovoltammetry 131

D

decavanadate 120
decay of alkyd resins 41
decomposition 121
deep-drawing 112, 137
deformation 120
degradation 19, 120, 123, 127, 131, 132
deicing salts 120
deionised water, DI 129, 149
delamination 20, 112, 115, 118, 122, 130, 133, 134
di- and tetrafunctional siloxanes 69
dichromate 20
dielectric 19, 113, 131
diffusion 16, 19, 21, 122, 134
difunctional siloxane 69
diisocyanate 123, 156
dimethyl formamide 110
discoloration 101, 106, 127, 147
domestic cleaner 127
double bonds 57
drilling 110

dripping edge 113
dry ice 156
dual cure systems 104
duroplastic 101, 112, 123
dye 127, 143

E

ε-caprolactam 101
EB curing 105
economic 102
EIS, electrochemical impedance spectroscopy 113, 118-122, 131, 132
elasticity 112, 121
electrical resistance, resistivity 18, 131
electroactive conducting polymers
ECP 85
electrochemical 122, 124
electrochemical corrosion cell 72
electrochemical potential 15, 17, 20, 124, 130, 131, 137
electrochemical process 15
electrodeposition 53
electrode potential 17
electro-dip primer 18, 129, 141, 146
electro-galvanised 114, 116, 117, 134, 138, 141
electrolyte 16-21, 112, 113, 120, 122, 131-133, 138, 143
electromagnetic spectrum 102
electromotive series of metals 72
electron 15-17, 21, 122, 130, 131
electron beam (EB) 101
electro-sheet 137
ELV, End of Life Vehicles Directive 148
energy saving 101
environmental efficiency 102
epichlorohydrine 48, 50
epoxidated oils 51
epoxy amine 113, 134
epoxy, epoxies 19, 21, 47, 111, 116, 119-121, 124, 156
epoxy-polyamine 121
epoxy/polyester 143
equivalent circuit 131
Erichsen 111, 119, 120
ester 121, 123
ether 100, 121
ethoxysilanes 67
EU Directive 148
European Coil Coating Association, ECCA 135
extender pigments 89

F

façade 113
fatty acid chain, decay 42
ferrate 18
filiform corrosion, FFC 123, 124, 128, 129
filler 20, 156
film forming agent 38
film forming mechanism 40
film forming temperature 99
final parameters 41
fingerprint resistance 147
flange, flanging 18, 110
flash rust 155, 156
flexibility 19, 21, 155
Florida test 135
fluoride 142, 147, 148
fluorinated polymer 156
food can 110, 111
foodstuff 127
formic acid 157
frame profile 143
furniture 137

G

Galfan 113, 120, 132, 138, 141
Galvalume 120, 138, 141
galvanised steel 113, 120-123, 127, 129, 132, 138, 142-144, 150
galvanising 16, 17, 137, 145, 147
garage door 143
German Automobile Association, VDA 118, 129
gibbsite 138
glass transition temperature, T_g 19, 112
gloss 21, 109, 120, 127, 140
graphite 116
green chromate 142, 150
green pigments 88

H

haematite 137
halloysite 187, 190
hang-on part 116
hardness 99
heavy machinery 17
hem flanging 117
hexafluoro silicic acid 142
hexafluoro titanate 147
hexafluoro titanic acid 142
hexafluoro zirconate 124
honing 110
hopeite 141
hot-dip galvanised steel 116, 119, 132, 138, 140, 141
humidity 127-129, 133-135, 155, 156
hybrid anticorrosive inhibitors 84
hydrocarbon 146
hydrochloric acid, hydrogen chloride 121, 123, 128
hydrofluoric acid 139
hydrogen bridge bond 99, 111, 122

hydrogen peroxide 147
hydrolysis 68, 120–122
hydrolytical decay 42
hydroperoxide 121
hydrotalcite 17, 19, 20, 119
hydrothermal sealing 123
hydrothermal stress 110, 121
hydroxide 17
hydroxy chloride 17
hydroxyl amine 141
hydroxy silane 146

I

immersion 127, 129, 140–142, 149, 150
impact, impact test 110, 112, 118–120
induction 140
insertion 57
intercoat adhesion 109
interstice 19, 99, 131
intramolecular cyclisation 45
ion exchange 120
ion exchange pigments 84
IR, infrared 118, 130, 140, 156
iron, Fe 17, 137, 140, 141, 156
iron oxide 122, 137, 156
iron phosphate, iron phosphating 140–143, 146, 156
iron phosphide 83
iron tannate 156
isocyanate 100, 156
isocyanate reaction 56
isocyanic acid 55

K

kaolin 90
ketone 123

L

lactams 101
lactones 101
layer-by-layer, LBL 163
layered double hydroxides 177
lead pigments 74
lighting 137
lime 137, 148, 149
lithium silicates 63
long oil 51
lubrication 147

M

magnesium phosphate 141
magnetite 137
maleic anhydride modified alkyd resins 40
malonic acid 101
mandrel 111
manganate 18
manufacturing 53
manufacturing of acrylic resins 54
manufacturing processes 41
marine 19
mechanical cleaning 137, 156
mechanical hardness 101
mechanical resistance 21, 99
medium oil 51
mercapto 111
metal/coating interface 15, 131
metallic state 15
metal soap 122
metal working 137
metasilicate 139
metavanadate 119
methyl ethyl ketone 110
mica 90
micaceous iron oxide 21, 73, 85, 156
microcapsules 184, 185
Mil Spec 143
mixed-metal construction 117
modification of epoxy esters 52
modified orthophosphates 78, 81
modified polyphosphates 80
moisture 104
molybdate 18, 20, 83, 141
monofunctional siloxanes 68
mould 120
multi-functional acrylates 104
mutagenic 148

N

nano 19, 143
nanocomposites 177, 180
nanocontainers 159, 173, 178, 183–192
nanomaterials 177
native oxide 17, 18, 122, 132
natural rubber 44
NCO free polyurethanes 59
nitrate 147
nitric acid 147
nitrite 141
nitro compound 141
noble metal 131, 139
non-ferrous metal 147
non-ionic type 59
non-resistant coatings 101
no-rinse 113, 141, 144, 150
Norrish Type I reaction 103
N-substituted carbamic acid 121
Nyquist plot 132

O

offshore 155
organically modified zinc phosphate 79
organic corrosion inhibitors 80
organic film former 38
organic inhibitors 81
organosilanes 65
orthophosphates 80
outdoor exposure 124, 134
oxazoline groups 60
oxidation 72, 121
oxidative drying 43
oxide 122, 123
oxidic ore 137
oxy chloride 17
oxygen, O, oxygen reduction 16-21, 103, 120-123, 134, 137, 156
oxy-hydroxide 132

P

paint stripper 157
PAni, polyaniline 20
passivation 16, 73
peel test 111, 113
permanganate 18
phase angle plot 120
phenolic resin 123
phenol modified alkyd resins 40
phospho-molybdate 124, 146
phosphonate 146
phosphophyllite 141
phosphoric acid 141-143, 156
photochemical 120
photochemical resistance 127, 129
photo-chemistry 101
photo electron spectroscopy, XPS 17, 144
photoinitiator 102
pickling 124, 137, 139, 156
pipeline construction 155
pitting 17
plasma discharge lamps 107
plasma vapour deposition, PVD 146
plastisol 21
polyacids 43
polyacrylate 156
polyalkylene 17
polyaniline, PAni 20, 170
polyaspartics 61, 63
polyelectrolyte multilayers 163-165
polyepoxy backbone 51
polyester 21, 41, 111, 113, 116-124, 143, 144, 156
polyester-melamine 121
polyester-urethane 121
polyether amine 61
polymer 19, 21, 38, 99, 112, 122, 123, 144, 145
polyols 43
polyphosphates 80
polypyrrole 167, 168
polysiloxane 68, 146, 156
polyurea 61
polyurethane 19, 21, 61, 111, 116, 121, 124, 156
polyvinyl chloride, PVC 21, 111, 113, 121
polyvinylidene difluoride, PVDF 124
post-rinse 140, 141
potassium silicate 63
powder paint, powder coating 17, 124, 143
preprimed 110
pretreatment 18, 109, 111-118, 123, 124, 132, 139-151
primer 17-21, 109, 121, 123, 124, 139, 140, 147
primer-pretreatment 118, 120, 147
profiling 110
prohesion 114, 129
protic 111
protic compounds 100
punch-hole 113
PVC, polyvinyl chloride 21, 111
PVDF, polyvinylidene difluoride 124

Q

QUV-A 121
QUV-B 118

R

radiation curing systems 101
radical 123
radical anion 123
railway vehicles 111
Raman 130
raw materials 41
REACH, registration, evaluation, authorisation and restriction of chemicals 142, 148
red lead 74, 155
red-ox 15, 18, 130, 131, 133
red pigments 88
reduction 72
relative humidity 113
repair coating 155
resin 38
retortability 127
rinsing 141, 145
RoHS, Regulation on Hazardous Substances 148
roll-coater 118, 150, 151
roller shutter 143
rolling 124, 137
roll profiling 137
roofing 135, 137, 143

rust 137, 155, 156
rust converter 156

S

sacrificial anode 17
sacrificial pigments 73, 86
salt spray 118-121, 127-129, 135, 143
SAM, self-assembling molecule, self-assembled monolayers 146, 159, 161, 163
sanding 156
saponification 121, 122
Scanning Kelvin Probe, SKP 114, 132, 133
scanning vibrating electrode technique, SVET 132, 160, 179
scratch resistance 139
screw fastening 110
scribe 124, 127, 128
sealants 106
seam 18
seam sealing 116
self-healing 18, 159, 164, 178, 183-192
self-polymerisation 50
semiconductor 16
shear test 111
shipbuilding 17, 110, 155
short oil 51
shot-blasting, grit-blasting 156
shrinkage 101, 104
silane 146
silicas 89
silicate 89, 137, 138
silicone modified alkyd resins 40
siloxane 68, 70, 146
SKP, Scanning Kelvin Probe 114, 115, 121, 122, 132-134
slag 137
sodium bisulphite 148
sodium chloride 114, 119, 127
sodium dihydrogen phosphate 141
sodium hydroxide 138
sodium silicates 63
solarisation 134, 135
solar radiation 129
sol-gel 162, 179-182
solvent 99, 100-102, 106, 107, 120, 147, 149, 155-157
solvent cleaning 132, 157
spinel 140
spot welding 117
spray, spraying 140-142, 149, 150, 156, 157
spray-cell 150
spray pretreatment 113
stainless steel 137, 147, 150
standard hydrogen electrode 130
standard potential 115
steel 17, 20, 113, 117, 122, 127-129, 135, 137, 138, 140-145, 150, 156
stone chipping 110, 111
strontium chromate 20, 77, 120
substrate 15-21, 109, 116, 118, 122, 127, 129, 131, 133, 134, 137-149
substrate adhesion 47
sulfidic ore 137
sulphate 73, 89
sulphur dioxide 120, 129, 134
sulphuric acid 143, 147
sun blinds 113
supramolecular system 63
surface tolerant coating 155
surfactant 141
swelling 19, 120, 122, 133
synergy 79, 80, 84
synthesis of polyaspartics 63

T

talc 89
tannic acid, tannin 156
T-bend 111, 114, 118, 120
temperature-sensitive substrates 101
temperature, thermal resistance 129
tetraalkoxysilanes 65
tetraethyl orthosilicate 65, 67
thermal decomposition, thermal destruction 120
thermogravimetric analysis, TGA 119
thermoplastic 99
thermoplastic acrylic resins 55
thermosetting acrylic resins 55
thiol 100
titanium dioxide 19, 87
titanium fluoro complex 113
titanium phosphate 141
topcoat 20, 21, 118, 121-124, 139, 140, 147
toxic to reproduction 148
tree gum 120
tri-cation phosphating 117, 141, 146
trifunctional siloxanes 69
trigger 183, 187, 188, 189, 192
trivalent chromium 141, 142, 148
tropical test 129
tungstate 18

U

ultra-high pressure water jet 155
ultrasound 160, 179, 193
ultraviolet, UV 21, 101, 120, 121, 127, 129, 134, 135, 156
under-curing 101
urea 121
urethane 121, 123

urethane modified alkyd resins 40
utility vehicles 111
UV equipment 107
UV lamps 106
UV initiated decay 42
UV photopolymerisation 102
UV polymerisation 103
UV resistance 121

V

valence band 16
vanadate 18, 20, 120
van-der-Waals force 111, 112
VDA, German Automobile Association 114, 118, 129
vehicle 38
V-groove panel 113
vinylated alkyd resins 40
vinylation 40
viscosity 99, 150
VOC, volatile organic compound 149
voltage 130, 131

W

wash primer 156
waste water 144, 148, 149, 150
water 18–21, 99, 100, 101, 113, 121, 133, 134, 140, 144, 150, 155
waterborne 118, 150, 156
waterborne alkali silicates 64
waterborne polysiloxanes 72
water engineering 17, 110, 155
water resistance 127, 129, 155, 156
water uptake 19, 113, 122, 133
weathering 118, 127, 137
weathering resistance 129
WEEE, Waste Electrical and Electronic Equipment Directive 148
welding 117
wettability 147, 155
wetting 19, 157
wheel 127, 146
white metal 147
wide spectrum anticorrosives 81
Williamson reaction 48
wollastonite 74, 90
wuestite 137

X

XPS, photo electron spectroscopy 17

Y

yellowing 101
yellow pigments 88

Z

zeolite 20
zinc aluminium phosphate 79
zinc, Zn 17, 18, 115, 116, 119, 124, 131, 132, 134, 135, 137, 138, 142, 145, 148, 150, 156
zincate 122
zinc borate 83
zinc calcium phospho-molybdate 119
zinc carbonate 137
zinc chromate 119, 156
zinc cyanamide 84
zinc dust 74, 83, 86
zinc flakes 86
zinc hydroxide 119, 137
zinc hydroxyphosphite 82
zinc oxide 16, 122, 137, 140
zinc phosphate, zinc phosphating 20, 77, 78, 119, 141, 143, 146, 156
zinc phosphating 141, 143
zinc phospho-molybdate 79, 119
zinc potassium chromate 77
zinc-rich primer 73, 83
zinc tetrahydroxy chromate 77
zirconium fluoro complex 113
zirconium oxide 146

www.ingramcontent.com/pod-product-compliance
Lightning Source LLC
LaVergne TN
LVHW081321300926
849161LV00008B/1841
9783866309111